W0261867

Proceedings of the German-Italian Symposium

Applications of Mathematics in Technology

March 26–30, 1984 Rome
(Under the auspices of the C.N.R.-D.F.G. agreement)

Edited by Prof. Dr. V. BOFFI, University of Bologna, Italy and Prof. Dr. H. NEUNZERT, University of Kaiserslautern, W.-Germany

1984. 484 pages. 16,2×23,5 cm. ISBN 3-519-02611-2. Paper DM 82,–

Contents

FLUID DYNAMICS

Karl, L. E. Nickel: Minimal Drag for Wings with Prescribed Lift, Roll Moment and Yaw Moment or How to Fight Adverse Yaw / C. Cercignani: Evaporation and Condensation: Conflicting Results from Two Different Models / R. Rautmann: Three Dimensional Flows: Models and Problems / G. P. Galdi: The Rotating Benard Problem: A Nonlinear Energy Stability Analysis / A. M. Anile; G. Russo: A Geometric Theory for the Propagation of Weak Shock Waves / A. Quarteroni: Spectral Methods for Flow Problems / C. Canuto: The Use of Spectral Methods for Exterior Problems / G. Benfatto; C. Marchioro; M. Pulvirenti: Vortex Methods in Planar Fluidodynamics / L. Triolo: Particle Models for Macroscopic Equations / E. Krause: Computation of Flows with Large Vortices / E. Martensen: Approximation of a Rarefaction Wave by Discretization in Time / V. Franceschini: Numerical Methods for Studying Periodic and Quasiperiodic Orbits in Dissipative Differential Equations / C. Tebaldi: Transitions to Turbulence in Truncated Navier-Stokes Equations / M. Dobrowolski; K. Thomas: On the Use of Discrete Solenoidal Finite Elements for Approximating the Navier-Stokes Equation / U. Bulgarelli; V. Casulli; M. Rosati: Numerical Stability for the Solution of Navier-Stokes and Euler Equations

INVERSE PROBLEMS

A. Fasano; M. Primicerio: Freezing in Porous Media — A Review of Mathematical Models / F. Natterer: Some Non-Standard Radon Problems / A. K. Louis: Fast Scanning Geometries in X-Ray Computerized Tomography / P. Colli Franzone: Inverse Problems in Electrocardiology / E. Schock: Regularization of Ill-Posed Equations with Selfadjoint Operators / F. Ebersoldt: Chain Systems in n-Compartment Analysis

MATHEMATICAL METHODS IN REACTOR TECHNOLOGY

A. Pignedoli: Transformational Methods for the Equations of the Reactor Theory / J. Batt: The Present State of the Existence Theory of the VLASOV-POISSON- and VLASOV-MAXWELL-System of Partial Differential Equations in Plasma Physics / R. Illner: On the Global Existence Problem for the Spatially Inhomogeneous Boltzmann Equation / N. Bellomo; R. Monaco: Molecular Gas Flow for Multicomponent Gas Mixtures: Some Discrete Velocity Models of the Boltzmann Equation and Applications / J. Wick: Numerical Aspects of Particle Simulation in the Plasma-Physical Case/ G. Spiga: Nonlinear Problems in Particle Transport Theory / G. Dukek; T. F. Nonnenmacher: Similarity Solutions of the Nonlinear Boltzmann Equation Generated by Lie Group Methods / W. Velte: Bounds for Critical Values and Eigenfrequences of Mechanical Systems

 **B. G. Teubner Stuttgart**

Proceedings of the Third German-Italian Symposium

Applications of Mathematics in Industry and Technology

June 18–22, 1988 Siena

(Under the auspices of the C.N.R. - D.F.G. agreement)

Edited by
Prof. Dr. Vinicio Boffi
University of Bologna, Italy

Prof. Dr. Helmut Neunzert
University of Kaiserslautern, W.-Germany

 B. G. Teubner Stuttgart 1989

CIP-Titelaufnahme der Deutschen Bibliothek

German Italian Symposium Applications of Mathematics in Industry and Technology:
Proceedings of the ... German Italian Symposium Applications
of Mathematics in Industry and Technology. – Stuttgart :
Teubner.
    1 angezeigt u.d.T.: German Italian Symposium Applications of
    Mathematics in Technology ‹01, 1984, Roma›: Proceedings of the
    German Italian Symposium Applications of Mathematics in
    Technology
NE: Applications of mathematics in industry and technology

3. June 18 – 22, 1988, Siena. – 1989
    ISBN 978-3-519-02628-0   ISBN 978-3-322-96692-6 (eBook)
    DOI 10.1007/978-3-322-96692-6

Gesamtherstellung: Präzis-Druck GmbH, Karlsruhe
Umschlaggestaltung: M. Koch, Reutlingen

PREFACE

This volume presents the proceedings of the third German-Italian Symposium on the Applications of Mathematics in Industry and Technology sponsored mainly by the research foundations of Germany and Italy, DFG and CNR. We had not published the proceedings of the second symposium held in Germany, but are now convinced that a publication would be appreciated by many scientists. The aim we had in mind organizing this symposium and communicating its scientific content was essentially the same as in the previous symposia - we were led by the same scientific and cultural (maybe even political) ideas.

The scientific idea is simple: Mathematics becomes a more and more important tool in technology, business and organisation - but mathematics does not play a corresponding role. The reason is that neither the public opinion nor the mathematicians themselves are really aware of this fact. To show the richness and efficiency of this "industrial raw material" mathematics - at least in some domains - was one subject of the conference.

In planning the structure a decision had to be made: If one wishes to present a huge variety of applications of mathematics in very different fields, the participants may loose the track, cannot find any connection with their own field of research in too many lectures; the symposium desintegrates like a puzzle into many pieces, which are not easy to be put together again. On the other hand, if one concentrates on only one subject the conference will become just one of the thousands of special workshops, and we cannot see a special need for increasing this number. Moreover, one would completely lose the point of view of a generalist, which, as we believe, is essential for a good "industrial mathematician" - practical problems almost never fit perfectly into only one mathematical box.

Our solution was a compromise, a compromise between too much generalism, where one knows "nothing about everything" and too much specialism, knowing "everything about nothing". We tried to gather fields, which are not directly connected but where each may profit from the ideas of the other. One of the fields we had chosen was fluid dynamics - a very old, almost "classical" field of industrial mathematics - and the other was system and control theory, which originated (and still lives) in classical engineering. Only two fields - less generality than before - but still a chance for mutual fertilization.

Another difference to the first and second symposium arose from the question: Whom do we want to convince? Industrialists, politicians - but it is not very likely that they will attend the conference. Young mathematicians, open for new ideas and new stimulations, maybe even seeking a new path for their future mathematical life - one can be more optimistic for an at least modest success with this target group. Therefore we invited 6 experts from each country to give survey lectures, additionally again 6 younger but advanced colleagues to report on their recent results and last not least 2 times 15 young scientists to listen, to discuss, to become interested. This is the "educational" aspect of this symposium.

But there was certainly also a cultural aspect: We believe strongly that all participants, from each side of the Alps can profit a lot from experiencing the culture on the other side. By culture we mean not only the beauty of towns (here we are not in equilibrium), we mean the way of life, the attitudes with respect to music and poetry, to family and state, to church and ecology, to dinners and football - and last not least the (different) views on mathematics and industry. It is our personal experience that one profits a lot in visiting each other, discussing with each other - one learns about mathematics, but also about the possibilies how to look on other, not necessarily scientific things. Italy and Germany are different, but not too far to become unintelligible - if one is willing to put some effort into it. This is, in an inner European exchange, why we believe that a European cooperation may be in some respects more rewarding than contacts with overseas.

Siena and the Toscana, university, city and country were excellent hosts - the most wonderful medieval piazza 100 m from the conference hall, the songs in the rooms of a contrada, the quiet evening in a certosa were precious supplements to a scientifically exciting, sometimes even exhausting symposium. We want to express our gratitude especially to the University of Siena, to its rector Prof. Berlinguer and to Prof. Millucci, to the Consiglio Nazionale Delle Ricerche and the Deutsche Forschungsgemeinschaft.

Vinicio Boffi, Bologna   December 1988
Helmut Neunzert, Kaiserslautern

TABLE OF CONTENTS

I. FLUID DYNAMICS

C. Cercignani:
Boltzmann Equation and Rarefied Gas Dynamics 9

S. Rionero:
Nonlinear Stability of Fluid Motions: The Lyapunov
Direct Method and its Applications to Natural Convection
in a Mixture 22

R. Rannacher:
Numerical Analysis of Nonstationary Fluid Flow 34

A. Quarteroni; G.S. Landriani:
Iteration by Subdomains in Numerical Fluid Dynamics 54

M. Wohlfahrt:
The Extended Lifting Line Theory for Systems of Sails 77

S. Oggioni; F. Premuda; G. Spiga:
Scattering Kernel Formulation of Nonlinear Extended
Kinetic Theory 97

G. Dziuk:
A Horizontally Twodimensional Climate Model 113

F. Gastaldi; A. Quarteroni:
On the Coupling of Hyperbolic and Parabolic Systems:
Analytical and Numerical Approach 123

G.P. Galdi; M. Padula:
New Contributions to Nonlinear Stability of the Magnetic
Benard Problem 166

D. Meinköhn:
A Criticality Concept for Reaction-Diffusion Systems far
from Thermodynamic Equilibrium 179

R. Monaco:
On the Approximation of Continuum Stochastic Systems by a
Discrete Stochastic System: A Problem with Moving
Boundary 189

A. Frezzotti:
Numerical Computation of Rarefied Gas Flows 199

G. Mulone:
On the Non-Linear Stability of Parallel Shear Flows 209

P. Dai Pra; M. Pavon:
A Rigorous Onsager-Machlup Formulation of
Nonequilibrium Thermodynamics 219

A.M. Gennai; C. Padovani:
Constitutive Equations for Masonry-Like Materials 229

G. Perrotta:
AMICS: A Multifunctional Assistent for State
Accounting Queries 239

II. SYSTEM AND CONTROL

L. Arnold:
Towards a Theory of Nonlinear Stochastic Systems 248

G. Picci:
Stochastic Aggregation 259

A.K. Louis:
Inverse Problems in Medicine 277

S. Beghelli; R.P. Guidorzi; U. Soverini:
Dynamical System Identification from Noisy Data 288

B. Aulbach:
Linearization Based on Eigenvalue Estimates 299

G.B. Di Masi; W.J. Runggaldier:
An Adaptive Linear Approach to Nonlinear Filtering 308

C.A. Clarotti; W.J. Runggaldier:
Stochastic Filtering in a Reliability Frame 317

H.J. Pesch; P. Rentrop:
Numerical Solution of Asymptotic Two-Point Boundary
Value Problems with Application to the Swirling Flow
over a Plane Disk 327

G. Casalino; R. Minciardi:
On Equilibrium Points of the Variational Adaptive
Control Scheme 339

BOLTZMANN EQUATION AND RAREFIED GAS DYNAMICS

Carlo Cercignani, Milano

Summary: A brief survey of the role played by the Boltzmann equation in rarefied gas dynamics, together with a review of recent results on the mathematical problems related to the same equation.

1 Introduction

Flight in the upper atmosphere must face the problem of a decrease in the ambient density with increasing height. This density reduction would alleviate the aerodynamic forces and heat fluxes that a flying vehicle would have to withstand. However, for virtually all missions, the increase of altitude is accompanied by an increase in speed; thus it is not uncommon for spacecraft to experience its peak heating at considerable altitudes, such as, *e.g.*, 70 km. When the density of a gas decreases, there is, of course, a reduction of the number of the number of molecules in a given volume and, what is more important, an increase in the distance between two subsequent collisions of a given molecule, till one may well question the validity of the Euler and Navier–Stokes equations, which are usually introduced on the basis of a continuum model which does not take into account the molecular nature of a gas. It is to be remarked that the use of those equations can also be based on the kinetic theory of gases [1-3], which justifies them as asymptotically useful models when the mean free path is negligible. According to kinetic theory the basic description of the evolution of a not-too-dense gas is in terms of a function of position $\underline{x}$, velocity $\underline{\xi}$, time t , $f=f(\underline{x},\underline{\xi},t)$, which gives the probability density of finding a molecule at position $\underline{x}$, with velocity $\underline{\xi}$, at time t. The usual quantities such as density ρ, bulk velocity $\underline{v}$, stresses p_{ij} (including pressure), temperature T, heat flux $\underline{q}$, are obtained as moments of the basic unknown f, through simple formulas such as:

$$(1.1) \qquad \rho(\underline{x},t) = \int f(\underline{x},\underline{\xi},t)d\underline{\xi} \ ; \qquad \rho\underline{v}(\underline{x},t) = \int \underline{\xi} f(\underline{x},\underline{\xi},t)d\underline{\xi} \ ;$$

$$p_{ij}(\underline{x},t) = \int c_i c_j f(\underline{x},\underline{\xi},t)d\underline{\xi} \ , \ \text{where} \ c_i = \xi_i - v_i \ ;$$

$$p=\rho RT(\underline{x},t)=\frac{1}{3}\int c^2 f(\underline{x},\underline{\xi},t)d\underline{\xi} \ ; \underline{q}(\underline{x},t) = \frac{1}{2}\int \underline{c}c^2 f(\underline{x},\underline{\xi},t)d\underline{\xi} \ .$$

where the subscripts take the values from 1 to 3 and R is the gas constant.

These formulas, without any further statement, are sufficient to show that the Navier Stokes equations must become invalid on a sufficiently small scale. In fact, a well-kown inequality shows that Eqs. (1.1) imply

$$(1.2) \qquad |P_{ij}| \le \frac{P_{ii} + P_{jj}}{2} \le 3p$$

Take now the component p_{12} in a simple shear flow, such as plane Couette flow; according to the Navier-Stokes constitutive relations:

$$(1.3) \qquad P_{12} = - \mu\frac{\partial u}{\partial y}$$

where $u=v_1$ is the x component of the velocity (the only one which is different from zero) and μ is the viscosity coefficient. Hence:

$$(1.4) \qquad |\mu\frac{\partial u}{\partial y}| \le 3p$$

which indicates that the velocity gradient cannot be higher than $3p/\mu$ (to say the least!) for the Navier-Stokes equations to be valid. Since the viscosity of a dilute gas is independent of the density, while the pressure is proportional to the density (for a given temperature), we see that in rarefied conditions the restriction on $\partial u/\partial y$ becomes quite severe. As a matter of fact, according to kinetic theory, Eq. (1.4) has a simple significance in molecular terms, because the mean free path λ of a molecule turns out to be related to μ by:

$$(1.5) \qquad \lambda = \frac{\mu}{p}\sqrt{2RT} \qquad .$$

Hence Eq. (1.4) may be rewritten as follows:

$$(1.6) \qquad \left|\frac{\partial u}{\partial y}\right| \leq \frac{3}{\lambda}\sqrt{2RT} \qquad .$$

In other words, the velocity gradient cannot be larger than a certain amount which is of the order of the ratio between the thermal speed and the mean free path. In order to appreciate this point, we remark that the mean free path is about 1 meter at an altitude of about 100 km. This explains why the Navier Stokes equations cannot be used for a rarefied gas and/or high speeds. Hence we must resort to the full apparatus of kinetic theory if we want to deal with high altitude flight.

2   <u>The Boltzmann equation</u>.

The natural tool for describing the behaviour of a rarefied gases is an integrodifferential equation called the Boltzmann equation [1-3], which rules the evolution of the distribution function, $f = f(\underline{x},\underline{\xi},t)$ and reads as follows:

$$(2.1) \qquad \frac{\partial f}{\partial t} + \underline{\xi}\cdot\frac{\partial f}{\partial \underline{x}} + \underline{X}\cdot\frac{\partial f}{\partial \underline{\xi}} = Q(f,f)$$

where

$$(2.2) \qquad Q(f,f) = \frac{1}{m}\iint (f'f'_* - ff_*)B(\theta,|\underline{\xi}-\underline{\xi}_*|)d\underline{\xi}_*\,d\theta d\varepsilon$$

Here $B(\theta,|\underline{\xi}-\underline{\xi}_*|)$ is a kernel associated with the details of the molecular interaction, m the molecular mass, f', f'_*, f_* are the same thing as f, except for the fact that the argument $\underline{\xi}$ is replaced by $\underline{\xi}'$, $\underline{\xi}'_*$, $\underline{\xi}_*$. The latter is an integration variable having the meaning of the velocity of a molecule colliding with the molecule of velocity $\underline{\xi}$, whose evolution we are following, while $\underline{\xi}'$ and $\underline{\xi}'_*$ are the velocities of two molecules entering a collision which will bring them into a pair of molecules with velocities ζ and ζ_*. θ and ε are two angles defining the direction of approach of two colliding molecules.

Eq. (2.1) has been the object of many studies on both physical and mathematical grounds, since Boltzmann proposed it in 1872, but in recent times it has become a practical tool in the hands of aerospace engineers dealing with upper atmosphere flights. In fact the Boltzmann equation is capable of describing the behaviour of gas from the continuum regime of a not-too-dense gas, for which the Navier-Stokes equations also apply, to the free-molecular regime of noninteracting particles. Between these two extreme regimes lies the transition regime, which cannot be described either by means of Navier-Stokes equations or a gas of noninteracting particles.

Eq. (2.1) must be solved with suitable initial and boundary conditions. The latter, in particular, describe the interaction of the gas molecules with the solid surfaces bounding the region where the gas moves.

3   Rarefaction regimes.

The two basic similarity parameters are the Knudsen number Kn and the molecular speed ratio S ; they are defined as the ratio of the mean freepath λ to a length L characteristic of the geometry of the flow and the ratio of the bulk speed u to the thermal speed C related to the temperature T and the gas constant R by

$$(3.1) \qquad C = \sqrt{2RT}$$

The Knudsen number and the speed ratio are related to the Mach and Reynolds numbers, Ma and Re, in the following way:

$$(3.2) \qquad Ma = \sqrt{2/\gamma}\, S \quad ; \quad Kn = \sqrt{2\gamma}\, Ma/Re$$

The breakdown of the description of the gas as a continuum described by the Navier-Stokes equations follows from the fact that the latter require a sufficiently large number of collisions per unit volume and unit time. This may easily lead to the necessity of using the Boltzmann equation for the flow past a solid body. In this case, in fact, an important macroscopic length is the thickness of the boundary layer δ and hence an appropriate Knudsen number is $Kn_\delta = \lambda/\delta$; when this number becomes larger than, say, 0.01, the effects occurring in a thin layer near the wall having a thickness of the order of a mean free path (Knudsen layers) will influence the behaviour in the entire viscous layer of thickness δ. A velocity slip u_s and a temperature jump $T_s - T_w$ between gas and wall will appear; these jumps are partly due to a real jump at the wall (microscopic slip and temperature jump) and partly to a quick change through the Knudsen layer.

An additional effect showing up is a significant thickening of the bow shock in front of a vehicle moving at supersonic speed with respect to the gas. In fact the shock wave thickness is of the order of 6λ and hence negligible when the mean free path λ is negligible.

The slip velocity is given by

$$(3.3) \qquad u_s = \zeta \left(\frac{\partial u}{\partial n} \right)_w$$

where ζ is the so called slip coefficient. ζ was shown by Maxwell [4] to be of the order of the mean free path; in fact by an approximate calculation he found $\zeta = (\sqrt{\pi}/2)\lambda$, by assuming that all the molecules are diffused with a Maxwellian distribution. This result was improved upon by several authors who showed that ζ is actually about 15% larger (see, e. g., [1-2]).

This simple example already shows that gas surface interaction influences the flow field development and the local aerodynamic actions on the body. It is to this interaction that one can trace back the origin of the draf and lift exerted by the gas on a solid body and the heat transfer between a gas and a solid wall.

The study of gas-surface interaction is an interdisciplinary subject related to molecular physics, surface physics and gas kinetics. From a physical point of view one must relate the distribution function of the reflected molecules to the distribution function of the incident ones. There exist several gas surface interaction models; the simplest one was proposed by

Maxwell [4] in 1979. He assumed that a fraction α of the incident molecules is diffused according to a Maxwellian distribution and the remaining fraction $1-\alpha$ is specularly reflected. A more complicated model was proposed by M. Lampis and myself [5] in 1970.

The influence of the gas surface interaction is particularly felt when the Knudsen number is large (few collisions). In particular, the limiting behaviour when Kn $\to \infty$ (free-molecular flow) depends on just the geometrical shape of the body and the gas surface interaction. When the effects of the intermolecular collisions can be treated as a perturbation of the free molecular results, we say that we are in the nealy free-molecular regime. Between this regime and the slip regime there is the so called *transition regime* where both gas surface interaction and intermolecular collisions are important. One of the striking features of this regime in the flow past a body is that the bow shock wave and the boundary layer merge. Hence the name merged layer regime used sometimes by the aerospace engineers to denote the transition regime.

4   Solving the Boltzmann equation.

How does one handle the already complicated Boltzmann equation with similarly complicated boundary conditions? Attempts began in the late 1950's and early 1960's. One of the first field to be explored was that of the "simple flows", such as Couette and Poiseuille flows in tubes and between plates; here it turned out that the equation to be solved is still formidable and various approximation methods where proposed. Some of these were perturbation methods: for large or small Knudsen numbers or about an equilibrium solution (Maxwellian). The first two approaches gave useful results in the limiting regimes, while the third method led to studying the so called Linearized Boltzmann Equation, which produced predictions which are in a spectacular agreement with experiment and have shed considerable light on the basic structure of transition flows, whenever nonlinear effects can be neglected [1-3]. This gave confidence to further use of the Boltzmann equation for practical problems. Other problems which were treated with the linearized equation were the half space problems which are basic in order to understand the structure of Knudsen layers and to evaluate the slip and temperature jump coefficients.

A particularly interesting problem is related to the structure of a shock wave; this is not a discontinuity surface as in the theory of compressible Euler equation but a thin layer (having, usually, a thickness of the order of a mean free path). In the case of a normal shock wave, one can imagine it in the entire space without boundaries; finding the shock wave structure means solving the Boltzmann equation when the solution (which depends on one space coordinate, say x, and the three velocity components ξ_i (i=1,2,3), but not on time and the other two coordinates) tends to two different Maxwellians when x tends to $+\infty$ and $-\infty$. The two Maxellians have the following shape:

$$(4.1) \qquad f_o^\pm = \rho^\pm (2\pi RT^\pm)^{-3/2} \exp[-(\underline{\xi} - u_{\underline{i}}^\pm)^2/(2RT^\pm)]$$

where the superscripts $\pm$ refer to the downstream and upstream state, respectively.

An early approach that was moderately successful in dealing with the shock wave problem was the Mott-Smith method [6]. The method postulates that there is a bimodal distribution, i.e. a linear combination of the two Maxwellians defined in Eq. (4.1):

$$(4.2) \qquad f = \nu f_o^+ + (1-\nu)f_o^-$$

Here $\nu=\nu(x)$ is a function that goes from 0 to 1 through the shock. Eq. (4.2) is easily shown to be compatible with the balance of mass, momentum and energy provided the constant values $\rho^\pm$, $u^\pm$, $T^\pm$ satisfy a set of compatibility conditions, which are nothing other than the Rankine-Hugoniot relations, familiar from the ideal fluid theory of shock waves. In order to determine $\nu(x)$ several procedures have been presented none of which is very satisfactory, since they are essentially arbitrary. A rational basis to the method is available only in the case of an infinitely strong shock. The results for several physical quantities, including the thickness of the shock are, however, considerably more accurate than the Navier-Stokes values for other than low Mach numbers. In addition to the unsatisfactory status from a mathematical point of view, the Mott-Smith approach suffers the further drawback of being restricted to the shock structure problem.

The most well known analytical (or semi-analytical) solutions for the Boltzmann equation are obtained through the so called BGK (Bhatnagar, Gross and Krook [7]) model, which is the simplest among the so called model equations or kinetic models. They differ from the Boltzmann equation because the collision integral is replaced by another, simpler term, which retains only the qualitative and average properties of the true collision operator. The idea behind this replacement is that a large amount of detail of the two-body interaction (which is contained in the collision term) is not likely to influence significantly the value of many experimentally measured quantities; thus it is expected that the fine structure of the collision operator can be replaced by a blurred image of it. The BGK model is characterized by the fact that the collision operator $Q(f,f)$ is replaced by:

$$(4.3) \qquad J(f) = \nu(\Phi - f)$$

where ν does not depent on $\underline{\xi}$ but is proportional to ρ and may depend on T as well (it has the physical meaning of a collision frequency), while Φ is the so-called local Maxwellian:

$$(4.4) \qquad \Phi = \rho(2\pi RT)^{-3/2} \exp[-(\underline{\xi} - \underline{v})^2)/2RT]$$

where ρ, $\underline{v}$, T are not given $a\ priori$ but are related to the unknown f through Eqs.(1.1). Thus the expression (4.3) is rather complicated in terms of f.

The kinetic models have been very useful in obtaining approximate solutions and forming qualitative ideas on the solutions of practical problems, but in general do not provide us with detailed and precise answers to the sort of question that is posed by the space engineer. Various numerical procedures exist which either attempt to solve for f by conventional techniques of numerical analysis or efficiently by-pass the formalism of the integrodifferential equation and simulate the physical situation that the equation describes (Monte Carlo methods). Only recently proofs have been given that these partly deterministic, partly stochastic games provide solutions that converge (in a suitable sense) to solutions of the Boltzmann equation. There appear to be very few limitations to the complexity of the flow fields that this approach can deal with. Chemically reacting and ionized flows can and have been analysed by these methods.

5   The mathematical theory of the Boltzmann equation.

The purely mathematical aspects of the Boltzmann equation began to be investigated in the thirties by the famous Swedish mathematician T. Carleman [8],who provided an existence proof for the purely initial value problem with homogeneous data (i. e. data independent of $\underline{x}$). The same problem was revisited by Arkeryd [9] in 1972; he provided solutions in a (weighted) L^1 space, rather than in a (weighted) L^∞ space. Solutions depending on the space variables are much more difficult to handle, if we do not restrict our attention to solutions existing only locally in time but look for solutions existing for an arbitrarily long time interval; the first results were obtained by several Japanese authors [10-12] and referred to solutions close to a (homogeneous) Maxwellian distribution. Then Illner and Shinbrot [13] provided an existence proof for a global solution close to vacuum; their assumptions were later relaxed by Bellomo and Toscani [14], while Toscani [15] has recently considered solutions close to a nonhomogeneous Maxwellian (which, however, must be a solution of the Boltzmann equation).

I proved [16] existence for a very particular case with data arbitrarily far from equilibrium, the so-called affine homoenergetic flows. Arkeryd, Esposito and Pulvirenti [17] proved existence for solutions close to a homogeneous solution (different from a Maxwellian).One should also mention the important paper of Arkeryd [18]who proved an existence theorem in the context of non-standard analysis.

Quite recently, R. DiPerna and J.P. Lions [19] provided an

existence theorem (without uniqueness) for the general case of inhomogeneous data; their proof is quite clever and makes use of a compactness lemma by Perthame, Golse, Sentis and Lions [20] to overcome the difficulties met by other authors.

It should be realized that any equation similar to the Boltzmann equation but having a little more of compactness in the dependence upon the space variables is rather easy to deal with; this was shown by Morgenstern [21] in the 1950' and Povzner [22] in the 1960'; they introduced mollifying kernels in the collision term of the Boltzmann equation, producing 8-fold and 6-fold integrations, respectively, in place of the original 5-fold integration. In the 1970' W. Greenberg, P. Zweifel and myself [23] indicated that a theorem of existence and uniqueness can be proved if the particles can sit only at discrete positions on a lattice.

While the initial value problem for the Boltzmann equation has received a great deal of attention, comparatively less work has been done for the steady problems, which, after all, are those of paramount interest for the space engineer. These problems, in a linearized form, were satisfactorily dealt with in the 1960's, with the exception of the important half-space problems that were completely treated only in the last few years. An early result for the nonlinear problem in a slab with data close to equilibrium was obtained by Pao [24] in 1967, by a suitable use of a previous result of mine [25] dealing with the corresponding linearized problem. Much later Ukai and Asano [26] were able to treat the small Mach number flow past a solid body.

The case of data arbitrarily removed from equilibrium was not considered till R. Illner, the late M. Shinbrot and myself [27] wrote a paper on the slab problem; for technical reasons we took the molecular velocities to be discrete and obtained existence for arbitrarily large data and domains. The treatment has been extended to the case of a rectangle [28] (but only for a very particular discrete velocity model) and to a half space (in collaboration with M. Pulvirenti [29]).

Before ending this survey, I would like to mention another

fascinating problem related to the Boltzmann equation: the problem of justifying the equation itself. The equation is deceivingly simple to derive, but a stringent analysis shows that it is not so easy to justify the steps; actually, for a long time it was thought that the Boltzmann equation could not be given any mathematical status (except, of course, by postulating it), because its irreversible features were thought to be in contradiction with the reversible physical model upon which it was based. Then H. Grad [30] pointed out that one should consider it to apply in the limiting case of a gas of infinitely many particles of vanishing diameter σ in such a way that the product $N\sigma^2$ (where N is the particle number) remains finite (the Boltzmann-Grad limit). In 1972 I was able to show [31] that Grad's conjecture was formally consistent; i.e. if all the required limits existed and appropriate existence and uniqueness and existence theorems applied to the limiting equations, the Boltzmann equation could be justified.

A few year later O. Lanford [32] was able to show that all this applied rigorously for a finite time interval (of the order of 1/4 of a mean free time); more recently R. Illner and M. Pulvirenti [33] have presented a validity proof valid for arbitrarily long times, provided the data are close to the vacuum solution. The recent theorem of DiPerna and Lions [19] has raised new hopes for a general treatment of the question, but the matter is not so easy.

<u>References</u>

[1] Cercignani, C.: *The Boltzmann equation and its applications*, Springer, New York (1988).

[2] Cercignani, C.: *Mathematical Methods in Kinetic Theory*, Plenum Press, New York (1969).

[3] Kogan, M. N.: *Rarefied Gas Dynamics*, Plenum Press, New York, 1969.

[4] Maxwell, J. C.: Phil. Trans. Royal Soc.,I, Appendix (1879)

[5] Cercignani, C.; Lampis, M: Transport Theory and Statistical

Physics, **1**, 101 (1971).

[6] Mott-Smith, H. M.: Phys. Rev., **82**, 885 (1951).

[7] Bhatnagar, P. L.; Gross, E. P.; Krook,M.: Phys. Rev., **94**, 511 (1954).

[8] Carleman, T.: Acta Math., **60**, 91 (1933).

[9] L. Arkeryd, Arch. Rational Mech. Anal., **45**, 1 and 17 (1972).

[10] Ukai, S.: Proc. Japan. Acad. Ser. A, Math. Sci. **50**, 179 (1974).

[11] Nishida, T.; Imai, K.: Publ. Res. Inst. Math. Sci., Kyoto Univ., **12**, 229 (1977).

[12] Shizuta, Y.; Asano, K: Proc. Japan Acad. Ser. A, Math. Sci., **53**, 3 (1977).

[13] Illner, R.; Shinbrot, M.: Comm. Math. Phys., **95**, 217 (1984).

[14] Bellomo, N.; Toscani, G.: J. Math. Phys., **26**, 334 (1985).

[15] Toscani, G.: preprint.

[16] Cercignani, C.: Arch. Rational Mech. Anal., to appear (1988).

[17] Arkeryd, L.; Esposito, R.; Pulvirenti, M.: Commun. Math. Phys., **111**,393 (1988).

[18] Arkeryd, L.: Arch. Rational Mech. Anal., **86**, 85 (1984).

[19] DiPerna, R.; Lions, P. L.: Ann. Math., to appear (1988).

[20] Perthame, B.; Golse, F.; Sentis,R; Lions, P. L.: J. Funct. Analysis, to appear (1988).

[21] Morgenstern, D.: J. Rational Mech. Anal., **4**. 533 (1955).

[22] Povzner, A. Ya.: Mat. Sbornik, **58**, 65 (1962).

[23] Cercignani, C.; Greenberg, W.; Zweifel, P.: J. Stat. Phys., **20**, 449 (1979).

[24] Pao, Y. P.: J. Math. Phys., **9**, 1893 (1968).

[25] Cercignani, C.: J. Math. Phys., **8**, 1653 (1967).

[26] Ukai, S.; Asano, K.: Arch. Rational Mech. Anal., **84**, 249 (1983).

[27] Cercignani, C.; Illner, R.; Shinbrot, M.: Duke Math. Journal, **55**, 889 (1987).

[28] Cercignani, C.; Illner, R.; Shinbrot, M.: Comm. Math. Phys. , **114**, 697 (1988).

[29] Cercignani, C.; Illner, R.; Pulvirenti, M.; Shinbrot, M.: J. Stat. Phys:, **52**, 885 (1988).

[30] Grad, H.: Comm. Pure and Appl. Math., 2, 33 (1949).

[31] Cercignani, C.: Transport Theory and Statistical Physics, 2, 211 (1972).

[32] Lanford, O.: in *Proceedings of the 1974 Battelle Rencontre on Dynamical Systems*, J. Moser, Ed., Lecture Notes in Physics, 35, 1, Springer, Berlin (1972).

[33] Illner, R.; Pulvirenti, M.:, Comm. Math. Phys., 105, 189 (1986).

Dipartimento di Matematica
Politecnico di Milano
Piazza Leonardo da Vinci, 32
20133 Milano (Italy)

NONLINEAR STABILITY OF FLUID MOTIONS : THE LYAPUNOV DIRECT METHOD AND ITS APPLICATIONS TO NATURAL CONVECTION IN A MIXTURE

SALVATORE RIONERO

Introduction

As is well known, general ideas of stability are fundamental to our perception to the universe. We can form mental images only of those phenomena and structures which recognizably persist or repeat themselves in time.

Our aim is to give an account of the theory of fluid stability and of the Italian school contribution to the nonlinear stability of fluid motions via the Lyapunov direct method.

PART I - Recall of stability

1.1 The method of linear stability. Main results.

Let H be a Hilbert space endowed with a scalar product $(\cdot,\cdot)$ whose associated norm is $\|\cdot\|$. Let us consider in H the following initial value problem:

$$(1.1) \qquad \begin{cases} u_t + Lu + N(u) = 0 \\ u(x,0) = u_o \end{cases} , \qquad u=u(x,t), \quad x \in \mathbb{R}^3, \quad t \in \mathbb{R}^+,$$

with

$$(1.2) \qquad \begin{cases} L = \text{linear operator} \\ N = \text{nonlinear operator such that } N(0) = 0. \end{cases}$$

The study of linear stability of the solution u=0 (basic solution) consists in neglecting the nonlinear term N(u) and in studying the behaviour in time of the perturbation u to the basic solution u=0 through the linear equation

$$(1.3) \qquad u_t = Lu \ .$$

Assuming (1.3) autonomous, one can look for asolution of the kind

$$(1.4) \qquad u = \hat{u}^{\langle}(x) \ e^{-\sigma t}$$

where σ is _apriori_ a complex parameter. Substituting (1.4) back into (1.3), one is led to solve the spectral problem

$$(1.5) \qquad \begin{cases} -\sigma\hat{u} = L \ \hat{u} \\ \hat{u}(x,0) = \hat{u}_o \end{cases} \ .$$

Definition 1 (linear stability). The basic solution is said to be linearly stable if

$$(1.6) \qquad re(\sigma) > 0 \quad \forall \ \sigma \ .$$

Definition 2 (linear instability). The basic solution is said to be linearly unstable if

$$(1.7) \qquad \exists \ \bar{\sigma} \ : \qquad re(\bar{\sigma}) < 0 \ .$$

Under suitable assumptions on L, [5], the spectrum consists of an (at most) denumerable number of eigenvalues $\{\sigma_n\}$, ($n \in \mathbb{N}$), with finite (algebraic and geometric) multiciplities and, moreover, such eigenvalues can cluster only at infinity and can be ordered in the following way:

$$(1.8) \qquad re\,(\sigma_1) \leq re\,(\sigma_2) \leq \cdots\cdots re\,(\sigma_n) \leq \cdots\cdots$$

Then, setting

$$(1.9) \qquad S = re\,(\sigma_1)\,,$$

one has

$$(1.10) \qquad S > 0 \;\;\Rightarrow\;\; \text{linear stability.}$$

In general, S will depend on the basic solution through dimensionless (positive) parameters R, T, ... and - in the case of periodic perturbations - on the wave numbers $\$\sigma_i$ associated with them. The values R_c, T_c, ... for which $S = 0$, are called critical values of R, T, ... , respectively.

1.2 - <u>Nonlinear stability</u>

<u>Definition 3 (nonlinear stability). The basic solution is said to be nonlinearly stable iff</u>

$$(1.11) \;\; \forall\, \varepsilon > 0,\; \exists\, \delta_\varepsilon > 0 : \|u_o\| < \delta_\varepsilon \;\Rightarrow\; \|u\| < \varepsilon \;\; \forall\, t \geq 0.$$

<u>Definition 4 (asymptotical stabiltiy) - The basic solution is said to be conditionally asymptotically stable iff is stable and moreover</u>

$$\exists\, \delta_1 > 0 : \qquad \|u_o\| < \delta_1 \;\Rightarrow\; \lim_{t\,\infty} \|u\| = 0.$$

<u>Iff $\delta_1 = \infty$, the basic solution is said to be unconditionally asymptotically stable.</u>

Following the modern version of the <u>energy method</u> , [2], [7], [11], [15], from (1.1) one obtains :

$$(1.12) \qquad \frac{1}{2}\frac{d}{dt}\|u\|^2 = (N(u),u) + (L(u),u)\,.$$

Therefore, if

25

$$(1.13) \qquad \exists\ \lambda > 0 \ : \quad \lambda \leq -\ \frac{(L(u),u)}{\|u\|^2}\ ,$$

then it follows

$$(1.14) \qquad \frac{1}{2}\frac{d}{dt}\|u\|^2 \leq \left[\ \frac{(N(u),u)}{\|u\|^2} - \lambda\ \right]\|u\|^2$$

and hence $\mu < \lambda$ implies

$$(1.15) \qquad \|u\|^2 \leq \|u_o\|^2\ e^{-2(\lambda-\mu)t}\ ,$$

where

$$(1.16) \qquad \mu = \max_{u}\ \frac{(N(u),u)}{\|u\|^2}\ .$$

Theorem 1 - The condition

$$(1.17) \qquad \mu < \lambda$$

assures the unconditional asymptotical exponential stability
of the basic solution.

Proof. See (1.15).

Of course μ will depend on the basic solution through the
same dimensionless (positive) parameters R, T, ... on which
depends S and the biggest values R_c^* , T_c^*, ... for which

$$\mu < \lambda$$

are the critical values of the nonlinear stability of R, T,
..., respectively. In general one has

$$R_c^* < R_c\ ,\quad T_c^* < T_c,\ \ldots \qquad .$$

For instance - in the case of Navier - Stokes equation,
when $\|u\| = L_2$- norm, the critical vales R_c^*, T_c^*, ... of
nonlinear stability are quite below the corresponding values
of linear stability. Only in some cases one has $R_c^* = R_c, T_c^* =$

T_c, ... and therefore the <u>coincidence</u> <u>between</u> <u>the</u> <u>linear</u> <u>and</u> <u>nonlinear</u> <u>stability</u> <u>conditions</u> (this happens, for instance, in the normal Bénard problem).

In general the following questions arise:

i) How do the critical numbers R_c^*, T_c^*, ... of nonlinear stability depend on the choice of the norm $\|u\|$ in H ?

ii) Is it possible to choose $\|u\|$ in such a way as to reach the results of linear stability ?

In order to answer these questions let us recall that in the case of ordinary differential equations the <u>Lyapunov</u> <u>direct</u> <u>method</u> holds. In fact, for instance, for an autonomous system

$$(1.18) \qquad \begin{cases} \dfrac{du}{dt} = F(u) & u = (u_1, u_2, \ldots, u_n) \\[2ex] u(0) = u_o & F(0) = 0 \end{cases}$$

the following Lyapunov theorem holds:

<u>Theorem II</u> - <u>Let</u> $V = V(u)$ <u>be</u> <u>positive</u> <u>definite.</u> <u>Then,</u> <u>if</u> <u>along</u> <u>the</u> <u>solutions</u> <u>to</u> (1.18) <u>one has</u>

$$(1.19) \qquad \dfrac{dV}{dt} \leq 0 \ ,$$

<u>the</u> <u>solution</u> $u = 0$ <u>is</u> <u>stable.</u>

The function V is called <u>Lyapunov</u> <u>function</u>.

In the case of <u>partial</u> <u>differential</u> <u>equations</u>, a general Lyapunov theorem does not exist and it has to be proved case to case. Consequently, the questions i) - ii) can be transformed to the following ones:

iii) Is it possible to chhose a Lyapunov function in such a way that theorem II holds and, moreover, the results of linear stability are reached ?

iv) Which are the guidelines that one has to follow in choosing the Lyapunov function ?

Actually now there are two guide-lines on how to choose the Lyapunov function.

a) <u>Mathematical</u> <u>guide-line</u>

This guide-line (Galdi [4]) is based on a deep analysis of the operators L and N and on the possibility of splitting L in symmetric and skew-symmetric parts.

b) <u>Physical</u> <u>guide-line</u>

This guide-line (Rionero, Mulone [8], [9], [12], [13], [14]) is based on introducing field variables X_1, X_2, ... which represent physical causes inhibiting or promoting the instability and on the use of balances

$$f_{ij} = X_i - c_{ij}X_j \quad (c_{ij} = const)$$

between field variables representing opposite causes. Then V is split into two parts

$$V = V_o + V_1$$

where V_o - depending on the field variables and on the balances - has to dominate the linear problem while V_1 has to dominate the nonlinear terms.

In the following sections we shall give an application of this guide-line to the stability of a mixture in a rotating layer.

Part II - Stability of a mixture in a rotating layer via the Lyapunov direct method.

2.1 Preliminaries

Let us consider an infinite horizontal layer of a mixture of two fluids. Let $Oxyz = (O,i,j,k)$ be a cartesian coordinate system with the z-axis pointing vertically upwards and let the mixture be confined between the planes $z=0$ and $z=d$, $d>0$, with assigned temperatures and concentrations. Moreover let the layer be:

  i) rotating about the vertical axis with angular velocity Ω

 ii) heated from below with constant gradient of temperature $\alpha > 0$

iii) salted from above with constant gradient of concentration $\beta > 0$.

Here we study in the Oberbeck-Boussinesq scheme , the stability of the rest state $m_o = (\hat{v}=o, \hat{T}=-\alpha z+\hat{T}_o, \hat{C}=-\beta z+\hat{C}_o, \hat{p})$, where v, T, C and p are the velocity, temperature, concentration and the pressure fields.

Indicated by

$u=(u,v,w)$	perturbation to the velocity field
ϑ	perturbation to the temperature field
γ	perturbation to the concentration field
$P_r = \nu/k$	Prandtl number
$P_c = \nu/k_c$	Schmidt number

$$R^2 = \frac{g\alpha_T \alpha d^4}{\nu k} \qquad\qquad \text{Rayleigh number(for the temterature)}$$

$$C^2 = \frac{g\alpha_c \beta d^4}{\nu k_c} \qquad\qquad \text{Rayleigh number (for the concentra-}$$

tion)

$$T^2 = \frac{4\Omega^2 d^4}{\nu^2} \qquad\qquad \text{Taylor number,}$$

the dimensionless equations for a perturbation (u,ϑ,γ) to m_o are

$$(2.1) \qquad \begin{cases} u_t + u\cdot\nabla u = -\nabla p + (R\vartheta - C\gamma)k + \Delta u + Tu\times k \\[4pt] P_r\vartheta_t + P_r u\cdot\nabla\vartheta = Rw + \Delta\vartheta \\[4pt] P_c\gamma_t + P_c u\cdot\nabla\gamma = -Cw + \Delta\gamma \\[4pt] \nabla\cdot u = 0 \end{cases}$$

where $(x,t) \in \mathbb{R}^2 \times [0,1]\times[0,\infty)$, under the initial and boundary conditions

$$(2.2) \qquad u(x,0)=u_o(x), \quad \vartheta(x,0)=\vartheta_o(x), \quad \gamma(x,0)=\gamma_o(x)$$

$$(2.3) \qquad \begin{cases} w(x,t)=\vartheta(x,t)=\gamma(x,t)=0, \\[4pt] u_z(x,t)=v_z(x,t)=0, \quad \text{on } z=0, z=1, t\geq 0 \ . \end{cases}$$

We assume that the perturbation fields are periodic functions of x and y of periods $\dfrac{2\pi}{a_x}$, $\dfrac{2\pi}{a_y}$, $(a_x>0, a_y>0)$ and we denote by Ω_1 the periodicity cell

$$\Omega_1 = [0,\frac{2\pi}{a_x}]\times[0,\frac{2\pi}{a_y}]\times[0,1].$$

We also require the "average velocity condition":

$$(2.4) \qquad \int_{\Omega_1} u\, d\Omega_1 = \int_{\Omega_1} v\, d\Omega_1 = 0 \ .$$

2.2 <u>Lyapunov function choice</u>

As field variables we choose $w, \zeta = k \cdot \nabla \times u, \vartheta_z, \gamma_z$. Because the instability occurs at the onset of convection, it is easy to

understand that w, $-\vartheta_z$ represent causes promoting instability while ζ and γ_z represent causes inhibiting the instability. Consequently, we choose as possible <u>balances</u> the functions

$$(2.5) \qquad f_1 = \zeta - c_1 \vartheta_z \quad , \qquad f_2 = \zeta + c_2 \gamma_z \ ,$$

where the constants c_i (i =1,2) will be chosen opportunely later.

As Lyapunov function we choose

$$(2.6) \qquad V(t) = V_o(t) + bV_1(t) \ , \qquad b = \text{const.}$$

where

$$(2.7) \qquad V_o(t) = \frac{1}{2} [\ \|\nabla w\|^2 + a_1 \|\vartheta_z\|^2 + a_2 \|\gamma_z\|^2 + a_3 \|f_1\|^2 + a_4 \|f_2\|^2 \ ]$$

$$(2.8) \qquad V_1(t) = \frac{1}{2} [\ \|\nabla u\|^2 + P_r \|\nabla \vartheta\|^2 + P_c \|\nabla \gamma\|^2] \ .$$

The quantities a_i (i=1,2,..,4) are positive constants and $\|\cdot\|$ is the L_2-norm.

2.3 - <u>Nonlinear stability</u>

Let

$$(2.9) \quad \begin{cases} I'_o = -(R(\Delta_1 \vartheta, w) + C(\Delta_1 \gamma, w) + R\sigma_1(w_z, \vartheta_z) - C\sigma_2(w_z, \gamma_z) \\[2mm] D'_o = \|\Delta w\|^2 + \alpha_1 \|\nabla \vartheta_z\|^2 + \alpha_2 \|\nabla \gamma_z\|^2 \\[2mm] \Delta_1 = \dfrac{\partial^2}{\partial x^2} + \dfrac{\partial^2}{\partial y^2} \quad , \quad (f,g) = \displaystyle\int_{\Omega_1} fgd\Omega_1 \quad , \\[3mm] H_1 = \{ \ w, \vartheta, \gamma \in {}^5 C \text{ which are periodic in } x \text{ and } y \\ \qquad\qquad \text{and satisfying the boundary conditions}\} \\[2mm] M = \sup_{H} \dfrac{I'_o}{D'_o} \quad , \end{cases}$$

with

σ_1 , σ_2 , α_1 , α_2 constants which depend on the basic solution.

Then the following theorem holds

Theorem III - Let

$$(2.10) \qquad \begin{cases} 0<M<1 \\ V(0)<A_o^{-2} \end{cases} \qquad .$$

Then there exists a positive constant α_o such that

$$(2.11) \qquad V(t) \leq V(0) \exp\{-\pi^2 \alpha_o (1-M)[1-A_o V(0)^{1/2}]t\} \quad ,$$

where

$$(2.12) \quad A_o = (8/b)^{1/2} c_o \{ 1 + [(P_r)^{1/2} + (P_c)^{1/2}]/2 + \frac{2}{b}(1+a_3+a_4) +$$

$$+ (\frac{1}{b(1-M)})^{1/2} [\frac{a_1}{(\alpha_1)^{1/2}} + \frac{a_2}{(\alpha_2)^{1/2}} + \frac{a_3 c_1}{(\alpha_1)^{1/2}} (2^{1/2}+c_1) +$$

$$+ \frac{a_4 c_2}{(\alpha_2)^{1/2}} (2^{1/2}+c_2) + 2^{1/2} (\frac{a_3 c_1}{(\alpha_1)^{1/2}} + \frac{a_4 c_2}{(\alpha_2)^{1/2}})]\} \quad ,$$

c_o is a positive computable constant depending on the periodicity cell,[3], [6], b, c_1,c_2 are positive constants depending on the basic solution, (see [14]).

From theorem III, by solving the maximum problem $(2.9)_6$, it is possible to see that for $P_r \geq 1$, $P_c \geq 1$, $T^2 \in [0,80\pi^4]$ the nonlinear results are coincident with the linear ones, [1], [10].

The values of the critical Rayleigh numbers R_c^{2*} are shown in table I. They are compared with those of linear stability R_c^2 .

<u>Table I – Critical Rayleigh numbers</u>

<u>against Taylor numbers T^2 in the case $P_r \geq 1, P_c \geq 1$.</u>

T^2	R_c^{*2}	R_c^2
0	657.511	657.511
10	677.1	677.1
50	748.3	748.3
10^2	826.3	826.3
5×10^2	1275	1275
10^3	1676	1676
2×10^3	2299	2299
5×10^3	3670	3670
7792	4675	4675
10^4	5366	5377
10^5	18151	21310
10^6	58636	92220
10^7	186679	4.147×10^5
10^8	591592	1.897×10^6

<u>References</u>

[1] Chandrasekhar, S.:*Hydrodynamic and hydromagnetic stability*, Oxford: Clarendon Press, 1961.

[2] Davis, S.H.; von Kerczeck, C.,*Arch. Rational Mech. Anal.*, <u>52</u> (1973).

[3] Galdi, G.P., *Arch. Rational Mech. Anal.* <u>87</u>,(1985).

[4] Galdi, G.P.: Nonlinear Energy methods in fluidmechanics (*in this volume*).

[5] Galdi, G.P.; Rionero, S. : *Weighted Energy methods in Fluid Dynamics and Elasticity*, Lecture Notes in Mathematics 1134, Springer - Verlag, 1985.

[6] Galdi, G.P.; Straughan B., *Proc.Soc. Lond. A* 402 (1985).

[7] Joseph, D.D.: *Stability of fluid motions* (2 vols.) Springer Tracts in Natural Phylosophy, vols. 27 and 28. Berlin: Springer - Verlag, 1976.

[8] Mulone, G.: On he nonlinear stability of parallel shear flows (*in this volume*).

[9] Mulone, G.; Rionero, S.:On the non-linear stability of the rotating Bénard problem via the Lyapunov direct method *J. Mat.Anal App.*,(*to appear*).

[10] Pearlstein,A.J., *J.Fluid Mech.*,103 (1981).

[11] Rionero, S., *Ann. Mat. Pura App.*, 78 (1968).

[12] Rionero, S.: On the choice of the Lyapunov function in the fluid motion stability, *lecture given at the meeting* "Energy stability and convection",Capri (1986).

[13] Rionero, S.; Mulone, G.:A nonlinear stability analysis of the magnetic Bénard problem via the Lyapunov direct method, *Arch.Rational Mech.Anal*,(*to appear*).

[14] Rionero, S.; Mulone, G.: On the stability of a mixture in a rotating layer via the Lyapunov second method, *ZAMM*, (*to appear*).

[15] Serrin, J., *Arch.Rational Mech.Anal.*,3 (1959).

Dipartimento di Matematica ed Applicazioni
"R.Caccioppoli"dell'Università di Napoli
via Mezzocannone 8,   80134 NAPOLI (Italy)

NUMERICAL ANALYSIS OF NONSTATIONARY FLUID FLOW

Rolf Rannacher

Summary. This lecture serveys several aspects of the numerical approximation of the nonstationary Navier-Stokes equations. In particular, some essential criteria are discussed for choosing appropriate time stepping schemes. These are the "smoothing property" and the "global regularity property". The importance of these principles for computing flows, at least in the range of moderate Reynolds-numbers, can be supported by a rigorous theoretical analysis. For flows with higher Reynolds-numbers, additional numerical damping is necessary.

1. Discretization of the Navier-Stokes Problem

The flow of a viscid incompressible fluid is described by the well-known initial-boundary value problem of the Navier-Stokes equations,

$$u_t - \nu\Delta u + (u\cdot\nabla)u + \nabla p = f \ , \qquad \nabla\cdot u = 0 \ , \quad \text{in} \ \ \Omega\times(0,T) \ ,$$

(1)

$$u\big|_{\partial\Omega} = g \ , \qquad u\big|_{t=0} = a \ .$$

Here, Ω is a bounded region in $\mathbb{R}^d$, d=2 or d=3 , with sufficiently regular boundary $\partial\Omega$, and $u=u(x,t)$ and $p=p(x,t)$ are the unknown velocity and pressure, respectively. The prescribed data are the viscosity, ν , the volume force, f , the boundary velocity, b , and the initial velocity, a . As usual, the density of the fluid has been normalized.

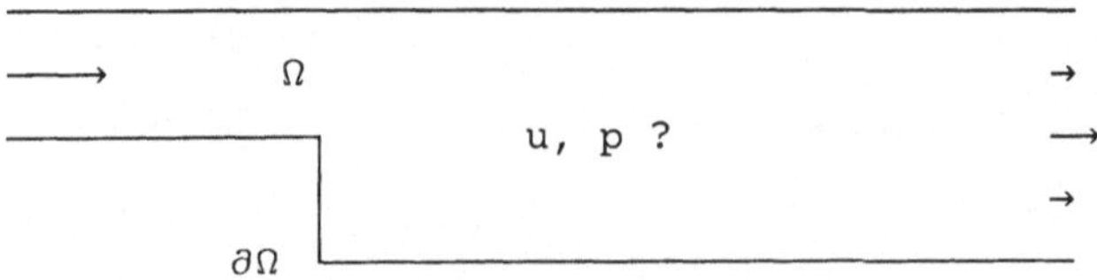

<u>Spatial discretization</u>: We suppose that problem (1) is discretized w.r.t. the spatial variables by any one of the standard (at least) second order methods:

$$\text{finite difference methods,}$$
$$\text{finite volume methods,}$$
$$\text{finite element methods.}$$

The parameter of the spatial mesh size is denoted by h. Let $U=U(t)$ and $P=P(t)$ be the vectors of the nodal values of the discrete velocity and pressure, respectively. Further, let M (mass matrix), A (stiffness matrix), B, and $N(u)$ denote the corresponding discrete analogues of the operators 1, $-\nu\Delta$, ∇, and $u\cdot\nabla$. Then, the spatially semi-discrete Navier-Stokes problem reads

$$(2) \qquad M\dot{U} + AU + N(U)U + BP = F , \qquad B^T U = 0 , \qquad t\in[0,T) .$$

As usual, the nonhomogeneous boundary data are thought to be incorporated into the equation. The initial value $U(0)$ is assumed to be a proper approximation to a. Problem (2) is a large system of ODE's which, in view of the constraint on U, is of a non-standard type. For further details on the spatial discretization of problem (1) we refer to [3], [5; Part I], and the literature cited therein.

<u>Time discretization</u>. For the solution of problem (2) we may consider any of the well-known discretization methods for ordinary differential equations. In choosing a particular method some problems inherent to (2) must be observed:

stiffness (requires A-stable methods),
nonlinearity $N(U)U$ (requires iterative methods),
constraint $B^T U = 0$ (saddle point problem),
$\nu\ll 1$ (requires numerical damping).

Further, the choice should be based on a rigorous analysis of the schemes, which has to take into account the following basic goals in flow modelling:

steady state approximation ("pseudo-time" stepping method)
tracking of periodic solutions ("forced" and "free")
long time simulation ($T\gg 1$) .

2. Some Theoretical Considerations

We begin with some basic theoretical considerations about the
Navier-Stokes problem which are of particular relevance for the
performance of discretization methods. For technical details
and proofs we refer to [5], and the literature cited therein.

a) Existence of solutions:

In two, and, for "small data", also in three dimensions it is
well-known that the Navier-Stokes problem (1) has a uniquely
determined strong (or classical) solution $u(t)$, $p(t)$, which
exists for all $t \geq 0$ (see [11], [16]). For general data in three
dimensions, however, the *global* existence of a strong solution
is still an open problem. Hence, rather than requiring the
domain to be two-dimensional, or the data to be "small", we
like to make the *global* existence of a (unique) strong solution
an assumption. Then, in the case of uniformly bounded data,
this solution stays also bounded as $t \to \infty$, i.e., particularly
the velocity has bounded rotation,

$$(3) \qquad \sup_{t \geq 0} \| \nabla \times u(t) \| < \infty \ .$$

b) Behavior at $t=0$:

Suppose that the data a, g, and f are smooth for all $t \geq 0$.
Then the assumption that $u(t)$ and $p(t)$ are also smooth, as
$t \to 0$, implies that

$$\nabla \cdot [\nabla p] = \nabla \cdot [f + \nu \Delta u - u_t - (u \cdot \nabla)u] \ , \quad \text{in} \ \ \Omega \ ,$$

$$\downarrow \ (t \to 0)$$

$$(4) \qquad \Delta p_o = \nabla \cdot [f_o - (a \cdot \nabla)a] \ , \quad \text{in} \ \ \Omega \ ,$$

$$n \cdot [\nabla p] = n \cdot [f + \nu \Delta u - u_t - (u \cdot \nabla)u] \ , \quad \text{on} \ \ \partial\Omega \ ,$$

$$\downarrow \ (t \to 0)$$

$$(5) \qquad \partial_n p_o = n \cdot [f_o + \nu \Delta a - b_t \big|_{t=0} - (a \cdot \nabla)a] \ , \quad \text{on} \ \ \partial\Omega \ .$$

The "Neumann problem" (4) and (5) is uniquely solvable in the
Sobolev quotient-space $H^1(\Omega)/\mathbb{R}$.

By an argument, analogous to that which led to (5), one obtains that its solution p_o necessarily satisfies

$$(6) \qquad \partial_\tau p_o = \tau \cdot [f_o + \nu \Delta a - b_t|_{t=0} - (a \cdot \nabla)a] , \qquad \text{on } \partial\Omega ,$$

for any tangential unit vector τ on $\partial\Omega$. This amounts to a compatibility condition for the data of the problem. Because of its non-local character, it is unverifiable in practice, and is rarely satisfied, even if $f \equiv 0$, $b \equiv 0$, and $a \in C_o^\infty(\Omega)$, $\nabla \cdot a \equiv 0$. Hence any practical choice of the initial data for a nonstationary flow simulation (say from experimental data) necessarily leads to an incompatibility at $t=0$, meaning that, e.g.,

$$(7) \qquad \overline{\lim_{t \to 0}} \, \|\nabla u_t\|_{L^2} = \infty .$$

In general, without this non-local compatibility, there holds

$$(8) \qquad \overline{\lim_{t \to 0}} \, \{\|\Delta u\|_{L^2} + \|u_t\|_{L^2}\} < \infty .$$

Furthermore, the blow up of higher order norms of $u(t)$, as $t \to 0$, can be described in terms of time weighted estimates of the form

$$(9) \qquad \overline{\lim_{t \to 0}} \, \{\sqrt{t}\|\nabla u_t\|_{L^2} + t\|u_{tt}\|_{L^2}\} < \infty .$$

These observations raise the following fundamental question:

Under what conditions does any given (at least) second order discretization scheme possess a "smoothing property" which, even in the case of "rough" initial data, allows for an error estimate of the form

$$(10) \qquad \|U_n - u(t_n)\|_{L^o} \leq K(t_n)\left\{h^2 + k^2\right\} , \qquad t_n > 0 ?$$

Here the error constant $K(t)$ may blow up as $t \to 0$. The rigorous theoretical analysis of discretization schemes for the non-stationary Navier-Stokes problem should be based on estimates like (10). Up to date this has been accomplished only for the finite element Galerkin method (of orders $m \leq 5$) and for the Crank-Nicolson scheme; see [5, Parts III and IV]. The very technical proofs are entirely by energy methods.

c) <u>Behavior as $t \to \infty$</u>:

In general, the constant in (10) behaves like $K(t) \sim e^{\alpha t}$, as $t \to \infty$, because of the instability of the solution. This makes the error estimate (10) meaningless for large t . Global error bounds for "long time" calculations necessarily require the sta bility of the solution. For "small" data the global classical solution u is known to be "universally energy stable", i.e., any instant perturbation decays exponentially. For arbitrary large data, the observability of a (laminar) fow over long periods of time indicates that it may also possess a certain stabiliy property. The simplest notion of such a property is that of "(conditionally) exponential stability":

There exist positive constants δ, α, A , such that, for any $t_0 \geq 0$, and for any perturbation w_0 , $\|\|w_0\|\| < \delta$, the corresponding perturbed solution $v(t)$, starting at time t_0 with the initial value $v_0 = u(t_0) + w_0$, satisfies

$$(11) \qquad \|\|(v-u)(t)\|\| \; \leq \; Ae^{-\alpha(t-t_0)}\|\|w_0\|\| \; , \qquad t \geq t_0 \; .$$

Here, $\|\|\cdot\|\|$ stands for the L^2-norm or for the H^1-norm. Equivalent concepts are obtained w.r.t. Sobolev norms of any order, provided that the data are sufficiently smooth. In [; Part II] it is shown that the exponential stability of the solution u guarantees error estimates uniform in time of the form

$$(12) \qquad \sup_{t_n \geq 1} \; \|U_n - u(t_n)\|_{L^2} \; \leq \; K\{h^2 + k^2\} \; .$$

The proof is by induction w.r.t. time using the "local" smoothing error estimate (10) over a sequence of time intervals, of unit length, tending to infinity. For this it is essential that the scheme under consideration possesses an appropriate damping property which may be expressed in terms of the estimate:

$$(13) \qquad \|U_n\|_{H^2} \; \leq \; c\left\{\|U_0\|_{H^2} + \sup_{0 \leq \mu \leq n} \|U_\mu\|_{H^1} + K(g,f)\right\} \; .$$

The proof of (13) is based on the "global damping property" of a time stepping scheme (see Section 4, below).

As a by-product, the proof of (12) also yields the following result for the approximation of a stationary solution u_∞ of problem (1), (2) by the pseudo-time stepping method:

$$(14) \qquad \|U_n - u_\infty\|_{L^2} \;\leq\; K_1 \exp(-\alpha t_n)\Big\{\|U_0 - u_\infty\|_{L^2} + k^2\Big\} + K_2 h^2 \;.$$

Hence, the "real time" needed for reaching a stationary limit does not very much depend on the order of the scheme.

For describing real nonstationary fluid flow the above concept of exponential stability is certainly much to restrictive. In [5; Part II] and [6] weaker concepts of stability are introduced. One is that of "quasi-exponential stability (w.r.t. time shifts)" which is thought to apply to time periodic flows which are stable only w.r.t. a perturbation of the amplitude of the oscillation but not w.r.t. its phase. An important feature of this definition is that the time shifts s are required to be bounded in terms of the initial perturbation, $|s| \leq B\|w_0\|$. Some tests on a simple nonlinear dynamical system show that this proportional bound is crucial for a global approximation of optimal order ([7]) . In some case the observed flow may have an overall stable appearence, but with some local nonstationary fluttering (turbulence). One may model this situation by the concept of "contractive stability to a tolerance", which, roughly speaking, means that instantly small perturbations, $\|w_0\| < \delta$, decay below some number $\rho < \delta$, within a fixed span of time. Assuming this stability property for u , one obtains global error estimates "to a tolerance":

$$(15) \qquad \sup_{t_n \geq 1} \|U_n - u(t_n)\| \;<\; K(h^2 + k^2) + 2\rho \;.$$

In turn, if under certain additional conditions, the discrete solution $\{U_n\}$ exhibits this kind of stability, then this guarantees the existence (and stability) of a closely neighboring continuous solution. This result supports the belief that the results of numerical flow simulations actually represent solutions of the Navier-Stokes problem, i.e., may tell us something about real flows.

3. Some Preliminaries on Time Stepping Schemes

We recall some basic properties of time stepping schemes which are well-known from the numerical analysis of ordinary differential equations (see, e.g., [12], [2]). For this it is most instructive to look first at the scalar linear "test equation".

$$(16) \qquad \dot{x}(t) + \lambda x(t) = 0 , \qquad t \geq 0 ,$$

where $\lambda \in \mathbb{C}$, $\mathrm{Re}\,\lambda \geq 0$. A time stepping scheme applied to (16), with constant length k of the time step, generates a sequence of values $x_n \sim x(t_n)$, $t_n = nk$. The behavior of the scheme as $t \to \infty$, depending on the parameter λ , is usually characterized by the "amplification factor" $\omega = \omega(\lambda k)$. In particular, for one-step schemes there holds $x_n = \omega^n x_o$. In terms of ω , we can formulate the following desirable properties of time stepping schemes:

1) (local) stability: $\qquad |\omega| \leq 1$ (A-stability),

2) (global) regularity: $\qquad \overline{\lim_{\mathrm{Re}\lambda \to \infty}} |\omega| \leq 1 - \mathcal{O}(k)$,

3) smoothing property: $\qquad \overline{\lim_{\mathrm{Re}\lambda \to \infty}} |\omega| \leq 1 - \delta$ (strong A-stability),

4) non dissipative: $\qquad |\omega| \sim 1$, for $\mathrm{Re}\,\lambda = 0$.

The following schemes are given in a form which applies to the general linear evolution equation

$$(17) \qquad u_t + A(t)u = f(t) , \qquad A_n = A(t_n) , \quad f_n = f(t_n) .$$

Nevertheless, their approximation properties can be expressed in terms of the amplification factor $\omega(\lambda k)$. Carrying these conclusions over to a non-selfadjoint, non-autonomous, and possibly weakly non-linear problem (e.g., the Navier-Strokes problem) is the subject of a growing number of research papers; see [5], [17], and the literature cited therein.

(I) One-step θ-schemes:

$$[I+\theta k A_{n+1}]u_{n+1} = [I-(1-\theta)k A_n]u_n + \theta k f_{n+1} + (1-\theta)k f_n ,$$

$$\omega(z) = \frac{1-(1-\theta)z}{1+\theta z} .$$

The explicit Euler scheme ($\theta=0$) is only conditionally stable, ($k\leq 1/\lambda$) and strongly amplifies free oscillations, $|\omega(ik)|>1$. Its implicit counterpart ($\theta=1$) is strongly A-stable, $\omega(z)\to 0$ as $Re\lambda\to\infty$, but it tends to damp out free oscillations, $|\omega(ik)|<1$ ($|\omega|=.995$ for $k=.1$). Both schemes are of _first_ order. The Crank-Nicolson scheme ($\theta=1/2$) is of _second_ order, but has only a low damping property, $|\omega(z)|\to 1$ as $Re\lambda\to\infty$. However, free oscillations are very well preserved, $|\omega(ik)|=1$. In order to enhance the damping property of the Crank-Nicolson scheme the following strategies may be used ([14], [15], [5; Part IV]):

a) The implicit shift to $\omega = \frac{1+k}{2}$ yields $\varlimsup_{Re\lambda\to\infty} |\omega| \leq 1-k$.

b) Supplementing the scheme by implicit Euler steps, one per unit of time, recovers the full damping property.

Both modifications (a) and (b) preserve the second order accuracy of the scheme.

(II) <u>Backward differencing two-step scheme</u>:

$$[I+\tfrac{2}{3}kA_{n+1}]u_{n+1} = \tfrac{4}{3}u_n - \tfrac{1}{3}u_{n-1} + \tfrac{2}{3}kf_{n+1} .$$

This scheme has a truncation error of order two.

$$\omega(z) = \frac{2\pm\sqrt{1+2z}}{3+2z} , \qquad \varlimsup_{Re\lambda\to\infty} |\omega| = 0 ,$$

$$\omega(ik)| < 1 \qquad (|\omega|\sim.995 \quad \text{for} \quad k=.4) .$$

(III) <u>(Diagonally) implicit Runge-Kutta scheme</u>:

$$u_{n+1} = u_n + k[(1-\theta)K^{(1)}+\theta K^{(2)}] , \qquad \theta = 1-\sqrt{2}/2 \sim .29289$$

$$K^{(1)} = f_n - A_n[u_n+\theta kK^{(1)}] ,$$

$$K^{(2)} = f_{n+1} - A_{n+1}[u_n+(1-\theta)kK^{(1)}+\theta kK^{(2)}] .$$

This scheme also has a truncation error of order two.

$$\omega(z) = \frac{1-\theta'z}{(1+\theta z)^2} \qquad (\theta'=1-2\theta), \qquad \varlimsup_{Re\lambda\to\infty} |\omega| = 0 ,$$

$$|\omega(ik)| < 1 \quad (|\omega|\sim.998 \quad \text{for} \quad k=.8) .$$

(IV) **(Fractional step) θ-scheme:**

Choosing $\theta \in (0,1)$, $\theta'=1-2\theta$, and $\alpha \in [0,1]$, $\beta=1-\alpha$, the time step $t_n \to t_{n+1}$ is split into three substeps as follows:

$$(I+\alpha\theta kA_{n+\theta})u_{n+\theta} = (I-\beta\theta kA_n)u_n + \theta kf_n \ ,$$

$$(I+\beta\theta'kA_{n+1-\theta})u_{n+1-\theta} = (I-\alpha\theta'kA_{n+\theta})u_{n+\theta} + \theta'kf_{n+1-\theta} \ ,$$

$$(I+\alpha\theta kA_{n+1})u_{n+1} = (I-\beta\theta kA_{n+1-\theta})u_{n+1-\theta} + \theta kf_{n+1-\theta} \ .$$

For the special choice $\theta = 1-\sqrt{2}/2$, this scheme has truncation error of order two. If one takes $\alpha=(1-2\theta)/(1-\theta)$, $\beta=2\theta/(1-\theta)$, then the coefficient matrices in all substeps are the same.

$$\omega(z) = \frac{(1-\beta\theta z)^2(1-\alpha\theta'z)}{(1+\alpha\theta z)^2(1+\beta\theta'z)} \ , \qquad \overline{\lim_{\mathrm{Re}\lambda\to\infty}} |\omega| = \frac{\beta}{\alpha} \ ,$$

$$|\omega(ik)| < 1 \qquad (|\omega|\sim.9998 \quad \text{for} \quad k=.8) \ .$$

The damping properties of the various schemes listed above can easily be visualized at the following simple test problem,

$$x: [0,T] \to \mathbb{R}^4 \ , \qquad \dot{x}(t) + Ax(t) = 0 \ , \qquad x(0)=0 \ .$$

The (nondiagonal) 4×4-matrix A is constructed to have the eigenvalues

$$\lambda_1 = 10^\gamma \ , \quad \lambda_2 = i \ , \quad \lambda_3 = -i \ , \quad \lambda_4 = 1 \ .$$

Hence, the solution contains a "stiff" component for $\gamma\gg1$ (rapid exponential decay), and two periodic components ("free" oscillations).

The following two groups of plots show how the schemes manage to represent the "free" sine-oscillation in the "non-stiff" case $\gamma=1$, first for small step size $k\sim.05$, and then also for larger step size $k\sim.4$. The third group of plots shows the performance of the schemes for the "stiff" case $\gamma=4$, which is thought to model "rough" initial data. The actual error in the discretization is indicated by shading.

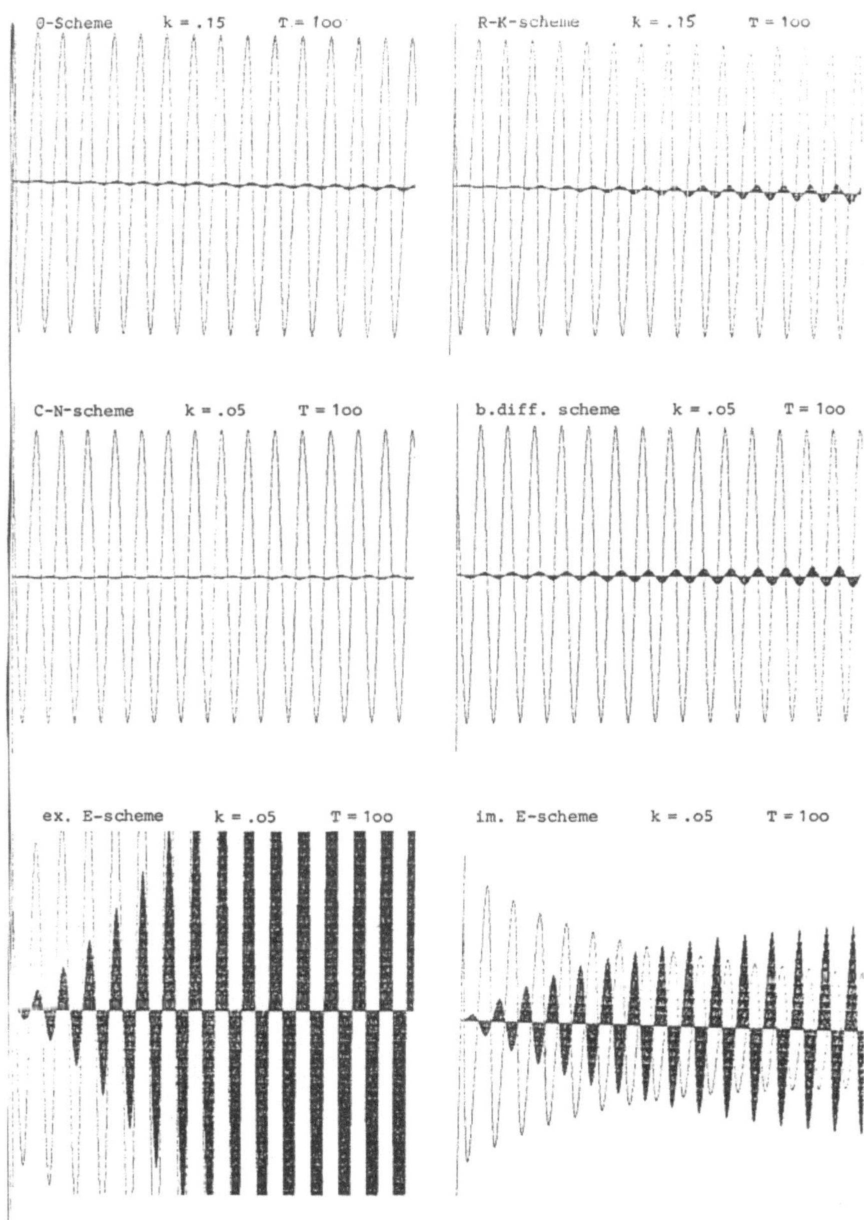
θ-Scheme   k = .15   T = 1oo
R-K-scheme   k = .15   T = 1oo
C-N-scheme   k = .o5   T = 1oo
b.diff. scheme   k = .o5   T = 1oo
ex. E-scheme   k = .o5   T = 1oo
im. E-scheme   k = .o5   T = 1oo

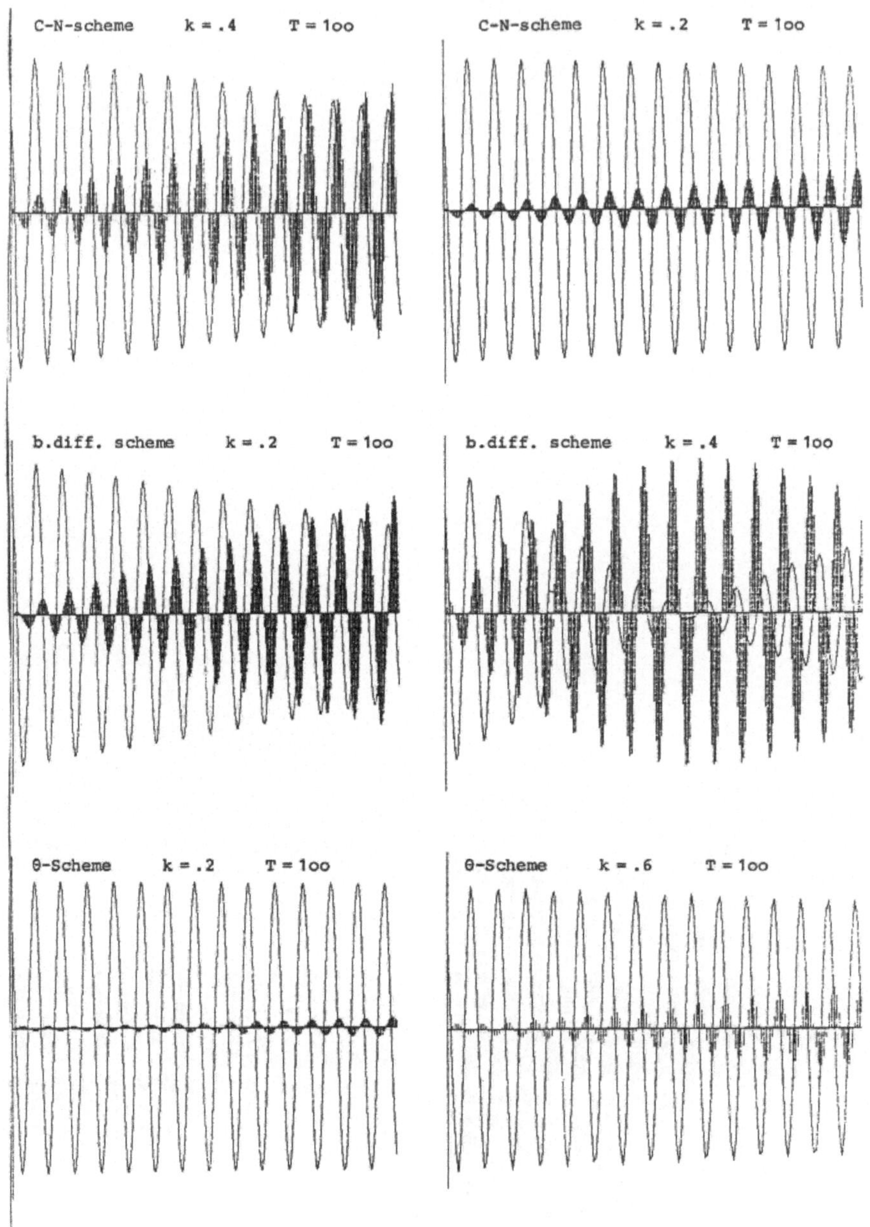

C-N-scheme   k = .4   T = 1oo
C-N-scheme   k = .2   T = 1oo
b.diff. scheme   k = .2   T = 1oo
b.diff. scheme   k = .4   T = 1oo
θ-Scheme   k = .2   T = 1oo
θ-Scheme   k = .6   T = 1oo

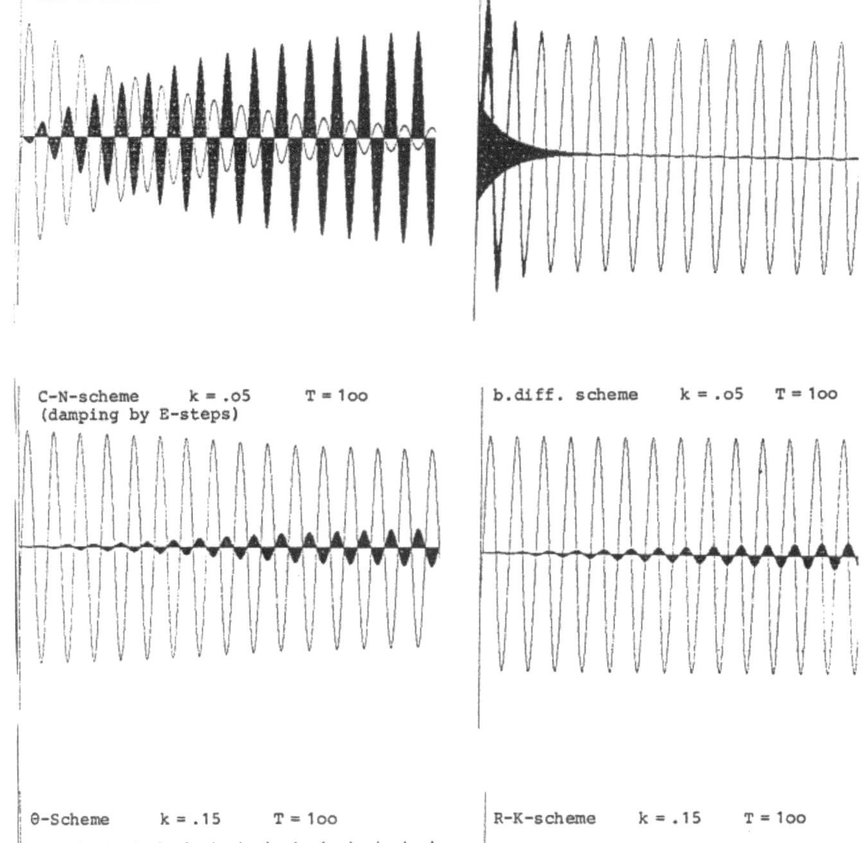
im. E-scheme   k = .o5   T = 1oo
C-N-scheme   k = .o5   T = 1oo
C-N-scheme   k = .o5   T = 1oo
(damping by E-steps)
b.diff. scheme   k = .o5   T = 1oo
θ-Scheme   k = .15   T = 1oo
R-K-scheme   k = .15   T = 1oo

These results clearly show the superiority of the (fractional step) θ-scheme over all the other methods including the Crank-Nicolson scheme. It has good damping properties (for $1/2 < \alpha \leq 1$) and low numerical dissipation. Therefore, only the application of the θ-scheme to the Navier-Stokes problem will be discussed below, although the Crank-Nicolson scheme seems still to be the most frequently used method in practice.

4. The θ-Scheme for the Navier-Stokes Problem

The use of the (fractional step) θ-method for solving the non-stationary Navier-Stokes equations was propagated in a series of papers by R.Glowinski and his group (see, e.g., Bristeau/ Glowinski/Periaux [1], and Glowinski/Periaux [4]). This scheme has good damping properties and low numerical dissipation. In particular, it allows to separate the two major difficulties, "nonlinearity" and "incompressibility constraint".

Choosing again $\theta = 1 - \sqrt{2}/2$, $\theta' = 1 - 2\theta$, $\alpha \in (1/2, 1]$, $\beta = 1 - \alpha$, the θ-scheme applied to the Navier-Stokes problem reads as folows:

$$U_o \quad \text{starting value,} \quad B^T U_o = 0 \; ;$$

1st step: (generalized) Stokes problem

$$[M + \theta k \alpha A] U_{n+\theta} + \theta k B P_{n+\theta} = [M - \theta k \beta A] U_n + \theta k F_n - \theta k N(U_n) U_n \;,$$

$$B^T U_{n+\theta} = 0 \;,$$

2nd step: nonlinear problem (Burgers equation)

$$[M + \theta' k \beta A] U_{n+1-\theta} + \theta' k N(U_{n+1-\theta}) U_{n+1-\theta} =$$

$$= [M - \theta' k \alpha A] U_{n+\theta} + \theta' k F_n - \theta' k B P_{n+\theta} \;,$$

3rd step: (generalized) Stokes problem

$$[M + \theta k \alpha A] U_{n+1} + \theta k B P_{n+1} = [M - \theta k \beta A] U_{n+1-\theta} + \theta k F_{n+1} -$$
$$- \theta k N(U_{n+1-\theta}) U_{n+1-\theta} \;,$$

$$B^T U_{n+1} = 0 \;.$$

A straightforeward calculation shows that the truncation error of this discretization is of the order two. In fact, it seems to work very well for numerical flow simulation in the range of moderate Reynolds-numbers. Although, for the linear model problem (16), the theoretical analysis of the (fractional step) θ-scheme is easily carried out by spectral arguments, only very little is known about this method in the general (nonlinear) case. Neither its stability nor any kind of error estimate has apparently been given in the literature yet. In view of the excellent performance of this scheme at the test problem (16), it is strongly suspected that its nice properties carry over also to the situation of the Navier-Stokes equations.

<u>Solution of the partial problems</u>:

<u>Steps 1 and 3</u>: "fast" Stokes solver

$$(18) \qquad [M+\theta k\alpha A]U + \theta kBP = b , \qquad B^T U = 0 .$$

The elimination of the variable U results in the reduced problem

$$(19) \qquad \mathscr{A}P \equiv B^T[M+\theta k\alpha A]^{-1}BP = B^T[M+\theta k\alpha A]^{-1}b$$

This compact system may be solved by the conjugate gradient method. This is particularly attractive, as the condition number of the matrix $\mathscr{A}$ behaves only like $\mathrm{cond}(\mathscr{A})= O(h^{-1})$. The evaluation of $[M+\theta kA]^{-1}y$ in each iteration step requires the solution of several Poisson problems which, in turn, is achived efficiently by using multi-grid techniques. If, additionally, the cg-iteration is preconditioned by solving a Neumann problem for P in each step, then one even obtains the conditioning $O(1)$, independent of the size of the domain Ω . An alternative to this "nested" solution procedure would be the direct application of the multi-grid approach to the Stokes problem. Several promising attempts in this direction have already been made.

Step 2:   **Least-square-cg method (and/or Newton method)**

(20) $[M+\theta'k\beta A]U + \theta'kN(U)U = b$.

The least-square-cg iteration described in Glowinsi/Periaux [4]
also requires the "fast" solution of several Poisson problems
in each step. It seems to be a very robust method, which works
well for the range of moderate Reynolds-number. Alternatively,
one may use the Newton method in order to achieve a faster
convergence to the limit. This , however, needs very good
starting values at each time step and, therefore, is not as
robust as the cg-method. Furthermore, the Newton iteration
requires the "fast" solution of *nonsymmetric* Poisson problems
at each step.

5. Numerical Damping

In computing flows with higher Reynolds-numbers a certain
amount of numerical damping is necessary in order to suppress
spurious oscillations in the solutions. Traditional techniques
are the "(spatial) upwind discretization" of the transport
terms or the use of "artificial viscosity". Both methods (in
their original form) are only first order accurate and tend to
smear out sharp boundary layers and interior "shocks". Below,
two more sophisticated approaches are described which may prove
to even manage the stable transition for Re $\to \infty$.

a) Backward characteristic method:

This approach is also known as "transport-diffusion algorithm"
(Pironneau [13]). It may be interpreted as an upwind discretization in space-time. The upwind direction in each time step is
locally given by the "characteristic flow" $\chi(\tau)=\chi(x,t;\tau)$
which, for fixed $(x,t)\in\Omega\times[0,T]$, is determined through the
initial value problem

(21) $\frac{d\chi}{d\tau}(\tau) = u(\chi(\tau),\tau)$, $\chi(t)=x$.

Using this notation, the Lagrangian time derivative at $\tau=t$ is

49

$$(22) \qquad \frac{d}{d\tau} u(x(\tau),\tau)\Big|_{\tau=t} = \left\{ \frac{\partial u}{\partial t} + (u\cdot\nabla)u \right\}(x,t) \ .$$

On the basis of (22) the step $t_n \to t_{n+1}$ of the "characteristic backward Euler scheme" applied to the continuous Navier-Stokes problem has the form

$$(23) \qquad \frac{1}{k}\left\{ u\Big|_{\tau=t_{n+1}} - u\Big|_{\tau=t_n} \right\} = f_{n+1} + \nu\Delta u_{n+1} - \nabla p_{n+1} \ .$$

The combination of this time stepping method with the spatial discretization reads as follows:

<u>Step 1</u>: For all (mesh points) $x\in\Omega$ compute the flow $x_n(x;\tau)$ from the initial value problem

$$\frac{d}{d\tau} x_n(x;\tau) = U_n(x_n(x;\tau)) \ , \quad \tau\in[t_n,t_{n+1}], \quad x_n(x;t_{n+1})=x \ .$$

<u>Step 2</u>: Evaluate $x_n(x) = x_n(x;t_n)$, $x\in\Omega$.

<u>Step 3</u>: Solve the (generalized) Stokes problem

$$(M+kA)U_{n+1} + kBP_{n+1} = MU_n(x_n(\cdot)) + kF_{n+1} \ , \qquad B^TU_{n+1} = 0 \ .$$

(If $x_n(x;\tau_*)\in\partial\Omega$, for some $\tau_*>t_n$, this formula has to be modified taking into account the boundary conditions.)

This algorithm is conservative and unconditionally stable, as $\nu\to 0$. Further, assuming the solution u to be sufficiently smooth, the error behavior was shown to be

$$(24) \qquad \|U_n-u(t_n)\|_{L^2} = \mathcal{O}(h + k + \frac{h^2}{k}) \ ,$$

uniformly for $\nu>0$ (Pironneau [13]). However, there are still many open questions:

a) What is the effect of numerical integration in evaluating the integrals $(U_n(x_n(\cdot)),\varphi)_{L^2}$?

b) Can one work with finite elements which are "divergence free" only in some discrete sense?

c) The known proof of convergence requires a high degree of regularity of u (independent of ν !). Can one get rid of this impractical assumption?

The major practical problem is the efficient computation of the flow $x_n(\cdot)$ and the evaluation of $U_n(x_n(\cdot))$. The standard approach is to construct, f.e., from a piecewise linear "stream function" ψ_n a piecewise constant approximation $U_n'=\nabla\times\psi_n \sim U_n$ satisfying

$$(25) \qquad \|U_n'-U_n\|_{L^\infty} = \mathcal{O}(h) \ , \qquad \nabla\cdot U_n'= 0 \ (\text{piecewise}).$$

Then, a flow $x_n'(\tau)$ is computed from the equations

$$(26) \qquad \frac{d}{d\tau}x_n'(\tau) = U_n'(x_n'(\tau)) \ , \qquad x_n'(t_{n+1})=x \ .$$

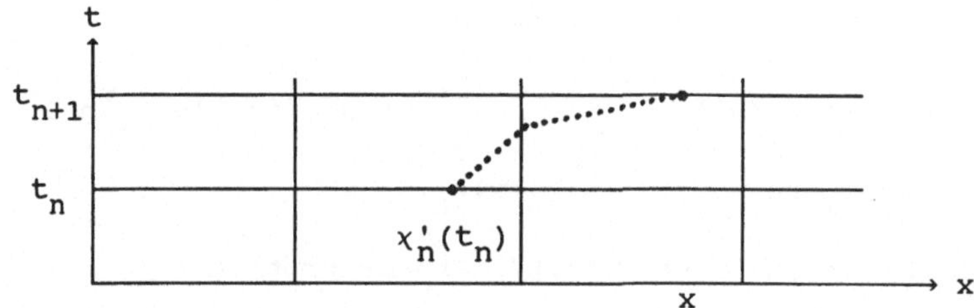

b) <u>Streamline diffusion method</u>:

The idea of (artificial) "streamline diffusion" is very simple. Instead of using a damping term acting in all directions, one would like to have the damping only in the flow direction, without necessarily reducing the order of consistency of the discretization. This approach was systematically developed and also analysed in a series of papers by Hughes, Johnson, e.al. (see Hughes/Franka/Mallet [8], Johnson [9], Johnson/Saranen [10], and the literature cited therein).

Consider first the scalar model equation

$$(27) \qquad u_t + (\underline{\beta}\cdot\nabla)u = f \qquad (\underline{\beta} \equiv \text{const.}) \ ,$$

together with appropriate boundary conditions. The standard Galerkin appoximation of equation (27), with piecewise linear finte elements, is numerically unstable for $h >\!>1/|\beta|$, and of only first order accuracy, $\mathcal{O}(h)$, as $h\to0$. Any sufficiently smooth solution of (27) automatically satisfies the equation

$$\nabla\cdot[\underline{\beta}u_t + \underline{\beta}(\underline{\beta}\cdot\nabla)u] = (\nabla\cdot\underline{\beta})f \; ,$$

and, consequently, with any $\delta>0$,

$$(28) \qquad u_t + (\underline{\beta}\cdot\nabla)u - \delta\nabla\cdot[\underline{\beta}u_t + \underline{\beta}(\underline{\beta}\cdot\nabla)u] = f - \delta(\nabla\cdot\underline{\beta})f \; .$$

The variational formulation of equation (28),

$$(u_t+(\underline{\beta}\cdot\nabla)u,\varphi) + \delta(u_t,(\underline{\beta}\cdot\nabla)\varphi) + \delta((\underline{\beta}\cdot\nabla)u,(\underline{\beta}\cdot\nabla)\varphi) +$$
$$(29)$$
$$+ \text{"boundary terms"} = (f,\varphi+(\underline{\beta}\cdot\nabla)\varphi) \; ,$$

contains a damping term (bold face) which acts only in the flow direction. For $\delta\sim h$, the standard Galerkin approximation of (29) is stable for all $h>0$, and has the improved order of convergence $O(h^{3/2})$. The resulting system of ODS's may then be solved by any time stepping scheme, e.g., the Crank-Nicolson scheme.

In contrast to steamline diffusion only "in space", described above, one may also introduce the damping simultaneously "in space-time" leading to variational equations of the form

$$(u_t+(\underline{\beta}\cdot\nabla)u,\varphi+\delta[\varphi_t+(\underline{\beta}\cdot\nabla)\varphi]) + \text{"boundary terms"}$$
$$(30)$$
$$= (f,\varphi+\delta[\varphi_t+(\underline{\beta}\cdot\nabla)\varphi]) \; ,$$

with time dependent test functions. This results in a Galerkin method which is global in the space-time region $\Omega\times[0,T]$. However, through the use of a "discontinuous" finite element discretization (piecewise linear) w.r.t. time, this scheme can be converted into a time marching procedure (see Johnson [9]). This method has good theoretical performance. But compared with the simpler diffusion "in space", it suffers from very high computational costs, since at least two solution vectors have to be carried at each time step. (For the two-dimensional compressible Euler equations the work ratio of the two approaches may be about 1:6 ; see Walter [17].)

The application of the streamline diffusion method to the diffusive Navier-Stokes problem is under progress (see Johnson/Saranen [10]). Here, the damping may be introduced through the use of the test function $\delta[\varphi_t+(u\cdot\nabla)\varphi+\nabla q]$. The major problem is the presence of the diffusive term $-\nu\delta(\Delta u,\ldots)_{L^2}$ which, in the conventional Galerkin approximation, requires the use of trial functions with an undesirably high degree of smoothness. The remedy could be to work with a "mixed" variational formulation of the Navier-Stokes problem having u, p, and $\omega \equiv \nabla\times u$ as unknowns. This approach admits some theoretical analysis but it still needs to be tested at practical problems.

<u>References</u>.

[1]   Bristeau,M.O.; Glowinski,R.; Periaux,J.: Numerical methods for the Navier-Stokes equations. Applications to the simulation of compressible and incompressible viscous flows, Research Report UH/MD-4, Febr. 1987, to appear in Computer Physics Report

[2]   Gekeler,E.: Discretization Methods for Stable Initial Value Problems, Springer, LNM Vol. 1044, 1984

[3]   Girault,V.; Raviart,P.A.: Finite Element Methods for Navier-Stokes Equations, Springer 1986

[4]   Glowinski,R.; Periaux,J.: Numerical methods for nonlinear problems in fluid dynamics, Proc. Intern. Seminar on Scientific Supercomputers, Paris, Feb. 2-6, 1987, North-Holland

[5]   Heywood,J.G.; Rannacher,R.: Finite element approximation of the nonstationary Navier-Stokes problem:

I. Regularity of solutions and second order spatial discretization, SIAM J.Numer.Anal. 19 (1982) 275-311;

II. Stability of solutions and error estimates uniform in time, SIAM J.Numer.Anal. 23 (1986) 750-777;

III. Smoothing propertyyy and higher order estimates for spatial discretization, SIAM J.Numer.Anal. 25 (1988)

IV. Error analysis for second order time discretization, to appear in SIAM J.Numer.Anal.

[6]   Heywood,J.G.; Rannacher,R.: An analysis of stability concepts for the Navier-Stokes equations, J.Reine Angew.Math. 372 (1986) 1-33

[7]   Heywood,J.G.; Rannacher,R.: On the global numerical approximability of dynamical systems, Research report, University of Saarbrücken, August 1986

[8] Hughes,T.J.; Franca,L.; Mallet,L.: A new finite element
     formulaton for computational fluid dynamics:

     I. Symmetric forms of the compressible Euler and Navier-
     Stokes equations and the second law of thermodynamics,
     Comp.Meth. in Appl.Mech.Eng. 54 (1986) 223-234;
     (Parts II-V, ibidem)

     VI. Convergence analysis of the generalized SUPG formula-
     tion for linear time dependent multidimensional advection-
     diffusive systems, Comp.Meth. in Appl.Mech.Eng. 60 (1987)

[9] Johnson,C.: Numerical Solution of Partial Differential
     Equations by the Finite Element Method, Cambridge Univer-
     sity Press 1987

[10] Johnson,C.; Saranen,J.: Streamline diffusion methods for
     the incompressible Euler and Navier Stokes equations.
     Math. Comp. 47 (1986) 1-18

[11] Ladyshenskaya,O.A.: The Mathamtical Theory of Viscous
     Incompressible Flow, Gordon and Breach 1969

[12] Lambert,J.D.: Computational Methods in Ordinary Differen-
     tial Equations, John Wiley 1973

[13] Pironneau,O.: On the transport-diffusion algorithm and its
     applications to the Navier-Stokes equations, Numer.Math.
     38 (1982) 309-332

[14] Rannacher,R.: Finite element solution of diffusion prob-
     lems with irregular data. Numer. Math. 43 (1984) 309-327

[15] Rannacher, R.: On the stabilization of the Crank-Nicolson
     scheme for long time calculations, Research Report, Univ.
     of Saarbrücken, September 1986

[16] Temam,R.: Theory and Numerical Analysis of the Navier-
     Stokes Equations, North-Holland 1977

[17] Thomee,V.: Galerkin-Finite Element Methods for Parabolic
     Problems, Springer, LNM Vol. 1054, 1984

[18] Walter,A.: Ein Finite-Elemente-Verfahren zur numerischen
     Lösung von Erhaltungsgleichungen, PhD thesis, University
     of Saarbrücken, Saarbrücken 1988

Prof. Dr. R. Rannacher
Institut für Angewandte Mathematik
Universität Heidelberg
Im Neuenheimer Feld 293/294
D-6900 Heidelberg, Fed.Rep. of Germany.

ITERATION BY SUBDOMAINS IN NUMERICAL FLUID DYNAMICS

A.Quarteroni (*) (**) and G.Sacchi Landriani (**)

INTRODUCTION

Domain decomposition methods are well designed to face fluid dynamical problems in complex geometries within parallel computational environements. The subdomain decomposition approach allows the use of fast accurate solvers for Euler or Navier-Stokes equations within subregions of simple form. In the case of external compressible flows, a zonal subdomain method makes it possible the coupling of the compressible Navier-Stokes equations in the vicinity of the body with the inviscid Euler equations in that partion of the domain where viscous effects are negligible. From the physical point of view, consistency of the subdomain problem with the original one is ensured by enforcing suitable transmission of information between adjacent subregions (continuity of the velocity field, of the normal stresses,...) Mathematically such an exchange of information is handled by proper "capacitance" or "influence" interface operators, whose numerical approximation by suitable iterative methods with preconditioning can effectively be implemented on a parallel architecture. The optimality of these iterative procedures is fulfilled whenever their rate of convergence does not decrease as the number of numerical unknowns increases.

In this report we review optimal iterative procedures for multidomain approximations of a collection of linear problems. Namely we consider elliptic equations, compressible Stokes equations, hyperbolic systems and coupled hyperbolic-parabolic systems. The presentation of the methodology on these model examples results in a relatively simple fashion, while its extension to their nonlinear counterparts is much more involved, though inspired to the same considerations.

For each problem we start giving an equivalence principle according to which the original problem is shown to be equivalent with a new one obtained in a multidomain approach. It is precisely on this alternative formulation that our iteration by subdomain procedures are carried out.

In section one we consider the two dimensional Helmholtz equation. We define the so called "Dirichlet-Neumann" iteration algorithm and we discuss its application in the frame of both finite element and spectral collocation methods.

In section two we deal with the incompressible linear Stokes equations. On its finite element approximation we generalize the previous Dirichlet-Neumann algorithm. We also propose a spectral collocation approximation whose numerical resolution by the Uzawa algorithm leads to a cascade of Helmholtz-type equations. In turns the solution of the latter problems can strongly benefit from the multidomain method presented in section one.

(*) Dipartimento di Matematica, Università Cattolica, Via Trieste 17, 25121 Brescia, Italy
(**) Istituto di Analisi Numerica del C.N.R., Corso C.Alberto 5, 27100 Pavia, Italy

In section three we deal with linear hyperbolic systems. In this case the iteration by subdomain procedure involves the characteristic form of the system at the interfaces between subdomains.

Finally section four is devoted to the coupling between hyperbolic and parabolic systems in different regions of the computational domain. This coupling reflects the more complicated mechanism of the interaction between Euler and Navier-Stokes equations. Such an interaction frequently occurs as a mathematical model to simulate the motion of a compressible viscous flow whenever the viscous effects can be neglected in a region of the computational domain.

1. ELLIPTIC PROBLEMS

In this section we consider an elliptic model problem. Let Ω be an open two dimensional domain with a Lipschitz continuous boundary $\partial\Omega$. If $f \in L^2(\Omega)$ we denote by u the solution of the boundary-value-problem

$$(1.1) \qquad \begin{cases} -\Delta u + \alpha_0 u = f \text{ in } \Omega \\ u = 0 \qquad\qquad \text{ on } \partial\Omega \end{cases}$$

with $\alpha_0(x) \geq 0, \forall\, x \in \Omega, \alpha_0 \in C^0(\overline{\Omega})$.

The variational formulation of this problem is

$$(1.2) \qquad u \in H_0^1(\Omega) : \int_\Omega \nabla u \nabla v \, dx + \int_\Omega \alpha_0 \, u \, v \, dx = \int_\Omega f \, v \, dx \qquad \forall\, v \in H_0^1(\Omega)$$

It is well known that (1.2) ha a unique solution. As usual, for each integer $k \geq 1$ we have denoted with $H^k(\Omega)$ the space of those functions of $L^2(\Omega)$ whose (distributional) derivatives of order less than or equal to k belong to $L^2(\Omega)$. Moreover, $H_0^1(\Omega)$ is the subspace of $H^1(\Omega)$ of those functions vanishing on $\partial\Omega$ (see, e.g., [LM]).

We non consider a multidomain formulation of the variational problem (1.2). Let Ω be partitioned into non intersecting subdomains Ω_i, i=1,...,M, as in fig.1.1, and denote by Γ the common boundary between Ω_i and Ω_{i+1}. We set $\Gamma = \bigcup_{i=1}^{M-1} \Gamma_i$ and we denote by Φ the space of functions defined on Γ which are the traces of functions of $H_0^1(\Omega)$ (it is known that $\Phi = \prod_{i=1}^{M-1} H_{00}^{1/2}(\Gamma_i)$, see [LM]).

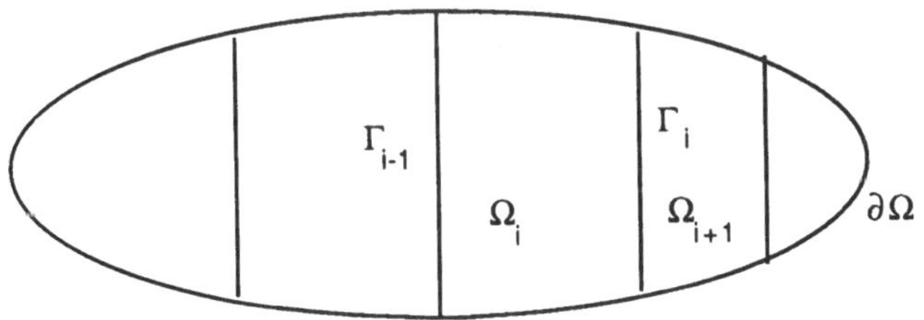

Fig.1.1 The partition of Ω

Problem (1.2) is equivalent to find $u \in H_0^1(\Omega)$ so that $u_i = u_{|\Omega_i}$, i=1,...,M are solutions to the problem

$$(1.3) \qquad \int_{\Omega_i} \nabla u_i \nabla v \, dx + \int_{\Omega_i} \alpha_0 \, u_i \, v \, dx = \int_{\Omega_i} f v \, dx \qquad \forall \, v \in H_0^1(\Omega_i) \qquad 1 \leq i \leq M$$

$$(1.4) \qquad u_i = u_{i+1} \quad \text{on } \Gamma_i \qquad 1 \leq i \leq M - 1$$

$$(1.5) \qquad u_i = 0 \quad \text{on } \partial\Omega \cap \partial\Omega_i \qquad 1 \leq i \leq M$$

(1.6)
$$\sum_{i=1}^{M} \left(\int_{\Omega_i} \nabla u_i \nabla \varphi_i^* dx + \int_{\Omega_i} \alpha_0 \, u_i \, \varphi_i^* dx \right) = \sum_{i=1}^{M} \int_{\Omega_i} f \, \varphi_i^* \, dx \qquad \forall \varphi \in \Phi \, , \, 1 \leq i \leq M$$

where φ_i^* is any H^1-extension of φ to Ω_i vanishing on $\partial\Omega \cap \partial\Omega_i$.

Remark 1.1 Under the current assumptions, (1.3) is equivalent to the differential equations

$$(1.7) \qquad -\Delta u_i + \alpha_0 u_i = f \text{ in } \Omega_i \qquad 1 \leq i \leq M$$

and (1.6) can be written in the form

$$(1.8) \qquad \frac{\partial u_i}{\partial \nu_i} = \frac{\partial u_{i+1}}{\partial \nu_i} \text{ on } \Gamma_i \, , \, 1 \leq i \leq M - 1$$

where ν_i is the outward unit vector to Ω_i.

We now build a sequence of functions that yields in the limit the solution to the problem (1.3)-(1.6). The approximation of such a sequence in the frame of a finite dimensional method (finite elements, spectral,...) will provide a domain decomposition method to approximate the solution of problem (1.1).

Let g^0 be an arbitrary function assigned on Γ (say, $g^0 \in \Phi$). The sequence $\{u_i^n, i = 1, ..., M\}$ is obtained as solution to the following problems (hereafter we are assuming for instance M odd).

For all *i* odd, i=1,...,M solve the Dirichlet problems:

$$(1.9) \qquad \int_{\Omega_i} \nabla u_i^n \, \nabla v \, dx + \int_{\Omega_i} \alpha_0 u_i^n \, v \, dx = \int_{\Omega_i} f \, v \, dx \qquad \forall \, v \in H_0^1(\Omega_i)$$

$$(1.10) \qquad u_i^n = 0 \quad \text{on} \quad \partial\Omega_i \cap \partial\Omega$$

$$(1.11) \qquad u_i^n = \begin{cases} \theta_{n-1}\, u_{i-1}^{n-1} + (1 - \theta_{n-1})\, u_i^{n-1} & \text{on } \Gamma_{i-1} \quad , i=3,...,M \\ \theta_{n-1}\, u_{i+1}^{n-1} + (1 - \theta_{n-1})\, u_i^{n-1} & \text{on } \Gamma_i \quad\ \ , i=1,...,M\text{-}2 \end{cases}$$

where θ_{n-1} is a suitable relaxation parameter.

Then for all $\underline{i\ \text{even}}$, i=2,...,M-1 solve the mixed Neumann-Dirichlet problems:

$$(1.12) \qquad \int_{\Omega_i} \nabla u_i^n\, \nabla v\, dx + \int_{\Omega_i} \alpha_0\, u_i^n\, v\, dx = \int_{\Omega_i} f\, v\, dx \quad \forall\, v \in H_0^1(\Omega_i)$$

$$(1.13) \qquad u_i^n = 0 \quad \text{on} \quad \partial\Omega_i \cap \partial\Omega$$

$$(1.14) \qquad \sum_{\substack{i=1 \\ i\ \text{even}}}^{M} \left(\int_{\Omega_i} \nabla u_i^n\, \nabla \varphi_i^*\, dx + \int_{\Omega_i} \alpha_0\, u_i^n\, \varphi_i^*\, dx \right) =$$

$$= -\sum_{\substack{i=2 \\ i\ \text{odd}}}^{M} \left(\int_{\Omega_i} \nabla u_i^n\, \nabla \varphi_i^*\, dx + \int_{\Omega_i} \alpha_0 u_i^n \varphi_i^*\, dx \right) + \sum_{i=1}^{M} \int_{\Omega_i} f\varphi_i^*\, dx \quad \forall\, \varphi \in \Phi$$

Here again equations (1.9) and (1.12) can be written in the strong form (1.7). Furthermore (1.14) is equivalent to require that:

$$(1.15) \qquad \begin{cases} \dfrac{\partial u_i^n}{\partial \nu_i} = \dfrac{\partial u_{i+1}^n}{\partial \nu_i} & \text{on } \Gamma_i \\[2mm] \dfrac{\partial u_i^n}{\partial \nu_i} = \dfrac{\partial u_{i-1}^n}{\partial \nu_i} & \text{on } \Gamma_{i-1} \end{cases} \qquad \text{for i even}$$

The following convergence result holds: for each given decomposition of Ω, there exists a non empty interval $[\theta', \theta'']$ with $\theta' > 0$ such that if $\theta_n \in [\theta', \theta'']$ we have

$$(1.16) \qquad \left[\sum_{i=1}^{M} ||u_i^{n+1} - u_i||_{H^1(\Omega_i)}^2 \right]^{1/2} \leq C\, k^{n+1}$$

where C and k are positive constants with $k < 1$.

The basic idea underlying (1.9)-(1.14) is to solve iteratively a sequence of Dirichlet problems in the odd domains and of mixed Neumann/Dirichlet problems in the even ones. This justifies the name "Dirichlet-Neumann procedure" which is frequently used to identify this method.

We briefly discuss below the application of this procedure to both finite element and spectral collocation approximations to the problem (2.2).

1.1 Finite Element Approximation (see [MQ1], [MQ2])

Let T_h be a regular decomposition (see [C]) of Ω into triangles T not crossing the interfaces $\Gamma_i, i = 1, ..., M - 1$. Define the conforming finite element space

$$(1.17) \qquad V_h = \{v \in C^0(\overline{\Omega}) : v_{|T} \in P_r(T) \quad \forall T \in T_h \ , \ v = 0 \text{ on } \partial\Omega\}$$

Here $P_r(T)$ denotes the space of polynomials on T whose global degree is less than or equal to r. The finite element approximation of problem (1.1) is

$$(1.18) \qquad u_h \in V_h : \int_\Omega \nabla u_h \nabla v \, dx + \int_\Omega \alpha_0 \, u_h \, v \, dx = \int_\Omega f \, v \, dx \quad \forall \, v \in V_h$$

We denote by $V_{h,i}, i = 1, ..., M$, the space of the restrictions to $\overline{\Omega}_i$ of the functions of V_h, by $V_{h,i}^0$ the subspace of $V_{h,i}$ of the functions vanishing on $\partial\Omega_i$ and by Φ_h the restriction of V_h to the subdomain interfaces Γ_i. The equivalence statement given in the previous section extends to the finite element problem as well. As a matter of fact, the problem (1.18) can be written in the form (1.3)-(1.6) by substituting the spaces $H_0^1(\Omega)$, $H_0^1(\Omega_i)$ and Φ with their corresponding finite element spaces $V_h, V_{h,i}^0$ and Φ_h, respectively. Analogously it is possible to achieve the solution u_h of the finite element problem (1.18) by an iterative procedure of the form (1.9)-(1.14). The convergence estimate (1.16) still holds, by substituting u_i^{n+1} and u_i with $u_{i,h}^{n+1}$ and $u_{i,h}$, respectively, with a constant C <u>independent of the discretisation parameter h</u>.

1.2 Spectral collocation approximations ([FQZ], [Q-SL1], [Q-SL2])

Let now Ω be a rectangle, and denote by $P_N(\Omega)$ the space of the algebraic polynomials of degree less than or equal to N with respect to each variable. If Ξ_N is the set of the Legendre (or Chebyshev) Gauss-Lobatto nodes in Ω (see e.g. [DR]) the spectral collocation approximation of problem (1.1) reads as follows (we are assuming here that the right hand side f of (1.1) be a continuous function).

Find $u_N \in P_N(\Omega)$ such that $u_N = 0$ on $\partial\Omega$ and

$$(1.19) \qquad -\Delta u_N + \alpha_0 u_N = f \quad \text{at each point of } \Xi_N \text{ internal to } \Omega \, .$$

A natural spectral multidomain formulation of problem (1.1) stemming from the spectral single-domain problem (1.19) is the following. One looks for polynomials $u_{N,i} \in P_N(\Omega_i)$, i=1, ...,M, which satisfy the differential equation (1.7) collocationwise at all internal collocation nodes of Ω_i, and such that the boundary and interface conditions (1.4), (1.5) and (1.8) hold. Clearly, the matching conditions (1.4) and (1.8) need only be satisfied at the collocation points induced on the interface boundaries.

Though the multidomain spectral solution fails to be equivalent to the single-domain one anymore (as apposite to the finite element case), it nevertheless enjoyes the same properties of stability and asymptotic accuracy.

If the normal derivative condition (1.8) is relaxed in a weak form similar to (1.6), then the overall spectral multidomain method can be formally interpreted variationalwise (as for the finite element case). Integrals need however to be replaced by Gaussian quadratures involving the collocation nodes. For both formulations (the former which is purely collocationwise, and the latter, which is of variational type), the above Dirichlet- Neumann iteration algorithm can be applied. At each step, a sequence of spectral subproblems with Dirichlet conditions need first to be solved within the odd subdomains. Then Neumann-like spectral subproblems are to be solved within the even subdomains. In both cases, a convergence estimate of the kind (1.16) can be proved with two positive constants C and k independent of the discretisation parameter N. The above remarks can be easily generalized to the case of a plurirectangular domain Ω partitioned into rectangles.

Remark 1.2 The iteration by subdomain algorithm introduced so far has been shown to be equivalent to a preconditioned iterative method for the solution of the influence system (see [MQ2] and [Q-SL1]). The influence matrix is nothing but the Schur complement of the matrix of the multidomain system with respect to the interface variables. In other words, it is precisely the matrix of the system of the interface unknowns which is obtained by the global system by the block Gaussian elimination.

Remark 1.3 The "Dirichlet-Neumann" iterative method was introduced in [FQZ] for spectral collocation approximations and, independently, in [BW] and [MQ1] for finite element approximations.

2. THE STOKES PROBLEM

In this section we consider multidomain approximations of the Stokes problem based on finite element and spectral collocation methods.

Given a two dimensional domain Ω and a force field $\mathbf{f} \in (L^2(\Omega))^2$ the continuous problem is the following ($\mathbf{u}$ is the velocity field, p is the pressure and $\nu > 0$ is the kinematic viscosity)

$$(2.1) \qquad \begin{cases} -\nu\Delta\mathbf{u} + \nabla p = \mathbf{f} & \text{in } \Omega \\ \nabla \cdot \mathbf{u} = 0 & \text{in } \Omega \\ \mathbf{u} = 0 & \text{on } \partial\Omega \end{cases}$$

It is well known that this problem has a unique solution (see e.g.[T]), moreover $\mathbf{u} \in (H^2(\Omega))^2$ and $p \in H^1(\Omega)$. The variational formulation of problem (2.1) reads as follows.

Find $\mathbf{u} \in (H_0^1(\Omega))^2$, $p \in L^2(\Omega)/\mathbf{R}$ such that

$$(2.2) \qquad \begin{cases} \nu \int_\Omega \nabla\mathbf{u} \cdot \nabla\mathbf{v}\, dx - \int_\Omega p\nabla \cdot \mathbf{v}\, dx = \int_\Omega \mathbf{f}\,\mathbf{v}\, dx & \forall \mathbf{v} \in (H_0^1(\Omega))^2 \\ \\ \int_\Omega \nabla \cdot \mathbf{u}\, q\, dx = 0 & \forall q \in L^2(\Omega)/\mathbf{R} \end{cases}$$

Let us now suppose that the domain Ω is partitioned into two non-intersecting subdomains Ω_1, Ω_2 and set $\Gamma := \partial\Omega_1 \cap \partial\Omega_2$.

For any function $\mathbf{v}$ defined in Ω we denote by $\mathbf{v}^i$ the restriction of $\mathbf{v}$ to Ω_i, $i = 1, 2$.

Defining the spaces

$$V = \{\mathbf{v} = (\mathbf{v}^1, \mathbf{v}^2) \in (H^1(\Omega_1) \times H^1(\Omega_2))^2 : \mathbf{v}^i = 0$$

60

$$\text{on } \partial\Omega_i \cap \partial\Omega, \ i=1,2, \text{ and } \mathbf{v} \text{ is continuous through } \Gamma\},$$

$$M = \{q = (q^1, q^2) \in (L^2(\Omega_1) \times L^2(\Omega_2))/\mathbf{R}\}$$

the problem (2.2) can be written in the following bi-domain formulation.
Find $\mathbf{u} \in V$ and $p \in M$ such that

$$(2.3) \quad \begin{cases} \displaystyle\sum_{i=1}^{2}(\nu \int_{\Omega_i} \nabla \mathbf{u}^i \cdot \nabla \mathbf{v}^i dx - \int_{\Omega_i} p^i \nabla \cdot \mathbf{v}^i dx) = \sum_{i=1}^{2} \int_{\Omega_i} \mathbf{f}^i \mathbf{v}^i dx \quad \forall \mathbf{v} \in V \\ \displaystyle\sum_{i=1}^{2} \int_{\Omega_i} \nabla \cdot \mathbf{u}^i q^i dx = 0 \hspace{4.5cm} \forall q \in M \end{cases}$$

It can be easily verified that solving (2.3) amounts to solve:

$$(2.4) \quad \begin{cases} -\nu\Delta\mathbf{u}^i + \nabla p^i = \mathbf{f}^i & \text{in } \Omega_i, i = 1,2 \\ \nabla \cdot \mathbf{u}^i = 0 & \text{in } \Omega_i, i = 1,2 \\ \mathbf{u}^1 = \mathbf{u}^2 & \text{on } \Gamma \\ \nu\frac{\partial \mathbf{u}^1}{\partial \mathbf{n}} - p^1\mathbf{n} = \nu\frac{\partial \mathbf{u}^2}{\partial \mathbf{n}} - p^2\mathbf{n} & \text{on } \Gamma \end{cases}$$

Here $\mathbf{n}$ denotes the outward normal unit vector to Ω_1. The last equation enforces the continuity of both normal and shear stresses across Γ.

Let us recall the Galerkin approximation of problem (2.2). We denote by h a discretization parameter and by H_h, L_h two finite dimensional spaces which approximate $(H_0^1(\Omega))^2$ and $L^2(\Omega)$ respectively. The discrete problem is the following. Find $\mathbf{u}_h \in H_h$ and $p_h \in L_h$ such that

$$(2.5) \quad \begin{cases} \nu \int_{\Omega} \nabla \mathbf{u}_h \cdot \nabla \mathbf{v}_h \, dx - \int_{\Omega} p_h \nabla \cdot \mathbf{v}_h \, dx = \int_{\Omega} \mathbf{f} \cdot \mathbf{v}_h \, dx \quad \forall \mathbf{v}_h \in H_h \\ \int_{\Omega} \nabla \cdot \mathbf{u}_h \, q_h \, dx = 0 \hspace{4.7cm} \forall q_h \in L_h \end{cases}$$

Clearly, in order to fix the value of the constant up to which the pressure field is defined, the space L_h is required to satisfy an extra condition (e.g., the functions of L_h attain a zero value at a given grid point, or their mean value over Ω is zero). Analogous schemes can be obtained by approximating the integrals appearing in (2.5) with suitable quadrature rules. This latter form includes both finite element and spectral approximations.

It is well known that problem (2.5) is well posed provided the couple of spaces (H_h, L_h) satisfies a suitable compatibility condition (which is also called the inf-sup condition, see [B]). In particular this condition implies that if $q_h \in L_h/\mathbf{R}$ and

$$(2.6) \qquad \int_{\Omega} q_h \nabla \cdot \mathbf{v}_h \, dx = 0 \text{ for any } \mathbf{v}_h \in H_h$$

then $q_h = 0$. This means that L_h contains no "spurious mode" of the pressure, i.e. any non-vanishing element q_h for which (2.6) holds. Various examples of spaces H_h and L_h which verify the compatibility condition can be found in the literature. We refer to [BF] for finite elements and to [CHQZ] for spectral methods.

We introduce now a finite dimensional approximation to the multidomain problem (2.3). For that, let V_h and M_h be suitable finite dimensional spaces representing respectively the approximate velocity and pressure fields. Then we look for $v_h \in V_h$ and $p_h \in M_h$ such that

$$(2.7) \quad \begin{cases} \sum_{i=1}^{2} (\nu \int_{\Omega_i} \nabla u_h^i \cdot \nabla v_h^i \, dx - \int_{\Omega_i} p_h^i \nabla \cdot v_h^i \, dx) = \sum_{i=1}^{2} \int_{\Omega_i} f^i \, v_h^i \, dx & \forall \, v_h \in V_h \\ \sum_{i=1}^{2} \int_{\Omega_i} \nabla \cdot u_h^i \, q_h^i \, dx = 0 & \forall \, q_h \in M_h \end{cases}$$

The super index i refers to the restriction upon Ω_i. Here again the integrals can be approximated by proper quadrature rules. Problem (2.7), which can be viewed as a Galerkin finite dimensional approximation to the multidomain problem (2.3), is a priori different from (2.5). In this form the numerical approximation of the Stokes problem looks more suited for a multidomain approach involving iteration by subdomain algorithms, as those presented above for elliptic problems. The couple of spaces (V_h, M_h) needs again to satisfy a compatibility condition. In particular the only element q_h of M_h satisying

$$(2.8) \quad \sum_{i=1}^{2} \int_{\Omega_i} q_h^i \nabla \cdot v_h^i \, dx = 0 \quad \forall \, v_h \in V_h$$

should be the null element, if the scheme (2.7) has to be free of spurious modes.

As an example we now consider two Galerkin approximation schemes for the solution of problem (2.3). The former is based on the finite element method and the latter on the spectral collocation method. In the case of finite elements we define an iteration-by- subdomain procedure to solve the discrete problem which generalizes the algorithm introduced in section one. For the case of the spectral scheme we consider a modification of the Uzawa method for solving the discrete problem. This allows the reduction of the Stokes problem to a sequence of Helmholtz problems for which the spectral version of the iteration by subdomain procedure (1.9)-(1.14) can be directly used.

2.1 Finite Element Approximation

In this section the discretization parameter h denotes the maximum of the diameters of the triangles of a regular decomposition T_h of Ω. As in the previous section we assume that the triangles of T_h do not cross the interface Γ. Let H_h and L_h be a family of finite element subspaces of $(H_0^1(\Omega))^2$ and $L^2(\Omega)$ respectively, such that the single domain compatibility condition holds. Then the discrete problem (2.5) is well posed and (u_h, p_h) converges to (u, p) when h tends to zero. We denote by $H_{h,i}$ and $L_{h,i}$, i=1,2, the spaces of the restrictions to $\overline{\Omega}_i$ of the elements of H_h and L_h. The space V_h is defined as follows

$$V_h = \{v_h = (v_h^1, v_h^2) \in H_{h,1} \times H_{h,2} : v_h^1 = v_h^2 \text{ on } \Gamma\}$$

while the pressure space M_h is the subspace of the functions of $L_{h,1} \times L_{h,2}$ vanishing at a given gridpoint of $\overline{\Omega}_i$. The bi-domain finite element approximation of (2.2) is then given by (2.7). It is proven in [MQ1] that if the functions of L_h are discontinuous (e.g. piecewise constant) then problems (2.5) and (2.7) are equivalent. Precisely we have that $u_h^i = u_{h_{|\Omega_i}}$ and $p_h^i = p_{h_{|\Omega_i}}$, i=1,2, where (u_h, p_h) is the standard finite element solution while (u_h^i, p_h^i), i=1,2, is the solution of the discrete bi-domain problem.

Let us denote bt Φ_h the space of the restrictions to Γ of the elements of V_h and by $H_{h,i}^0$, i=1,2, the subspace of $H_{h,i}$ of the functions vanishing on $\partial\Omega_i$. Problem (2.7) is equivalent to the following one.

Find $(u_{h,i}, p_{h,i}) \in H_{h,i} \times L_{h,i}$ i=1,2, such that

$$(2.1.1) \qquad \nu \int_{\Omega_1} \nabla u_h^1 \cdot \nabla v_h^1 \, dx - \int_{\Omega_1} p_h^1 \nabla \cdot v_h^1 \, dx = \int_{\Omega_1} f^1 v_h^1 \, dx \quad \forall v_h^1 \in H_{h,1}^0$$

$$(2.1.2) \qquad \int_{\Omega_i} \nabla \cdot u_h^i q_h^i \, dx = 0 \quad \forall q_h^i \in L_{h,i} \,, \, i=1,2$$

$$(2.1.3) \qquad u_h^1 = u_h^2 \text{ on } \Gamma$$

$$(2.1.4) \quad \nu \int_{\Omega_2} \nabla u_h^2 \cdot \nabla v_h^2 \, dx - \int_{\Omega_2} p_h^2 \nabla \cdot v_h^2 \, dx = \int_{\Omega_2} f^2 v_h^2 \, dx - \nu \int_{\Omega_1} \nabla u_h^1 \nabla \rho_h^1(v_{h_{|\Gamma}}^2) dx +$$

$$+ \int_{\Omega_1} p_h^1 \nabla \cdot \rho_h^1(v_{h_{|\Gamma}}^2) \, dx + \int_{\Omega_1} f^1 \, \rho_h^1(v_{h_{|\Gamma}}^2) \, dx \quad \forall v_h^2 \in H_{h,2}$$

where $\rho_h^1(v_{h,\Gamma}^2)$ is the finite element vector function which matches v_h^2 at all nodes of Γ, and vanishes at all other nodes of the finite element triangulations in $\overline{\Omega}_1$. The formulation (2.1.1)-(2.1.4) suggests the introduction of an iteration by subdomain procedure analogous to the one defined in section one for the Helmholtz problem. For any $g_h^0 \in \Phi_h$ we consider two sequences $(u_h^{i,n}, p_h^{i,n}) \in H_{h,i} \times L_{h,i}$, i=1,2 satisfying for each $n \geq 1$

$$(2.1.5) \qquad \begin{cases} \int_{\Omega_1} \nabla u_h^{1,n} \cdot \nabla v_h^1 \, dx - \int_{\Omega_1} p_h^{1,n} \nabla \cdot v_h^1 \, dx = \int_{\Omega_1} f^1 v_h^1 dx & \forall v_h^1 \in H_{h,1}^0 \\ \int_{\Omega_1} \nabla \cdot u_h^{1,n} q_h^1 \, dx = 0 & \forall q_h^1 \in L_{h,1} \\ u_h^{1,n} = g_h^{n-1} & \text{on } \Gamma, \end{cases}$$

$(2.1.6)$
$$\begin{cases} \nu \int_{\Omega_2} \nabla u_h^{2,n} \cdot \nabla v_h^2 \, dx - \int_{\Omega_2} p_h^{2,n} \nabla \cdot v_h^2 \, dx = \int_{\Omega_2} f^2 v_h^2 \, dx + \int_{\Omega_1} f^1 \rho_h^1(v_{h_{|\Gamma}}^2) dx - \\ -\nu \int_{\Omega_1} \nabla u_h^{1,n} \cdot \nabla \rho_h^1(v_{h_{|\Gamma}}^2) dx + \int_{\Omega_1} p_h^{1,n} \nabla \cdot \rho_h^1(v_{h_{|\Gamma}}^2) dx & \forall v_h^2 \in H_{h,2} \\ \int_{\Omega_2} \nabla \cdot u_h^{2,n} q_h^2 \, dx = 0 & \forall q_h^2 \in L_{h,2} \end{cases}$$

where we have set

$$(2.1.7) \qquad g^n = \theta_n u_h^{2,n} + (1 - \theta_n) g^{n-1} \text{ on } \Gamma, \, n \geq 1$$

and $\theta_n > 0$ is the relaxation parameter. If θ_n is chosen in a proper interval $[\theta', \theta'']$ then the sequences defined in (2.15)-(2.17) converge to the solution of problem (2.7). Furthermore the rate of convergence is independent of the discretization parameter h. For a detailed analysis of this method we refer to [MQ1].

Remark 2.1 (Influence Matrix) As for the elliptic problem considered in section 1, the algorithm (2.1.5)-(2.1.6) can be viewed as an effective preconditioned iterative procedure to solve the capacitance system (also called Schur complement system) associated with the multidomain finite element problem (2.1.1)-(2.1.4). In this case, the capacitance matrix is the one which handles the unknown values of the velocity field pertaining to the finite element nodes lying on the interface Γ (see [Q1]).

2.2 Spectral Collocation Approximation

Several spectral collocation methods have been proposed in the last years for the solution to the Stokes problem in a square within a single-domain framework. For a complete review we refer to [CHQZ], Ch.7. Here we consider the one proposed in [BMM] and [M], with the aim of generalizing it to a bi-domain case. We use the notation of figure 2.2.1.

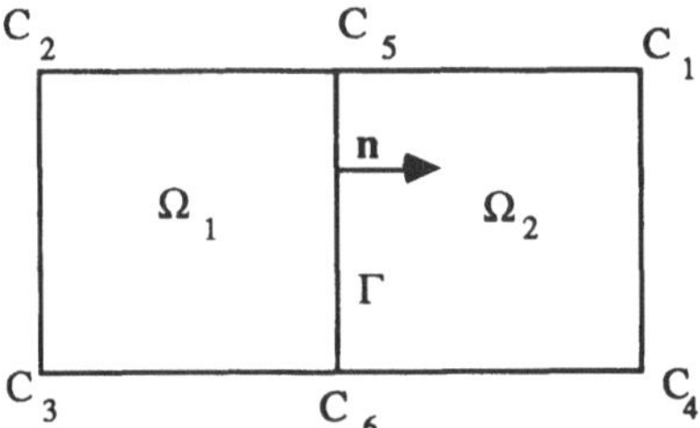

Fig.2.2.1 The decomposition of Ω

Let us set h := 1/N where N is a positive integer. We define the finite dimensional subspace V_h of V as follows.

$$V_h = \{v_h = (v_h^1, v_h^2) \in (P_N(\Omega_1) \times P_N(\Omega_2))^2 : v_h^i = 0 \text{ on } \partial\Omega_i \cap \partial\Omega, i=1,2, v_h^1 = v_h^2 \text{ on } \Gamma\}$$

We consider a discrete problem of the type (2.7) where the integrals are approximated by Gaussian quadrature rules. To this end let us define the discrete inner product on $P_N(\Omega_i)$, i=1,2:

$$(2.2.1) \qquad (w, z)_{N,i} = \sum_{k=0}^{N} \sum_{m=0}^{N} (wz)(x_k^i, y_m^i)\, \omega_k^i\, \omega_m^i, \quad \forall w, z \in C^0(\overline{\Omega_i})$$

Here (x_k^i, y_m^i), $k, m = 0, ...,N$, i=1,2, are the collocation points related to the Legendre-Gauss-Lobatto quadrature formula in $\overline{\Omega_i}$ and ω_k^i, ω_m^i are the corresponding weights (see e.g. [CHQZ] Ch.2).

We denote again by $(\cdot, \cdot)_{N,i}$ the inner product induced by (2.2.1) on the vector valued function space $(P_N(\Omega_i))^2$. The pressure space M_h is required to be a subspace of $P_N(\Omega_1) \times P_N(\Omega_2)$. Then the spectral collocation approximation to the problem (2.3) takes the following form.

64

$$(2.2.2) \quad \begin{cases} \mathbf{u}_h \in V_h, p_h \in M_h : \\[4pt] \displaystyle\sum_{i=1}^{2}\{(\nu\nabla u_h^i, \nabla v_h^i)_{N,i} - (p_h^i, \nabla\cdot v_h^i)_{N,i}\} = \sum_{i=1}^{2}(\mathbf{f}^i, v_h^i)_{N,i} \quad \forall v_h \in V_h \\[10pt] \displaystyle\sum_{i=1}^{2}(\nabla\cdot u_h^i, q_h^i)_{N,i} = 0 \qquad\qquad\qquad\qquad\qquad \forall q_h \in M_h \end{cases}$$

As mentioned in section 2 the spaces V_h, M_h must satisfy a suitable compatibility condition. In the current situation, the condition necessary to exclude the existence of spurious modes reads as follows (it is the counterpart of (2.8)). If $q_h \in M_h$ is such that

$$(2.2.3) \qquad \sum_{i=1}^{2}(q_h^i, \nabla\cdot v_h^i)_{N,i} = 0 \quad \forall v_h \in V_h$$

then q_h must be equal to zero. Unfortunately (2.2.3) has 12 independent non-vanishing solutions within $\mathbf{P}_N(\Omega_1) \times \mathbf{P}_N(\Omega_2)$. They are precisely the spurious modes (see [SV] for the proof):

$$(2.2.4) \quad \begin{cases} q_{1,h} = (L_0^1, L_0^2),\ q_{2,h} = (L_N^1(x), 0),\ q_{3,h} = (L_N^1(x)L_N^1(y), 0), \\[8pt] q_{4,h} = (L_N^1(y), L_N^2(y)),\ q_{5,h} = (0, L_N^2(x)),\ q_{6,h} = (0, L_N^2(x)L_N^2(y)), \\[8pt] q_{7,h} = (0, \chi_{h,C_1}^2),\ q_{8,h} = (\chi_{h,C_2}^1, 0),\ q_{9,h} = (\chi_{h,C_3}^1, 0), \\[8pt] q_{10,h} = (0, \chi_{h,C_4}^2),\ q_{11,h} = (\chi_{h,C_5}^1, -\chi_{h,C_5}^2),\ q_{12,h} = (\chi_{h,C_6}^1, -\chi_{h,C_6}^2) \end{cases}$$

Here we have denoted by $L_k^i(x)L_m^i(y)$, $k, m = 0, ..., N$, the Legendre polynomial of degree k in x and m in y defined in Ω_i, $i = 1, 2$. Moreover, for any collocation point $\mathbf{p}$ of $\overline{\Omega}_i$, i=1,2, we have denoted with $\chi_{h,\mathbf{p}}^i$ the element of $\mathbf{P}_N(\Omega_i)$ whose value is 1 in $\mathbf{p}$ and 0 at the other collocation points.

In order to render the pressure field free of spurious modes we must remove the functions (2.2.4) from $\mathbf{P}_N(\Omega_1) \times \mathbf{P}_N(\Omega_2)$. A possible choice of the new pressure space M_h is therefore the following.

$$(2.2.5) \qquad M_h = \{q_h = (q_h^1, q_h^2) \in \mathbf{P}_N(\Omega_1) \times \mathbf{P}_N(\Omega_2) : \sum_{i=1}^{2}(q_h^i, q_{k,h}^i)_{N,i} = 0, k = 1, ..., 12\}$$

Other choices of M_h are possible, however (see [SV]). We enphasize that with the choice (2.2.5), the solution $\mathbf{u}_h$ of (2.2.2) is exactly dvergence free in Ω for any value of the discretization parameter h. This property is important in general, and it is particularly relevant when the full nonlinear Navier-Stokes equations are considered.

Problem (2.2.2) can be written in the following collocation form (see [SV]): $\mathbf{u}_h \in V_h$, $p_h \in M_h$:

$$(2.2.6) \qquad -\nu\Delta u_h^i + \nabla p_h^i = f^i \text{ at the collocation points internal to } \Omega_i \ (i=1,2)$$

$$(2.2.7) \qquad \nabla \cdot u_h^i = 0 \text{ at the collocation points of } \overline{\Omega}_i \ (i=1,2)$$

(2.2.8)
$$\nu\frac{\partial u_h^1}{\partial n} - p_h^1 n - \nu\frac{\partial u_h^2}{\partial n} + p_h^2 n = -\sum_{i=1}^{2}(-\nu\Delta u_h^i + \nabla p_h^i - f^i)\,\omega_N^i \text{ at the collocation points of } \Gamma$$

Note that in (2.2.8) the right hand side is the residual of the momentum equation multiplied by ω_N^i (which behaves like N^{-2}), hence this condition asymptotically enforces the continuity of the normal and shear stresses across Γ.

For the solution of the problem (2.2) we suggest using the Uzawa method. The idea underlying this method is to decouple the pressure field from the velocity field. Precisely, applying first the inverse of the Laplacian operator (with Dirichlet homogenous boundary conditions) and then the divergence operator to the momentum equation one obtains

$$(2.2.9) \qquad \nabla \cdot \Delta^{-1}\nabla p = \nabla \cdot \Delta^{-1} f \text{ in } \Omega \ .$$

This equation does not require any boundary condition. Hence one can compute the pressure and then obtain the velocity field from the identity

$$(2.2.10) \qquad u = -\frac{1}{\nu}\Delta^{-1}(f - \nabla p)$$

We now define an operator A_h which is the spectral approximation of the operator $\nabla \cdot \Delta^{-1}\nabla$ appearing in (2.2.9). For any $q_h \in M_h$ let w_h^q be the solution of the problem

$$(2.2.11) \qquad w_h^q \in V_h : \sum_{i=1}^{2}(\nabla w_h^{q,i}, \nabla v_h^i)_{N,i} = \sum_{i=1}^{2}(q_h^i, \nabla \cdot v_h^i)_{N,i} \ \ \forall v_h \in V_h$$

Then $A_h q_h$ is defined by

$$(2.2.12) \qquad A_h q_h \in M_h : \sum_{i=1}^{2}((A_h q_h)^i, r_h^i)_{N,i} = \sum_{i=1}^{2}(\nabla \cdot w_h^{q,i}, r_h^i)_{N,i} \ \ \forall r_h \in M_h$$

For any continuous f let w_h^f be the solution of the problem

$$(2.2.13) \qquad w_h^f \in V_N : \sum_{i=1}^{2} (\nabla w_h^{f,i}, \nabla v_h^i)_{N,i} = \sum_{i=1}^{2} (f^i, v_h^i)_{N,i} \quad \forall v_h \in V_h$$

We define $F_h f \in M_N$ by

$$(2.2.14) \qquad \sum_{i=1}^{2} ((F_h f)^i, r_h^i)_{N,i} = \sum_{i=1}^{2} (\nabla \cdot w_h^{f,i}, r_h^i)_{N,i} \quad \forall r_h \in M_h$$

One can easily prove that problem (2.2.2) is equivalent to find $u_h \in V_h$ and $p_h \in M_h$ such that

$$(2.2.15) \qquad \sum_{i=1}^{2} ((A_h p_h)^i, r_h^i)_{N,i} = \sum_{i=1}^{2} ((F_h f)^i, r_h^i)_{N,i} \quad \forall r_h \in M_h$$

$$(2.2.16) \qquad u_h = \frac{1}{\nu} w_h^f + w_h^p \ .$$

Therefore the solution can be obtained by solving first the pressure equation (2.2.15), then computing the velocity field using (2.2.16). Since A_h is linear, continuous symmetric and positive definite, the use of the conjugate gradient method is especially indicated for the solution of (2.2.15). In this frame the matrix associated to A_h needs not to be computed; rather one simply needs to calculate $A_h q_h$ for a sequence of given $q_h \in M_h$. To this end problem (2.2.11) has to be solved, then $A_h q_h$ is obtained from (2.2.12).
Problem (2.2.11) is of elliptic type, as those considered in section one. Actually integrating by parts in (2.2.11) we are led to:

$$(2.2.17) \qquad w_h^q \in V_h : \sum_{i=1}^{2} (\nabla w_h^{q,i}, \nabla v_h^i)_{N,i} = \sum_{i=1}^{2} (g_h^i, v_h^i)_{N,i} \quad \forall v_h \in V_h$$

where g_h^i is the following vector ($\tilde{q}_h^i$ being the polynomial which equals q_h^i on Γ and vanishes at the collocation points of $\overline{\Omega}_i \setminus \Gamma$)

$$\left(-\frac{\partial q_h^i}{\partial x} - (-1)^i \frac{1}{\omega_N^i} \tilde{q}_h^i , \ -\frac{\partial q_h^i}{\partial y} \right)^t$$

In conlcusion the use of the conjugate gradient method within the Uzawa iterations yields a sequence of elliptic problems that can be efficiently solved by using the iteration by subdomain procedure described in section one.

3. HYPERBOLIC PROBLEMS ([Q2])

In this section we deal with the approximation of hyperbolic systems of first order. In order to keep our presentation plain, we confine ourselves to the case of a linear system in one-dimension. The case of a general nonlinear system of conservation laws in one or several dimensions is presented in [Q2]. For the sake of simplicity we consider a system of the form

$$(3.1) \qquad \begin{cases} \mathbf{w}_t + A\,\mathbf{w}_x + B\,\mathbf{w} = 0 & -1<x<1\,,\, 0 \le t \le T \\ \mathbf{w}(x,0) = \varphi(x) & -1 \le x \le 1 \end{cases}$$

where φ is a given initial data and A and B are 2×2 constant matrices. We assume that A is diagonalizable and has real eigenvalues λ_i, i=1,2, with $\lambda_1 > 0, \lambda_2 < 0$ (this is the case if, e.g., $A_{ii} = \alpha, A_{12} = A_{21} = 1, |\alpha| < 1$). In order to make (3.1) a well-posed problem, we need to enforce two boundary equations. Among several possible choices (which depend on the form of A) we require for instance that:

$$(3.2) \qquad w_1(-1,t) = w_1(1,t) = 0 \quad 0 \le t \le T \quad .$$

We suppose that φ is continuous and compatible (at $x = \pm 1$) with the boundary conditions (3.2), so that problem (3.1) has a unique continuous solution. Despite its apparently simple form, the problem (3.1)-(3.2) incorporhates most of the peculiar aspects of first order hyperbolic systems of more general form. In particular, it is a good model for the definition of a multidomain solution method. Actually eigenvalues of different sign of A ensure the coexistence of waves travelling along opposite directions, and this aspect needs to be properly faced within any multidomain approach. At first we note that if $\mathbf{w}^1$ and $\mathbf{w}^2$ are the restrictions of $\mathbf{w}$ to the space intervals $[-1,0]$ and $[0,1]$ at any time t, then $\mathbf{w}^1 \in (C^0([0,T] \times [-1,0]))^2$ and $\mathbf{w}^2 \in (C^0([0,T] \times [0,1]))^2$ are such that

$$(3.3) \qquad \begin{cases} \mathbf{w}_t^1 + A\,\mathbf{w}_x^1 + B\,\mathbf{w}^1 = 0 & -1<x<0\,,\, 0<t \le T \\ \mathbf{w}_t^2 + A\,\mathbf{w}_x^2 + B\,\mathbf{w}^2 = 0 & 0<x<1\,,\, 0<t \le T \\ w_1^1(-1,t) = w_1^2(1,t) = 0 & 0 \le t \le T \\ \mathbf{w}^1(0,t) = \mathbf{w}^2(0,t) & 0 \le t \le T \\ \mathbf{w}^1(x,0) = \varphi(x) & -1 \le x \le 0 \\ \mathbf{w}^2(x,0) = \varphi(x) & 0 < x \le 1 \end{cases}$$

Let T be the matrix which diagonalizes A, i.e.

$$A = T \wedge T^{-1}, \text{ with } \wedge = \text{diag}\,\{\lambda_1, \lambda_2\}$$

We denote by $\sigma_{ij}, i,j = 1,2$ the entries of T, and by $\mathbf{z}^i$, i=1,2, the <u>characteristic variables</u> defined by

$$(3.4) \qquad \mathbf{z}^i = T^{-1}\mathbf{w}^i, i = 1,2 \quad (\text{i.e., } z_k^i = \sum_{j=1}^{2} \sigma_{kj} w_j^i, k = 1,2)$$

68

z_1^i is the characteristic variable moving from left to right and z_2^i the one which moves along the opposite direction.

Our aim is to define an iteration by subdomain approach for the solution of problem (3.3). Assume for example that an implicit scheme is used for the time-integration of (3.3). Then at each time step one is left with a problem of the following type.

$$(3.5) \qquad \begin{cases} \tilde{A}\,\tilde{w}_x^1 + \tilde{B}\,\tilde{w}^1 = f^1 & -1 < x < 0 \\ \tilde{A}\tilde{w}_x^2 + \tilde{B}\tilde{w}^2 = f^2 & 0 < x < 1 \\ \tilde{w}_1^1(-1) = \tilde{w}_1^2(1) = 0 \\ \tilde{w}^1(0) = \tilde{w}^2(0) \end{cases}$$

where $\tilde{A}$ and $\tilde{B}$ are matrices which depend on A, B and of the time step Δt, while f^i, $i = 1, 2$, depend on the solution at previous time levels.

The following iteration-by-subdomain method can be used in order to solve (3.5). Given two initial values g_1^0 and g_2^0 at the n-th step of the procedure one solves the two independent problems

$$(3.6) \qquad \begin{cases} a) \;\; \tilde{A}\,\tilde{w}_x^{1,n} + \tilde{B}\,\tilde{w}^{1,n} = f^1 & -1 < x < 0 \\ b) \;\; \tilde{w}_1^{1,n}(-1) = 0 \\ c) \;\; (\sigma_{21}\,\tilde{w}_1^{1,n} + \sigma_{22}\,\tilde{w}_2^{1,n})(0) = g_2^{n-1} \end{cases}$$

and

$$(3.7) \qquad \begin{cases} a) \;\; \tilde{A}\,\tilde{w}_x^{2,n} + \tilde{B}\,\tilde{w}^{2,n} = f^2 & 0 < x < 1 \\ b) \;\; \tilde{w}_1^{2,n}(1) = 0 \\ c) \;\; (\sigma_{11}\,\tilde{w}_1^{2,n} + \sigma_{12}\,\tilde{w}_2^{2,n})(0) = g_1^{n-1} \end{cases}$$

where

$$(3.8) \qquad \begin{cases} g_1^n = (\sigma_{11}\,\tilde{w}_1^{1,n} + \sigma_{12}\,\tilde{w}_2^{1,n})\,(0) \\ g_2^n = (\sigma_{21}\,\tilde{w}_1^{2,n} + \sigma_{22}\,\tilde{w}_2^{2,n})\,(0) \end{cases}$$

According to this method, at the new step, we don't require that each physical variable matches continuously at the interface between subdomains, but rather that the characteristic variable inpinging one subdomain attains the same value enjoyed by the omologous characteristic variable in the complementary subdomain at the previous iteration. In other words at each step instead of requiring the continuity of the physical variables we require the one of the characteristic variables, in the respect of their direction of propagation. This allows the achievement of a fast convergence rate as shown by several numerical experiments and proven analytically (see [Q2]).

The problems (3.6)-(3.8) can be discretized in space by the spectral collocation method as follows. For each n, $\tilde{w}^{1,n}$ and $\tilde{w}^{2,n}$ are assumed to be algebraic polynomials of degree N, that satisfy (3.6 a) and (3.7 a), respectively at the $N-1$ Chebyshev collocation points internal to $[-1, 0]$ and $[0, 1]$. Both (3.6) and (3.7) need to be supplemented by two further equations, one at the boundary point ($x = -1$ for (3.6) and $x = 1$ for (3.7)), and the other at the interface point $x = 0$. These are precisely the "compatibility equations" which amounts to collocate at a given boundary point of the computational domain that linear combination of the differential equations corresponding to the outgoing characteristic variable. In the current context, the compatibility equations take the following form:

$$\sum_{k=1}^{2} \sigma_{2k}[\tilde{A}\,\bar{\mathbf{w}}_x^{1,n} + \tilde{B}\,\bar{\mathbf{w}}^{1,n} - \mathbf{f}^1]_k = 0 \quad \text{at } x = -1$$

$$\sum_{k=1}^{2} \sigma_{1k}[\tilde{A}\,\bar{\mathbf{w}}_x^{1,n} + \tilde{B}\,\bar{\mathbf{w}}^{1,n} - \mathbf{f}^1]_k = 0 \quad \text{at } x = 0$$

$$\sum_{k=1}^{2} \sigma_{2k}[\tilde{A}\,\bar{\mathbf{w}}_x^{2,n} + \tilde{B}\,\bar{\mathbf{w}}^{2,n} - \mathbf{f}^2]_k = 0 \quad \text{at } x = 0$$

$$\sum_{k=1}^{2} \sigma_{1k}[\tilde{A}\,\bar{\mathbf{w}}_x^{2,n} + \tilde{B}\,\bar{\mathbf{w}}^{2,n} - \mathbf{f}^2]_k = 0 \quad \text{at } x = 1$$

Here the notation $[\cdot]_k$ denotes the k-th component of the vector between brackets.
The adaptation to the discrete collocation method of the above iteration by subdomain procedure is straightforward. The extension to the case of decomposition by more than two subdomains is made accordingly. At each step, the boundary value problems to be solved within each subdomain are independent each other, and they can therefore be solved simultaneously. The interested reader is referred to [Q2], where the method is applied to several dimensions and to nonlinear hyperbolic systems.

4. COUPLING OF PARABOLIC AND HYPERBOLIC SYSTEMS
In this section we consider the problem of coupling elliptic and hyperbolic one dimensional systems in different zones of the computational domain. Let A and B be the matrices defined in section 3; we seek for a pair of two dimensional vector functions such that

$$(4.1) \qquad \mathbf{w}_t^1 - \nu\,\mathbf{w}_{xx}^1 + A\mathbf{w}_x^1 + D\mathbf{w}^1 - f^1 \quad \text{for} \quad -1 < x < 0$$

$$(4.2) \qquad \mathbf{w}_t^2 + A\mathbf{w}_x^2 + B\mathbf{w}^2 = f^2 \quad \text{for} \quad 0 < x < 1$$

$$(4.3) \qquad \mathbf{w}^1(-1) = 0$$

$$(4.4) \qquad w_1^2(1) = 0$$

As in (3.3), the initial values of $\mathbf{w}^1$ and $\mathbf{w}^2$ at the time $t = 0$ need to be prescribed. The above is a simple model example suggested from the general problem of coupling the Navier Stokes equations for viscous flows with the Euler equations in that part of the computational domain where viscous effects can be disregarded. The crucial issue is about the kind of matching conditions that should be enforced at the viscous-inviscid interface in order to get a solution which is physically consistent with the solution of the overall viscous problem.

Despite the estremely simplified form of problem (4.1)-(4.4) the mechanism of the interaction of $\mathbf{w}^1$ with $\mathbf{w}^2$ is not far different than the one pertaining to the physical situation described above. A more precise discussion about similarities and differences between the two problems is given in [GQ]. To solve this problem we must impose three further "interface conditions" at the point 0. As a matter of fact two conditions are needed to solve the parabolic problem in the domain $(-1, 0)$ and one condition is needed for the hyperbolic system defined in the domain $(0, 1)$. An analysis of the possible choices of the interface conditions can be found in [GQ]. In this work the authors suggest the following approaches.

<u>"Variational" interface conditions</u>

$$(4.5) \qquad \sigma_{11}w_1^1 + \sigma_{12}w_2^1 = \sigma_{11}w_1^2 + \sigma_{12}w_2^2 \ \text{ in } 0$$

$$(4.6) \qquad -\nu\,\mathbf{w}_x^1 + A\mathbf{w}^1 = A\mathbf{w}^2 \ \text{ in } 0$$

<u>"Nonvariational" interface conditions</u>

$$(4.7) \qquad \mathbf{w}^1 = \mathbf{w}^2 \ \text{ in } 0$$

$$(4.8) \qquad \sigma_{11}w_{1,x}^1 + \sigma_{12}w_{2,x}^1 = \sigma_{11}w_{1,x}^2 + \sigma_{12}w_{2,x}^2 \ \text{ in } 0$$

We recall that $\sigma_{ij}, i, j = 1, 2$ are the elements of the matrix T which diagonalizes A (see section 3). Hence the conditions (4.5) and (4.8) enforce the continuity of the characteristic variable outcoming from the parabolic region and of its first derivative respectively. Note that with the second approach both physical variables are continuous at the interface point (see eq. (4.7)); on the contrary, in the former approach continuity is only enforced for that characteristic variable impinging the parabolic subdomain. Both couples of interface conditions are obtained with a vanishing viscosity procedure. Precisely to obtain (4.5)-(4.6) one consider the fully parabolic system

$$(4.9) \qquad w_{\varepsilon,t}^1 - \nu\,w_{\varepsilon,xx}^1 + Aw_{\varepsilon,x}^1 + Bw_\varepsilon^1 = f^1 \ \text{ in } \ -1 < x < 0$$

$$(4.10) \qquad w_{\varepsilon,t}^2 - \varepsilon\,w_{\varepsilon,xx}^2 + Aw_{\varepsilon,x}^2 + Bw_\varepsilon^2 = f^2 \ \text{ in } 0 < x < 1$$

$$(4.11) \qquad w_{\varepsilon}^{1}(-1) = 0 \ , \ w_{1,\varepsilon}^{2}(1) = 0 \ , \ w_{2,\varepsilon,x}^{2}(1) = 0$$

(where ε is a real positive parameter) supplemented by the interface conditions

$$(4.12) \qquad w_{\varepsilon}^{1} = w_{\varepsilon}^{2} \ \text{in} \ 0 \ , \ \nu\, w_{\varepsilon,x}^{1} = \varepsilon\, w_{\varepsilon,x}^{2} \ \text{in} \ 0$$

It can be proved that if ε tends to zero the solution w_{ε}^{i} i=1,2 of problem (4.9)-(4.12) tends to the functions w^{i} for i=1,2 which verify (4.1)-(4.6). This approach is called "variational" because the problem (4.9)-(4.12) can be written in a variational form in the whole domain $(-1, 1)$. On the other hand one can consider, for any $\varepsilon > 0$, the couple w_{ε}^{i} for i=1,2 which verify (4.9)-(4.11) together with the interface conditions

$$(4.13) \qquad w_{\varepsilon}^{1} = w_{\varepsilon}^{2} \ \text{in} \ 0 \ , \ w_{\varepsilon,x}^{1} = w_{\varepsilon,x}^{2} \ \text{in} \ 0$$

In this case w_{ε}^{i} i=1,2 tends to a couple w^{i} i=1,2 which verify (4.1)-(4.4) and (4.7)-(4.8). Both approaches require an extra compatibility equation at the interface point $x = 0$ whenever the spectral collocation methods is applied to get a numerical solution. Again, an iteration by subdomain procedure can be used in order to solve a sequence of hyperbolic problems in $(0, 1)$ and parabolic ones in $(-1, 0)$. The details can be found in [GQ].

4.1 Numerical results

We present some numerical results obtained for the test problem (4.1)-(4.4) in the case where

$$B = I \ , \ A = \begin{pmatrix} 0 & 1 \\ 3 & 2 \end{pmatrix} \ f^{1} = f^{2} = 1$$

so that the eigenvalues of A are 3 and -1. All the results are obtained by the spectral collocation method using 41 collocation nodes within each subdomain. In all figures below the solid line represents the first component w_1 of the solution, while the second component w_2 is represented by the dashed line.

Figures 4.1, 4.2 and 4.3 refer to the case in which $\nu \equiv 1$ in $(-1,0)$. The results of fig.4.1 are obtained using the variational interface conditions (4.5), (4.6), while those of fig. 4.2 using the nonvariational ones (4.7), (4.8). Note that at the interface point $x = 0$ the curves in fig.4.1 exhibit a jump discontinuity, as predicted by the theory of [GQ]. Fig.4.3 shows the results obtained for the same problem by enforcing at the point $x = 0$ the differential equations (4.1) rather than the proper set of interface conditions. This (rather striking) result is a clear evidence that our approach about the setting of suitable interface conditions is necessary in order to garantee that the numerical solution of the truncated problem is consistent with the overall viscous solution. Fig.4.4 shows the difference between the solution of the perturbed problem (4.9), (4.10), (4.11) and (4.13) and that of the unperturbed problem (4.1)-(4.4), (4.7) and (4.8). Here $\varepsilon = 0.01$ and again $\nu = 0.1$.

Finally, fig.4.5 shows the jump in the solution corresponding to the use of variational interface conditions for different values of the viscous coefficient (precisely, $\nu = 0.01, 0.05, 0.1$). It turns out that the size of the jump is proportional to ν, so it disappears if the interface is fixed at a point where the viscosity is negligible.

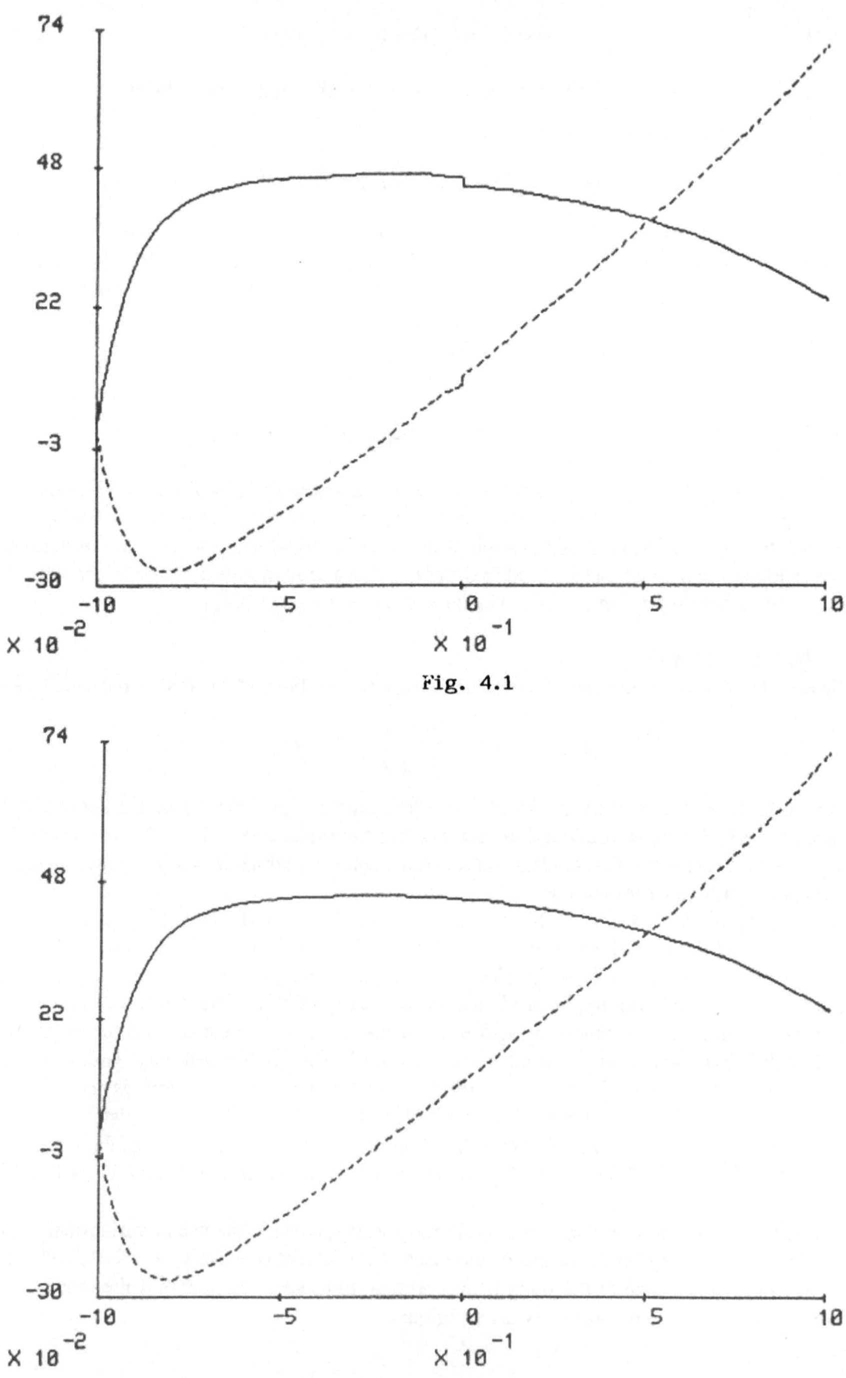

Fig. 4.1

Fig. 4.2

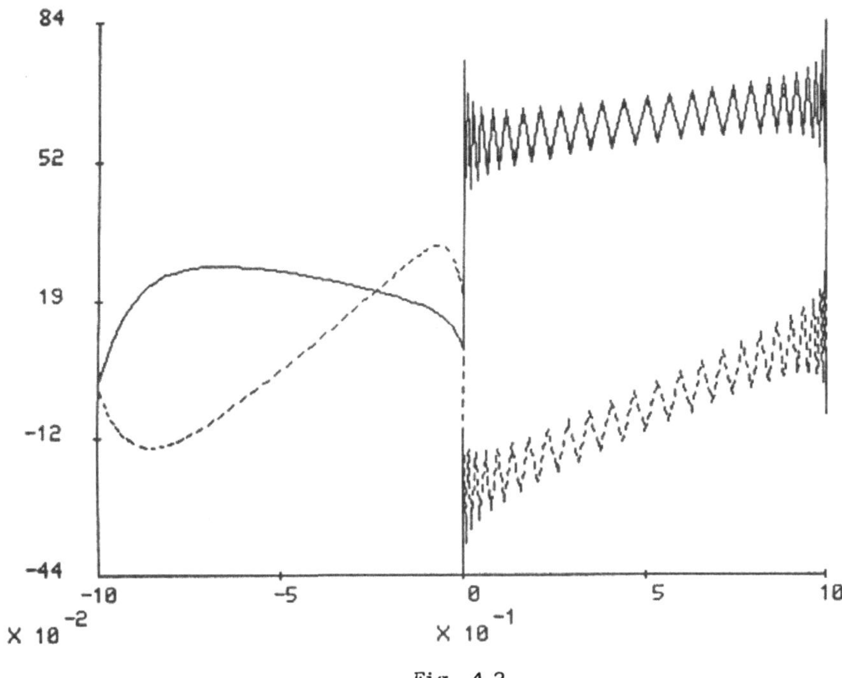

Fig. 4.3

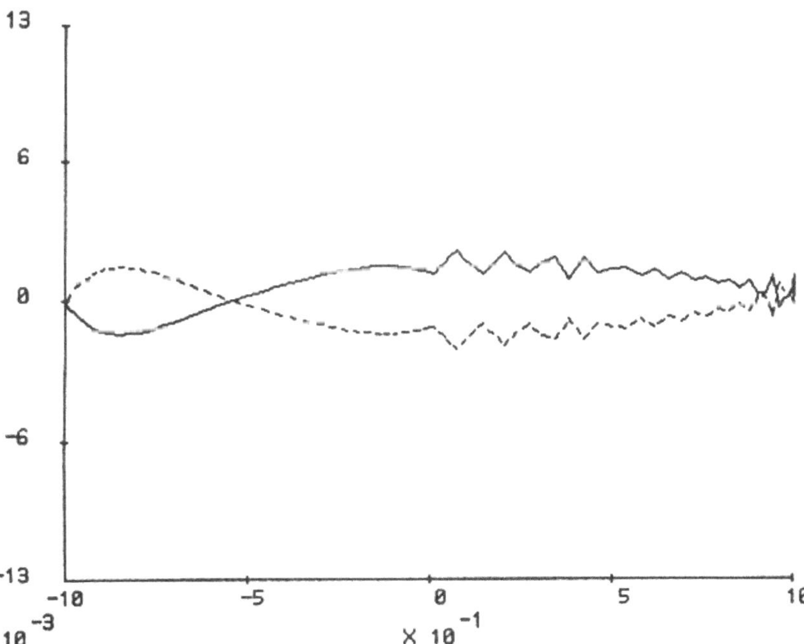

Fig. 4.4

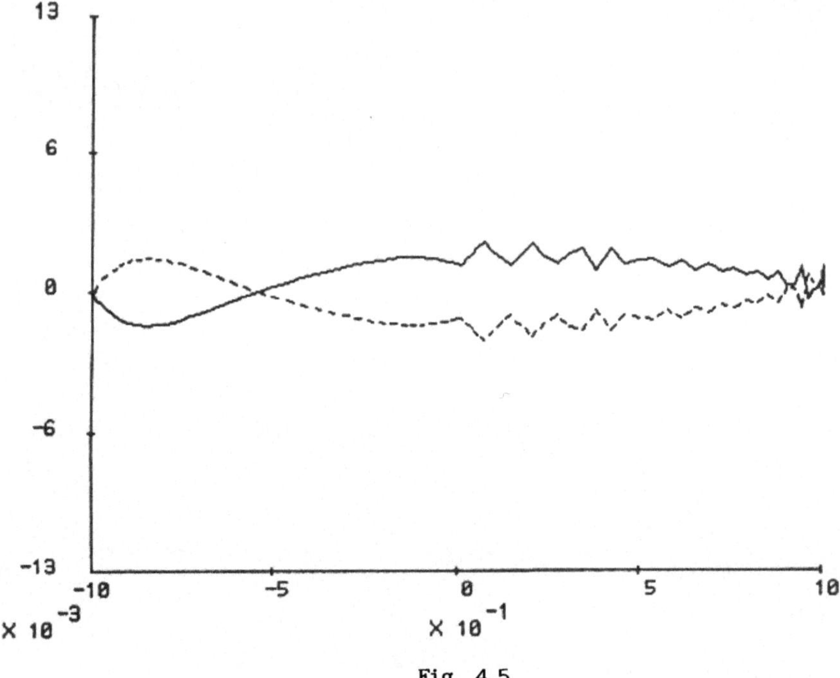

Fig. 4.5

REFERENCES

[LM] J.L.Lions, E.Magenes, <u>Nonhomogeneous boundary value problems and applications</u>, Vol.I Springer-Verlag, Berlin 1972.

[Q-SL1] A.Quarteroni, G.Sacchi-Landriani, Domain Decomposition Preconditioners for the Spectral Collocation Method, to appear in J.Scientif.Comput. 1988 (also ICASE Report n.88-11)

[Q-SL2] A.Quarteroni, G.Sacchi-Landriani, Parallel Algorithms for the Capacitance Matrix Method in Domain Decomposition, to appear in Calcolo 1988 (also I.A.N. Publication n.578)

[MQ1] D.Marini, A.Quarteroni, A Relaxation Procedure for Domain Decomposition Methods using Finite Elements, submitted to Numer. Math. 1988. (also I.A.N. Publication n.577, Pavia)

[FQZ] D.Funaro, A.Quarteroni, P.Zanolli, An Iterative Procedure with Interface Relaxation for Domain Decomposition Methods, to appear in SIAM J. Numer.Anal. December 1988 (also I.A.N. Publication n.530)

[C] P.G.Ciarlet, <u>The Finite Element Method for Elliptic Problems</u> North-Holland, Amsterdam,
1978.

[DR] P.J.Davis, P.Rabinowitz, <u>Methods of Numerical Integration</u>, Academic Press, New York, 1975.

[CHQZ] C.Canuto, M.Y.Hussaini, A.Quarteroni, T.A.Zang, <u>Spectral Methods in Fluid Dynamics</u>, Springer-Verlag, New York, 1988.

[BW] P.E.Bjorstad, O.B.Widlund, Iterative Methods for the Solution of Elliptic Problems on Regions Partitioned into Substructures, SIAM J. Numer. Anal., 23, pp.1097-1121, 1986.

[BMM] C.Bernardi, Y.Maday, B.Metivet, Calcul de la Pression dans la Resolution Spectrale du Problème de Stokes, La Recherche Aerospatiale, 1, pp.1-21, 1986.

[MP] Y.Maday, A.T.Patera, Spectral Element Methods for the Incompressible Navier-Stokes Equations, in <u>State of the Art Surveys in Computational Mechanics</u>, Noor A. Ed., ASME, 1987.

[SV] G.Sacchi Landriani, H.Vandeven, A Multidomain Spectral Collocation Method for the Stokes Problem, submitted to Numer.Math. 1988 (also I.A.N. Publication n.642, Pavia).

[GQ] F.Gastaldi, A.Quarteroni, On the Coupling of Hyperbolic and Parabolic Systems: Analytical and Numerical Approach, to appear in Applied Numerical Mathematics (also I.A.N. Publication n.619)

[MQ2] D.Marini, A.Quarteroni, An Iterative Procedure for Domain Decomposition Methods: A Finite Element Approach. In First Int. Conf. on <u>Domain Decomposition Methods for Partial Differential Equations</u>, R.Glowinski, G.H.Golub, J.Periaux, eds.SIAM, Philadelphia 1987.

[T] R.Teman, <u>Navier-Stokes Equations</u>, North-Holland, Amsterdam, 1977.

[B] F.Brezzi, On the Existence Uniqueness and Approximation of a Saddle-Point Problem Arising from Lagrange Multipliers, RAIRO Anal. Numer. 8-R2, pp.129-151, 1974.

[BF] F.Brezzi, M.Fortin, Book, to appear.

[Q1] A.Quarteroni, Domain Decomposition Algorithms for the Stokes Equations, in 2nd Int. Conf. on <u>Domain Decomposition Methods for Partial Differential Equations</u>, T.F.Chan et al., Eds.,
SIAM, Philadelphia, 1988 (to appear) (also I.A.N. Publication n.612)

[M] B.Metivet, <u>Resolution Spectrale des Equations de Navier-Stokes par une Methode de Sous-Domaines Courbes</u>, these Univ. Paris VI, 1987.

[Q2] A.Quarteroni, Domain Decomposition Approximations for Systems of Conservation Laws: Spectral Collocation Approximations, ICASE-Report, NASA Langley Research Center, Hampton, VA (in press).

THE EXTENDED LIFTING LINE THEORY FOR SYSTEMS OF SAILS

M. WOHLFAHRT

1. Introduction

This contribution contains some aspects of the Ph.D. thesis of the author [9]. In this thesis a theory for the aerodynamic and geometric properties of systems of sails is developed. It describes the feed-back between the aerodynamic forces and the tensions in the sail. The aerodynamic effects are described by Weissinger's extended lifting line theory [8]. In accordance with this one-dimensional theory the torsion of the sail-membrane is described by a one-dimensional equation following the idea of Nickel [3].

Even for the treatment of a boat with one single sail a theory for two symmetric sails with a gap is necessary. This is one reason why systems of sails are discussed in this work.

Prandtl formulated two main-problems [4] for the lifting line theory. Transferred to the sail, the first one raises the question, which sail(-geometry) produces a desired lift distribution. The second one is the inversion of the first one. Both problems are discussed with regard to solvability properties and to the numerical solution.

Examples for the cases of a hangglider and a boat with one sail are given. Optimal sails with minimal induced drag are especially interesting and are, therefore, treated too.

2. The extended lifting line theory for systems of wings

The idea of the lifting line theories (Prandtl [4]) is to replace the wing by one single lifting line. It is placed on one quarter of the chord length. The whole chordwise circulation is concentrated on this line. The concept of Weissinger's extended lifting line theory is to satisfy the downwash-condition at three quarters of the chord length.

We denote with

ℓ the chord length,

f   the chord/4-line,

Γ the circulation,

α the angle of attack,

V   the speed of the air.

Let the coordinate system be chosen as in Fig. 1. The wing reaches from y = a to y = b. Then the variables above are functions from I := [a,b] into $\mathbb{R}$. If we have a number N (N > 1) of wings, we identify the values of the μ-th wing by an upper index μ in brackets.

It is easy to generalize Weissinger's lifting line theory [8] to more than one wing. The downwash condition leads to

$$\sum_{\nu=1}^{N} \left(\frac{1}{2\pi} \oint_{I^{(\nu)}} \frac{\Gamma^{(\nu)'}(u)}{y-u} \, du + \frac{1}{4\pi} \int_{I^{(\nu)}} K^{(\mu,\nu)}(y,u) \Gamma^{(\nu)}(u) du \right)$$

$$= V(y)\alpha^{(\mu)}(y) \qquad (y \in I^{(\mu)}, \ \mu = 1(1)N), \qquad (2.1)$$

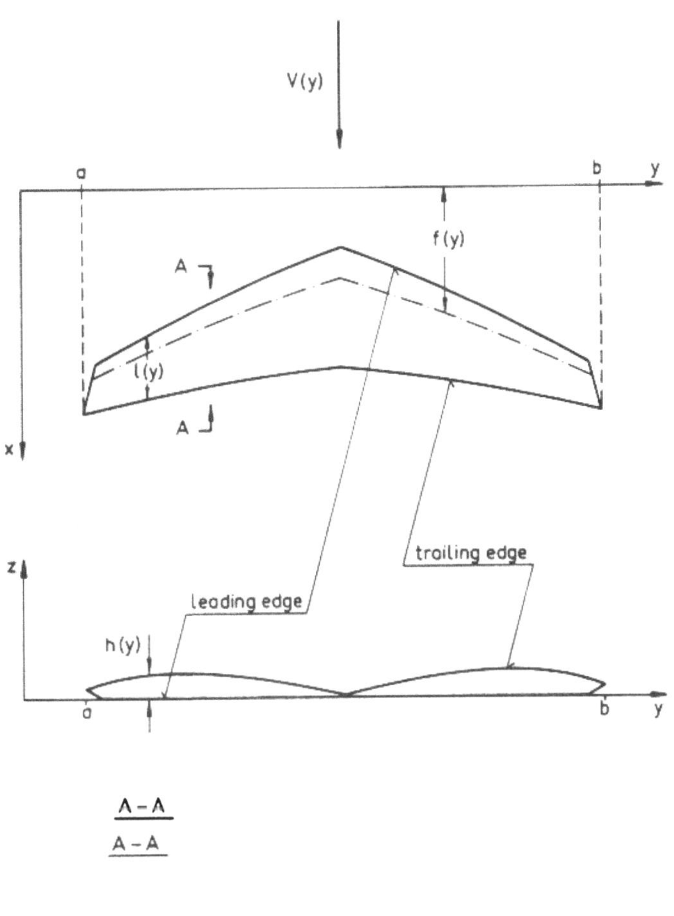

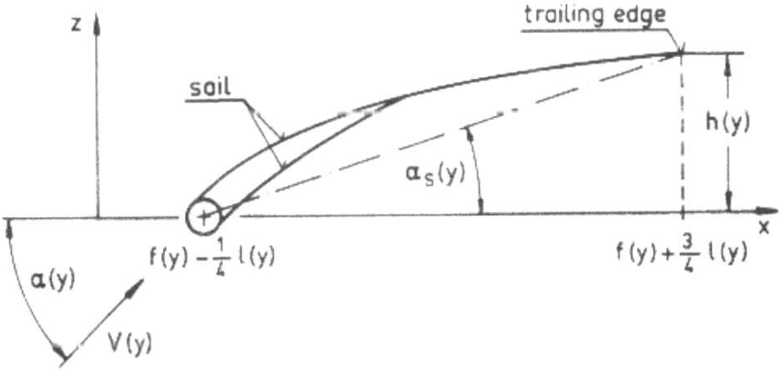

Fig. 1: A typical hangglider

where

$$K^{(\mu,\nu)}(y,u) := \frac{1}{(y-u)^2} \cdot$$

$$\left(1 - \frac{1+\varphi^{(\mu,\nu)}(y,u)}{((1+\varphi^{(\mu,\nu)}(y,u))^2 + \frac{4}{\ell^{(\mu)}(y)^2}(y-u)^2)^{1/2}}\right) \qquad (2.2)$$

with

$$\varphi^{(\mu,\nu)}(y,u) := 2\,\frac{f^{(\mu)}(y)-f^{(\nu)}(u)}{\ell^{(\mu)}(y)} \cdot \qquad (2.3)$$

3. Application of the extended lifting line theory to systems of sails

In accordance with the idea of Nickel's theory of sail-wings [3] we compute the (circulation-dependent) twist α_S of the sail and replace the angle of attack α by $\alpha - \alpha_S$ (see Fig.1). If h means the deformation of the trailing edge, then the sail angle α_S is approximately given by

$$\alpha_S := \frac{h}{\ell} \cdot \qquad (3.1)$$

Analogous to Nickel [3] we get

$$h''(y) = -\frac{\varrho V(y)}{\cos\beta S}\, d(y)\Gamma(y). \qquad (3.2)$$

Herein ϱ means the density of the air, β the sweep of the trailing edge, S the tension of the trailing edge, and $d(y)$ the position of the chordwise centre of pressure.

In the case of the **rigid profile** we have [6]

$$d(y) = \frac{1}{4} - \frac{V(y)\ell(y)}{2} \frac{c_{m_0}(y)}{\Gamma(y)} \tag{3.3}$$

with the coefficient c_{m_0} of the pitch moment. In the case of the **flexible airfoil**, however, d is independent of α, and

$$1/4 \leq d(y) \leq 1/2 \tag{3.4}$$

holds [7]. Therefore $d(y) \cdot \Gamma(y)$ depends in both cases linear on Γ.

According to the conditions for h at the ends of the trailing edge (e.g. $h(a) = h(b) = 0$) and to (3.2) we get the height h(y) of the so-called tunnel

$$h(y) = - \frac{\varrho}{\cos\beta S} \left(\int_I S(y,t)V(t)d(t)\Gamma(t) \, dt + s(y) \right) \tag{3.5}$$

with suitable functions S and s (see Satz 4.3.2 in [9]).
Thus we get the following equations for the circulations of a system of sails:

$$\sum_{\nu=1}^{N} \left(\frac{1}{2\pi} \oint_{I^{(\nu)}} \frac{\Gamma^{(\nu)'}(u)}{y-u} \, du + \frac{1}{4\pi} \int_{I^{(\nu)}} K^{(\mu,\nu)}(y,u)\Gamma^{(\nu)}(u)du \right)$$

$$= V(y)\alpha^{(\mu)}(y) + \frac{\varrho V(y)}{\cos\beta^{(\mu)}S^{(\mu)}\ell^{(\mu)}(y)} \cdot$$

$$\cdot \left(\int_{I^{(\mu)}} S^{(\mu)}(y,u)V(u)d^{(\mu)}(u)\Gamma^{(\mu)}(u)du + s^{(\mu)}(y) \right)$$

$$(y \in I^{(\mu)}, \quad \mu = 1(1)N) \tag{3.6}$$

These are equations in $L^p_*(I^{(\mu)})$, if $\Gamma^{(\mu)} \in \overset{\circ}{W}{}^{1,p}_*(I^{(\mu)})$,
$\alpha^{(\mu)} \in L^p_*(I^{(\mu)})$, $\ell^{(\mu)} \in L^p(I^{(\mu)})$, $\ell^{(\mu)}(y) > \varepsilon > 0$ ($\varepsilon \in \mathbb{R}$), V and
$c_{m_0}^{(\mu)}$ respectively $d^{(\mu)}$ are bounded for all $y \in I^{(\mu)}$ and
$\mu = 1(1)N$. They are equations in $C^{0,\alpha}(I^{(\mu)})$, if $\Gamma^{(\mu)} \in C^{1,\alpha}(I^{(\mu)})$,
$\ell^{(\mu)}$, $f^{(\mu)}$, V, $\alpha^{(\mu)} \in C^{0,\alpha}(I^{(\mu)})$, $\ell^{(\mu)}(y) > 0$ and $c_{m_0}^{(\mu)}$ respectively $d^{(\mu)}$ are bounded for all $y \in I^{(\mu)}$ and $\mu = 1(1)N$.
($L^p_*(a,b)$ denotes the space of summable functions $f:[a,b] \to \mathbb{R}$,
for which

$$\int_a^b |(f(t)|^p ((b-t)(t-a))^{\frac{p-1}{2}}$$

exists. $W^{s,p}_*(I)$ is the Sobolev-space of functions f, for which
$f^{(s)}$ is in $L^p_*(I)$. The circle over the W indicates $f(a)=f(b)=0$.
Finally $C^{s,\alpha}(I)$ is the space of functions f, for which $f^{(s)}$
satisfies a Lipschitz condition with exponent α.)

4. The first main problem

Prandtl formulated two main problems for the lifting line theory
[4]: The first one is, to determine the chord-length $\ell^{(\mu)}$ for
given $\Gamma^{(\mu)}$, $f^{(\mu)}$ and $\alpha^{(\mu)}$.

In the flexible case the determination of $\ell^{(\mu)}(y)$ from (3.6) for
fixed μ and $y \in I^{(\mu)}$ leads to a one-dimensional equation (i.e.:
there is no dependency from μ or y). For each μ and every $y \in I^{(\mu)}$
we have a new problem to solve, which is independent of the
others.

Theorem 4.1: Let (3.6) be an equation in $L_*^p (I^{(\mu)})$. If

$$\Gamma^{(\mu)}(y) > 0 \qquad (y \in I^{(\mu)}, \ \mu = 1(1)N), \tag{4.1}$$

if, furthermore,

$$h^{(\mu)}(a^{(\mu)}) \geq 0, \ h^{(\mu)}(b^{(\mu)}) \geq 0 \quad (\mu = 1(1)N) \tag{4.2}$$

and if, finally,

$$V(y)\alpha^{(\mu)}(y) > \sum_{\nu=1}^{N} \frac{1}{2\pi} \oint_{I^{(\nu)}} \frac{\Gamma^{(\nu)'}(u)}{y-u} \, du, \tag{4.3}$$

then $\ell^{(\mu)}(y)$ exists and is uniquely determined for all $y \in I^{(\mu)}$ and $\mu = 1(1)N$.

If in addition (3.6) is an equation in $C^{0 \cdot \alpha}(I^{(\mu)})$, then $\ell^{(\mu)} \in C^{0 \cdot \alpha}(I^{(\mu)})$ holds too.

Remark 4.2:

1) The condition (4.3) is also necessary!

2) In the most interesting case of the "elliptical" circulation the resulting chord-length for the straight rigid wing is <u>not</u> elliptical, as Prandtl's lifting line would suggest [4].

For the numerical solution of equation (3.6) we approximate the integrals with Multhopp's formula [2] for the Cauchy-Integral in (3.6) and with the trapezoidal rule. This leads again to independent problems for $\ell^{(\mu)}$. Existence and uniqueness can be proven. The assumptions are analogous to those in Theorem 4.1. For $\Gamma^{(\mu)} \in C^{2 \cdot \beta}(I^{(\mu)})$ $(0 < \beta < 1)$ convergence is proven.

<u>**Remark 4.3:**</u>

1) Higher convergence order can be achieved for $\Gamma^{(\mu)} \in C^{\eta \cdot \beta}(I^{(\mu)})$
   if $2 < \eta \leq 4$ or with $\eta \geq 4$ and with the Simpson-rule instead
   of the trapezoidal rule.

2) For $S^{(\mu)} \to \infty$ one gets the equations for rigid wings.

5. The second main problem

Prandtl's second main problem is the inversion of the first: For
given $\ell^{(\mu)}$, $f^{(\mu)}$ and $\alpha^{(\mu)}$ one looks for $\Gamma^{(\mu)}$. From this point of
view the equation (3.6) is a system of integro-differential
equations for $\Gamma^{(\mu)}$.

On the condition that equation (3.6) is an equation in L_*^p an
equivalent regularization is possible: The system of integrodifferential equations (3.6) is equivalent to a system of Fredholm
equations of second order. Therefore the Fredholm alternative
[10] holds true: If equation (3.6) has one solution at most,
then one solution exists. Thus one is interested in uniqueness-
conditions.

In the case of one single sail, several uniqueness-conditions
have been deducted. For $S = 0$ and $S \to \infty$ it is easy to prove uniqueness. Another approach is possible with a contraction criterion according to Schleiff [5]. The best results however were
obtained with the argumentation with positive definiteness:

<u>Theorem 5.1:</u> Let $h(a) \geq 0$ and $h(b) \geq 0$. Then there is at most
one non-negative solution of (3.6).

Theorem 5.2: Let (3.6) be an equation in L_*^2 , let V, d and ℓ be piecewise continuous, let h(a) = h(b) = 0 and assume

$$2\pi^2 e(y) \geq \left(\int_a^b e(u) \, du \right)^2 \left(\frac{1}{e(y)} \right)''$$

with $e(y) := \dfrac{\varrho V(y)}{\cos\beta S} d(y) \cdot \ell(y)$.

Then (3.6) has for every S at most one solution.

The numerical solution of (3.6) can be done similarly to Multhopp's method [2]. The integrals are approximated by replacing $\Gamma^{(\mu)}$ by trigonometric polynomials of order M. The resulting equation is satisfied at M points (collocation). This leads to a $(M \cdot N) \times (M \cdot N)$ system of linear equations for the functions $\Gamma^{(\mu)}$, each at M points of $I^{(\mu)}$.
For only one sail (N=1) and the L_*^2 case a convergence theorem has been given.

6. Examples

For hanggliders and sailboats with one sail some solutions to the first and to the second main problem are presented in what follows. The typical shape of a hangglider can be seen from Fig. 1. Typical data are: b = - a = 5 m and the wing area F = 16 m^2. In the middle of the sail h(0) = 0 always holds. Most hanggliders have a condition of the form $\dfrac{h'(b)}{h(b)} = - \dfrac{h'(a)}{h(a)} = C$ with a suitable C $\geq$ 0 at the ends of the sail. The weight of a

hangglider is about 1000 N. Therefore the total lift in
stationary flight has approximately the same value.

6.1 Hanggliders with optimal circulation

The elliptical circulation

$$\Gamma_{ell}(y) := \Gamma_0(1-(y/b)^2)^{1/2}$$

has for given total lift

$$A := \varrho V \int_a^b \Gamma(y)dy$$

the lowest induced drag

$$W_i := \frac{\varrho}{4\pi} \int_a^b \Gamma(y) \oint_a^b \frac{\Gamma'(u)}{y-u} \, du \, dy.$$

From $A = 1000$ N we get the value of Γ_0. For practical reasons α
is constant and the area of the new sail is unchanged $16m^2$.
For simplicity's sake we assume $c_{m_0} = 0$. Furthermore we choose
$V = 10m/s$, $f(y) := 0,4|y|$, $C = -0,5/m$ and $S = 200$ N, if nothing
else is mentioned. The number M in all computations is 63.

Influence of the parameter C

This parameter describes the behaviour of the trailing edge at
the wing tips. $C = \infty$ means $h(a) = h(b) = 0$; the trailing edge is
fixed. The opposite case $C = 0$ stands for a free-rotating rib on
the wing tip. If the rib on the wing tip is flexibly attached
then $0 < C < \infty$ holds true (see Fig. 6.1).

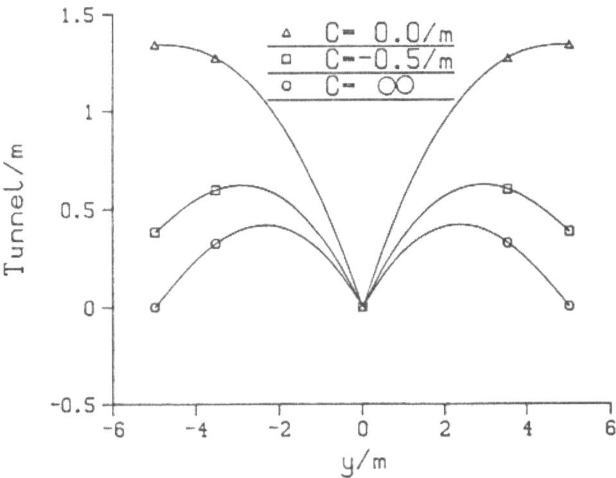

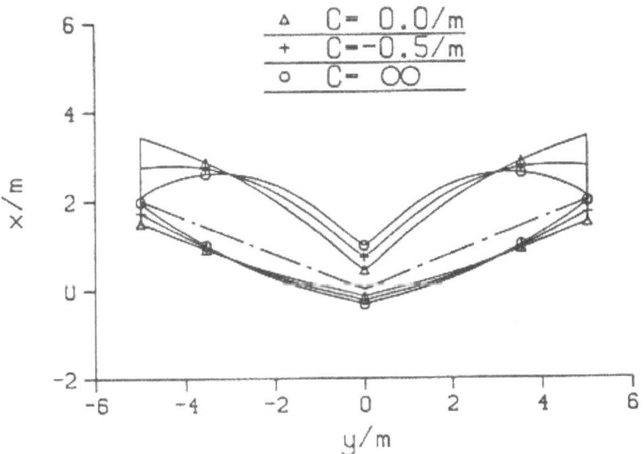

Fig. 6.1: Chord and tunnel for different C

Influence of the tension S

Decreasing tension S increases the tunnel h. This leads to stronger washout and requires more chord whereever the tunnel is high (see Fig. 6.2).

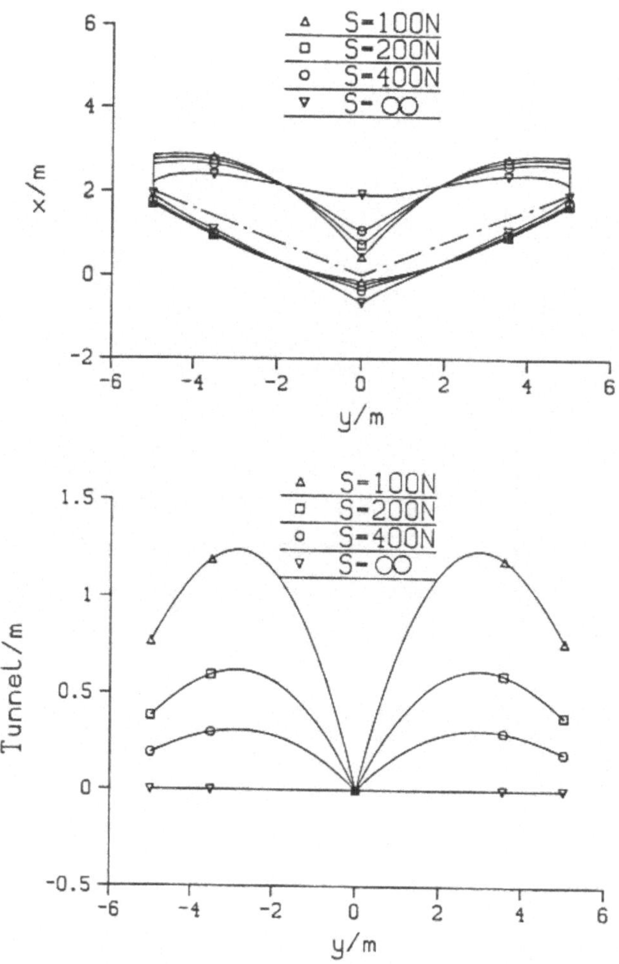

Fig. 6.2: Chord and tunnel for different tensions S

Rigid wings

Tension S → ∞ leads to a theory for rigid wings. Fig. 6.3.
illustrates the influence of the sweep-back on optimal chord-
distributions. The chord of the straight wing is not elliptical.

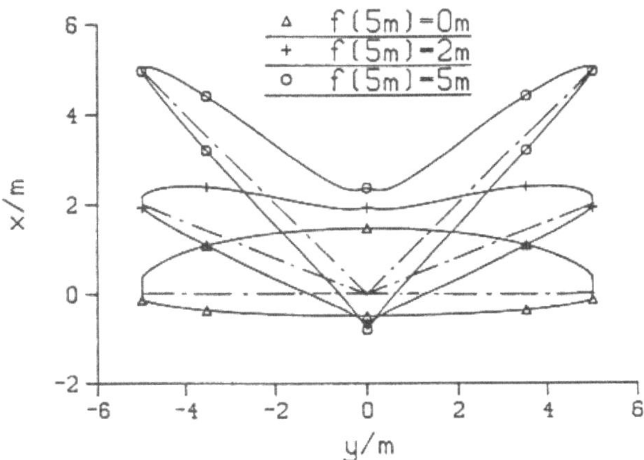

Fig. 6.3: Optimal chord distributions for swept rigid wings

6.2 Circulations of hanggliders

In this paragraph the solutions to the second main problem are
demonstrated for a conventional hangglider as shown in Fig. 1.
We assume $\ell(y) = 2,4$ m $- 0,32|y|$ and standard data as above. The
total lift A = 1000 N is prescribed and the (constant) angle of
attack α is chosen accordingly.

Influence of the parameter C

Fig. 6.4 shows the importance of the method of attachement of
the trailing edge at the wing tips. With decreasing C the qua-
lity of the circulation distribution gets worse.

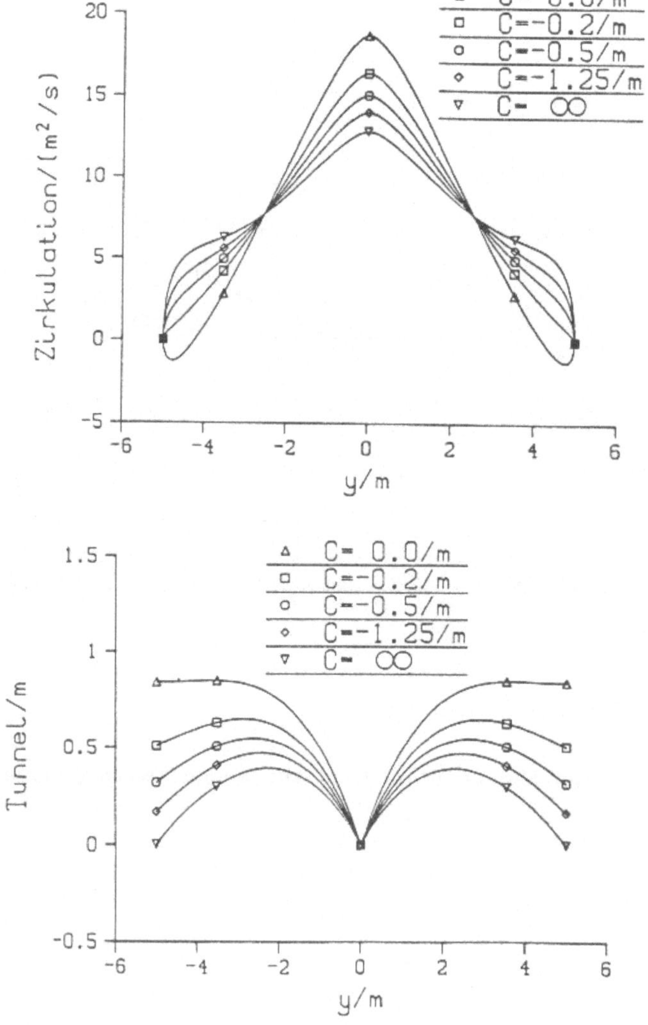

Fig. 6.4: Circulation and tunnel for different C

Influence of the tension S

Lower tension leads to a higher tunnel (see Fig. 6.5). Therefore there is a loss of circulation in the outer range of the sail. The circulation gets deformed and the induced drag is increased.

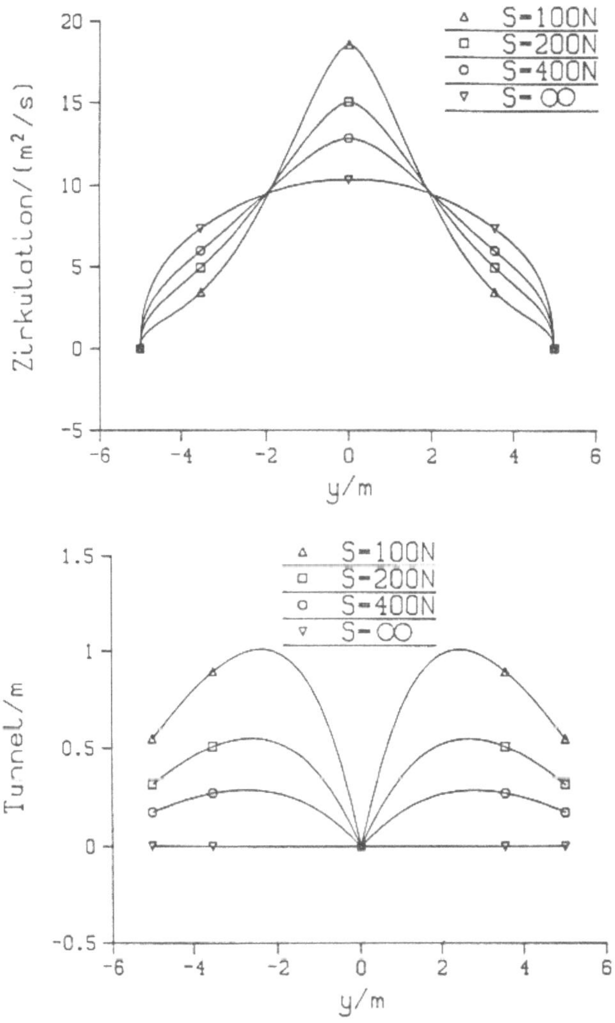

Fig. 6.5: Circulation and tunnel for different tensions S

Influence of the speed V

The tunnel does not vary greatly with the speed. But the circulation in the outer areas of the wing tends to zero (see Fig. 6.6).

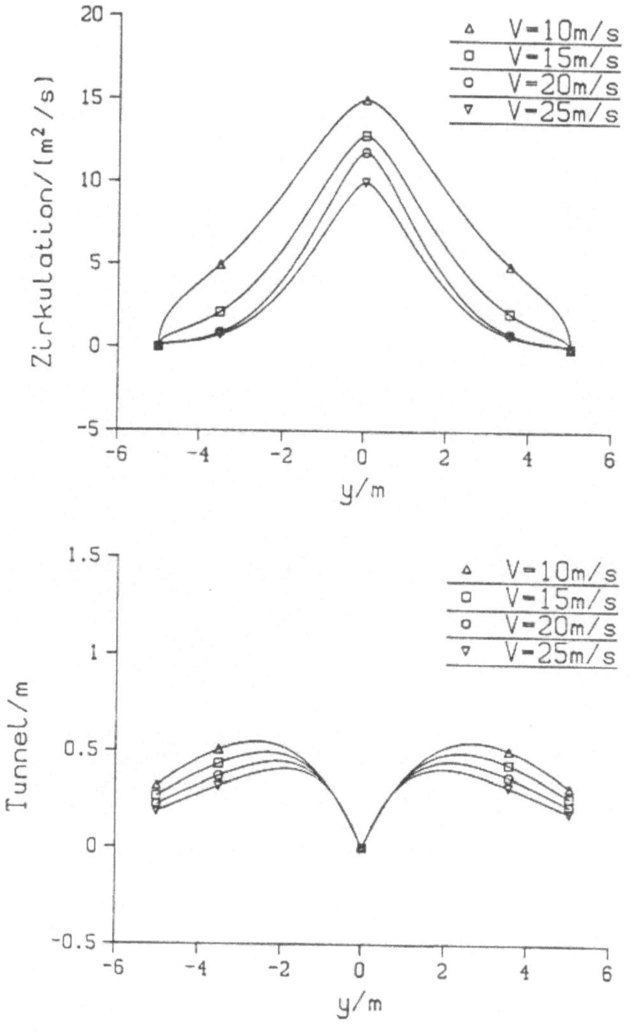

Fig. 6.6: Circulation and tunnel for different speed V

6.3 Optimal sails for sailboats with one sail

Even for sailboats the induced drag - and therefore the distribution of the circulation - is very important [1].

We denote the speed of the boat with V_S, the speed of the wind with V_w, and the course-angle between V_S and V_w with γ_w. Then we have from Fig. 6.7 the speed of the appearant wind

$$V = ((V_w \sin\gamma_w)^2 + (V_S + V_w \cos\gamma_w)^2)^{1/2}$$

and the angle of attack (against V_S)

$$\gamma = \arctan \frac{V_w \sin\gamma_w}{V_S + V_w \cos\gamma_w}$$

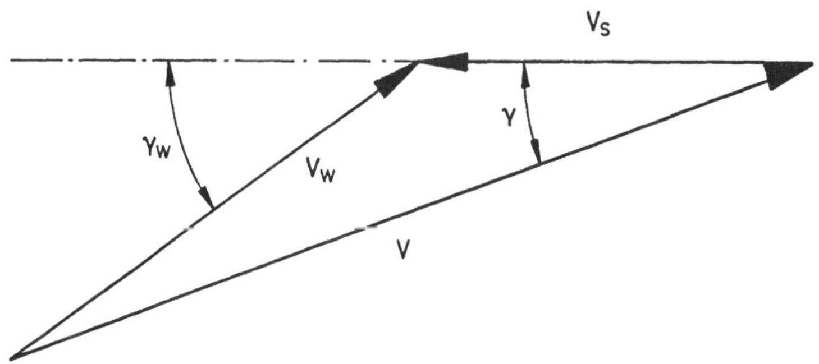

Fig. 6.7: Speed of wind and boat

The speed of the wind V_w varies with the distance y to the water surface (y=0). We assume that this wind-boundary-layer is given by (s. Marchaj [1])

$$V_w(y) = \frac{y}{0.87m + y} \cdot 7.3 \ \frac{m}{s} \ .$$

Therefore V and α also depend on y. We assume that the sail reaches from y = 0,6m to 6,6m, that its area is 10,2m^2 and that h(a) = h(b) = 0. The course angle γ_w = 38° and the boat speed V_S = 2,318m/s are measured values for a Finn-Dinghy (see Marchaj [1]). The angle of attack is α(y) := $\gamma(\eta)$-δ (δ position of the sail). δ is chosen in such a way, that the total lift is 350 N. Then for a straight chord/4-line some sails with elliptic circulation for different tensions S are shown in Fig. 6.8. For high tensions S the tips at y = a result from the boundary layer of the wind!

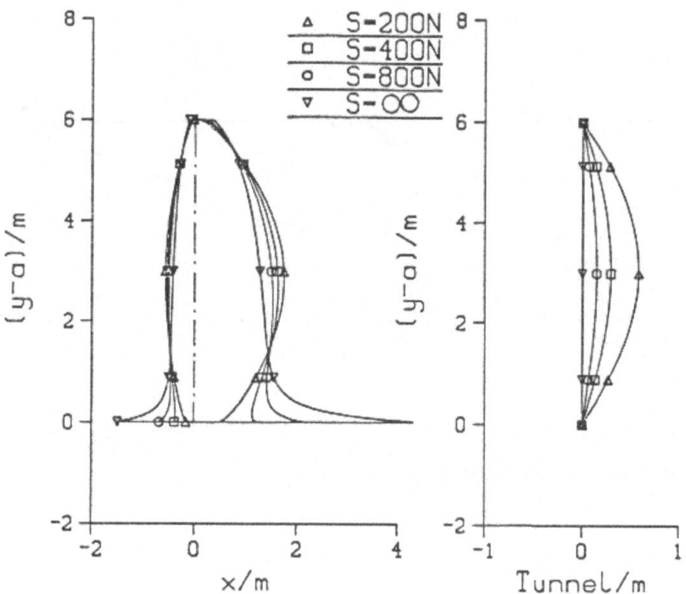

Fig. 6.8: Chord and tunnel for different tensions S

6.4 Circulation of a boat sail

Finally the effect of an increased tunnel on the circulation is
shown (see Fig. 6.9). The sail loses lift at the top and gains
(due to constant lift) at the bottom. The growing tunnel
increases the induced drag once again.

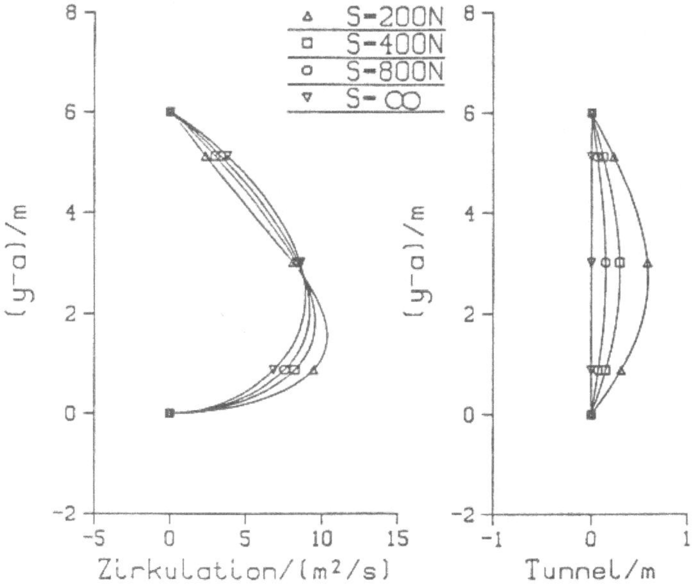

Fig. 6.9: Circulation and tunnel for different tensions S

7. Literatur

[1]   Marchaj, C.A.: Aerodynamik und Hydrodynamik des Segelns.
       Bielefeld: Verlag Delius, Klasing und Co. 1982.

[2]   Multhopp, H.: Die Berechnung der Auftriebsverteilung von
       Tragflügeln. Luftfahrtforschung 15 (1938) 153-169.

[3]   Nickel, K.: A Theory of Sail-Wings. Z. Flugwiss. Welt-
       raumforsch. 11 (1987) 321-328.

[4]   Prandtl, L.: Tragflügeltheorie I. Nachr. Ges. Wiss.
       Göttingen, math.-phys. Kl. (1918) 451-477.

[5]   Schleiff, M.: Über Näherungsverfahren zur Lösung einer
       singulären Integrodifferentialgleichung. ZAMM 48 (1968)
       477-483.

[6]   Schlichting, H.; Truckenbrodt, E.: Aerodynamik des Flug-
       zeuges. Berlin: Springer-Verlag 1959.

[7]   Voelz, K.: Profil und Auftrieb eines Segels. ZAMM 30
       (1950) 302-317.

[8]   Weissinger, J.: Über eine Erweiterung der Prandtl'schen
       Theorie der tragenden Linie. Math. Nachr. 2 (1949)
       46-109.

[9]   Wohlfahrt, M.: Anwendung der erweiterten Traglinien-
       theorie auf Systeme von Segeln. Dissertation, Albert-
       Ludwigs-Universität Freiburg i.Br., West-Germany (1988).

[10]   Zabreyko, P.P.: Integral equations - a reference text.
       Leyden: Noordhoff International Publishing 1975.

Michael Wohlfahrt

Institut für Angewandte Mathematik

Universität Freiburg

D-7800 Freiburg i.Br.

West-Germany

Scattering kernel formulation of nonlinear extended kinetic theory

S.Oggioni[+], F.Premuda[x], G.Spiga[*]

Summary.   In the frame of an extended kinetic theory for gas mixtures, the scattering kernels appearing in the elastic scattering collision terms are explicitly evaluated for any value of the mass ratio between the colliding particles. Their expressions are then used in order to obtain the relevant moment equations needed for the solution of the generalized transport problem. A Laguerre series expansion for the polynomial reconstruction of the distribution functions is finally derived, and the influence of different physical effects is briefly discussed.

1. Introduction

In a recent series of papers[14,7,17] an extended kinetic theory for gas mixtures, allowing for binary chemical or nuclear reactions, and including effects of background media and external sources, has been presented. Elastic scattering collisions have been treated in the frame of the so called scattering kernel formulation of the nonlinear Boltzmann equation[4]. Several meaningful simple cases have been solved either analytically or numerically[15,5,11,19,20] and in this respect an important role has been played by the moment method[16], coupled to group theoretical techniques for evaluating similarity solutions for the generating function of the moments[8] One of the main hindrances in the development of the theory is the lack of an explicit expression of the scattering kernel itself, which has been cast in analytical form only in the case of scattering between equal Maxwellian molecules in isotropic physical conditions and according to the Krook-Wu interaction model (constant collision frequency, differential cross section independent of the deflection in the center of mass system). The quite cumbersome task of bridging this gap is undertaken in this paper, where the scattering kernels for the isotropic Krook-Wu model are explicitly given for any mass ratio between the colliding particles. Furthermore, the moments of the scattering kernel with respect to the velocity after collision are evaluated, and the coefficients to be used in the set of moment equations are determined. Finally, it is shown that a polynomial reconstruction of the distribution functions f, in the presence of removal collisions and external sources, is possible in terms of suitable Laguerre polynomials. The Laguerre coefficients are in fact solutions of essentially the same set of equations derived for the ordinary moments. All goals achieved here are thus generalizations of previous results, available in the transport theory literature, to the case of a mixture of different gases, and/or to the presence of removal, background, and regeneration effects. The set of nonlinear, integrodifferential Boltzmann-like transport equations, homogeneous in space and isotropic in velocity, for a mixture of N species reads as

$$\frac{\partial f_i}{\partial t} + \sum_{j=1}^{N} C_{ij}\rho_j(t)f_i(v,t) -$$

$$- \sum_{j=1}^{N} C_{ij}^{S} \iint \pi_{ij}(v',w'\rightarrow v)f_i(v',t)f_j(w',t)d\underline{v}'d\underline{w}' + Q_i(v,t) \tag{1}$$

with $i=1,...,N$ and initial conditions $f_i(v,0)=f_i^{0}(v)$. With reference to the i-th species, symbols S and R are used to label scattering and removal collisions, respectively, Q_i is the intensity of the external source, $C_{ij}=C_{ij}^{S}+C_{ij}^{R}$ is the total (constant) frequency for collision with particles of the j-th species, and ρ_i denotes the number density

$$\rho_i(t) - \int f_i(v,t)d\underline{v} \; . \tag{2}$$

The scattering kernels π_{ij} are expressed as[1]

$$\pi_{ij}(v',w'\rightarrow v) - \tfrac{1}{2}\int_{-1}^{+1} P_{ij}(v',w',\mu\rightarrow v)d\mu \tag{3}$$

in terms of the Amaldi distributions, defined by

$$P_{ij}(v',w',\mu\rightarrow v) - \frac{(1+r_{ij})^2}{8\pi v} \; |r_{ij}\underline{v}'+\underline{w}'|^{-1}|\underline{v}'-\underline{w}'|^{-1} \tag{4}$$

for $||r_{ij}\underline{v}'+\underline{w}'|-|\underline{v}'-\underline{w}'||<(1+r_{ij})v<|r_{ij}\underline{v}'+\underline{w}'|+|\underline{v}'-\underline{w}'|$, and by zero otherwise, as follows from conservation of both momentum and kinetic energy in an elastic scattering collision between particles of species i and j, with mass ratio $r_{ij}=m_i/m_j$. The variable μ represents the cosine of the angle between the vector velocities before collision, $\underline{v}'$ and $\underline{w}'$. If necessary, one or more background species can be taken into account by considering the relevant distribution functions as known, and dropping the relevant equation from the set (1).

Selfgenerating processes could be easily added to Eq.(1), but will not be considered in this work. For later reference we define the ordinary and Laguerre moments of the distribution functions as

$$M_n^{\,i}(t) = \frac{1}{(2n+1)!!} \int v^{2n} f_i(v,t) \, d\underline{v} \tag{5}$$

and

$$\xi_n^{\,i}(t) = \frac{2^n n!}{(2n+1)!!} \int L_n^{(\frac{1}{2})}(\beta_i v^2) f_i(v,t) \, d\underline{v} \;, \tag{6}$$

the latter being needed in the Laguerre polynomial representation

$$f_i(v,t) = M(v,\beta_i) \sum_{n=0}^{\infty} \xi_n^{\,i}(t) \, L_n^{(\frac{1}{2})}(\beta_i v^2) \tag{7}$$

where the parameters $\beta_i = m_i/2KT_i$ will be chosen later on, and M denotes the normalized Maxwellian

$$M(v,\beta_i) = (\beta_i/\pi)^{3/2} \exp(-\beta_i v^2) \;. \tag{8}$$

In order to derive moment equations, the moments of the scattering kernel

$$I_n^{\,ij}(v',w') = \frac{1}{(2n+1)!!} \int v^{2n} \pi_{ij}(v',w' \rightarrow v) \, d\underline{v} \tag{9}$$

are also needed. In the present frame, any field particle distribution function can be adequately treated, and in particular the most meaningful case of a Maxwellian at a given temperature, enlarging thus the results of the papers referenced to above, where field particles free and at rest (delta like distributions) have mainly been considered. In the following sections the structure of π_{ij}, $I_n^{\,ij}$, of the moment equations for $M_n^{\,i}$ and $\xi_n^{\,i}$ are analyzed and discussed versus the basic parameters r_{ij}, in view of possible applications, especially to a binary mixture, or to a single species in a background medium.

2. **Evaluation of the scattering kernels**

In order to simplify notation, indices i and j will be dropped in this section, and it is implicitly understood that we are dealing with binary elastic scattering of a particle i of velocity $\underline{v}'$ impinging on a particle j of velocity $\underline{w}'$, the particles being given velocity $\underline{v}$ and $\underline{w}$, respectively, after collision. Conservation laws read as

$$r\,\underline{v} + \underline{w} = r\,\underline{v}' + \underline{w}' \qquad\qquad r\,v^2 + w^2 = rv'^2 + w'^2 \tag{1}$$

and imply $|\underline{v} - \underline{w}| = |\underline{v}' - \underline{w}'|$. Introduce the new variable

$$x = |\underline{v}' - \underline{w}'| = \left[v'^2 + w'^2 - 2v'w'\mu \right]^{\frac{1}{2}} \tag{2}$$

which is a monotonically decreasing function of $\mu\epsilon(-1,1)$, so that

$$|v' - w'| < x < v' + w' \ . \tag{3}$$

The integral in Eq.(1.3) can be performed by using x as integration variable. The condition for the integrand P to be nonvanishing, namely

$$\left| [(1+r)E - rx^2]^{\frac{1}{2}} - x \right| < (1+r)v < [(1+r)E - rx^2]^{\frac{1}{2}} + x, \tag{4}$$

where $E = rv'^2 + w'^2$, can be solved in terms of x. After some algebra, one gets the condition

$$|v - w| < x < v + w, \tag{5}$$

in which $w = (E - rv^2)^{\frac{1}{2}}$ as follows from Eq.(2.1), plus the constraint on v

$$v_m < v < v_M, \tag{6}$$

where upper and lower bounds are functions of r, v' and w', given by

$$v_m = \begin{cases} \dfrac{(1-r)v' - 2w'}{1 + r} & \text{for } r<1, \ v'>2w'/(1-r) \\[2mm] \dfrac{(r-1)v' - 2w'}{1 + r} & \text{for } r>1, \ v'>2w'/(r-1) \\[2mm] 0 & \text{otherwise} \end{cases} \tag{7}$$

and

$$v_M = \begin{cases} \dfrac{(1-r)v' + 2w'}{1 + r} & \text{for } r<1, \ v'<(1-r)w'/2r \\[2mm] \dfrac{(r-1)v' + 2w'}{1 + r} & \text{for } r>1, \ v'<(r-1)w'/2r \\[2mm] [E/r]^{\frac{1}{2}} & \text{otherwise}, \end{cases} \tag{8}$$

respectively. The sought scattering kernel becomes thus

101

$$\pi(v',w'\to v) = \frac{(1+r)^2}{4}\ \frac{1}{4\pi vv'w'}\ \int [(1+r)E-rx^2]^{-\frac{1}{2}}dx \tag{9}$$

where the integration domain is the intersection of the intervals (2.3) and (2.5), and the integrand is simply the derivative of $r^{-\frac{1}{2}}\sin^{-1}\{x[(1+r)E/r]^{-\frac{1}{2}}\}$. The problem is reduced then to determining actual upper and lower integration limits in (2.9), for fixed r,v', and w', with v varying according to Eq.(2.6). When $r=1$, the result of Ref.14 is easily recovered. When instead $r\neq 1$, all symmetries are broken, and one has to partition the problem into very many separate subcases, with analytical expressions for π different from each other. For a complete treatment of all possible cases, one may refer to the following quantities

$$v_1 = \frac{2w' - (1-r)v'}{1+r} \qquad v_2 = \frac{2w' + (1-r)v'}{1+r}$$

$$w_1 = \frac{2rv' + (1-r)w'}{1+r} \qquad w_2 = \frac{2rv' - (1-r)w'}{1+r} \tag{10}$$

where $|w_i|=(E-rv_i^2)^{\frac{1}{2}}$, $i=1,2$. They come from the comparison of $|v-w|$ to $|v'-w'|$, and of $v+w$ to $v'+w'$, whereas the inequalities $|v-w|\leq v'+w'$ and $|v'-w'|\leq v+w$ are always in order; in fact, $|v_1|$, $|v_2|$ and v' correspond to the values taken by v when π changes from one explicit expression to another. In any event, the scattering kernel is always given in terms of inverse sine functions, and, on the basis of the integration limits, there are only four possibilities, namely:

A) $|v'-w'|\leq|v-w|\leq v+w\leq v'+w'$,

B) $|v-w|\leq|v'-w'|\leq v'+w'\leq v+w$,

C) $|v'-w'|\leq|v-w|\leq v'+w'\leq v+w$,

D) $|v-w|\leq|v'-w'|\leq v+w\leq v'+w'$.

Results are listed below after rearrangement according to the addition theorem.

$$\pi^A(v',w'\to v) = \frac{(1+r)^2}{4r^{\frac{1}{2}}}\ \frac{1}{4\pi vv'w'}\ \sin^{-1}\left[\frac{2r^{\frac{1}{2}}v}{E}\ (E-rv^2)^{\frac{1}{2}}\right] \tag{11a}$$

$$\pi^B(v',w'\to v) = \frac{(1+r)^2}{4r^{\frac{1}{2}}}\ \frac{1}{4\pi vv'w'}\ \sin^{-1}\left[\frac{2r^{\frac{1}{2}}v'w'}{E}\right] \tag{11b}$$

$$\pi^C(v',w'\rightarrow v) = \frac{(1+r)^2}{4r^{\frac{1}{2}}} \frac{1}{4\pi vv'w'} \sin^{-1}\left[\frac{r^{\frac{1}{2}}}{E} [w_2v+v_2 (E-rv^2)^{\frac{1}{2}}]\right] \qquad (11c)$$

$$\pi^D(v',w'\rightarrow v) = \frac{(1+r)^2}{4r^{\frac{1}{2}}} \frac{1}{4\pi vv'w'} \sin^{-1}\left[\frac{r^{\frac{1}{2}}}{E} [w_1v+v_1 (E-rv^2)^{\frac{1}{2}}]\right]. \qquad (11d)$$

Where appearing, w has again to be understood as $(E-rv^2)^{\frac{1}{2}}$ and E as $rv'^2+w'^2$. The final expression for the scattering kernel as a function of v for fixed values of r, v' and w' follows then from a patient and careful examination of all 30 subcases to be singled out. Results are sketched in Table 1, where the capital letter refers to the scheme (2.11), and the v_i's, i=1,2, are given by Eq.(2.10). Notice that either v_1 or v_2 may become negative, but only their moduli appear in the table. Indeed, it is worth remarking that, when r<1, v_2 is always positive and greater than v_1; furthermore, $v_M=v_2$ as soon as $w_2<0$, and $v_m=-v_1$ as soon as $v_1<0$. The same is true, when r>1, provided v_1 and v_2 are interchanged. The value of v_M, where occurring, turns out to be given by $(E/r)^{\frac{1}{2}}$; it is meant throughout that v_m (v_M) always coincides with the minimum (respectively, maximum) value of v reported in the picture. Of course, π vanishes when v is outside the interval shown for each case. For instance, if 1/3<r<1 and w'<v'<w'/r, then the scattering kernel is given by π^A for $0=v_m<v<v_1$ and $v_2<v<v_M=(E/r)^{\frac{1}{2}}$, by π^D for $v_1<v<v'$, by π^C for $v'<v<v_2$, and by zero otherwise.

3. <u>Moments of the scattering kernels</u>

This section is devoted to the evaluation of the integrals (1.9), and again it is implicitly understood that binary elastic scattering of a particle i against a particle j is being dealt with, so that the labels i and j will be dropped. On using once more Eq.(2.9), one gets, after a permissible inversion of integration order

$$I_n(v',w') =$$

$$= \frac{2^{n-2} n!}{(2n+2)!} \frac{1}{(1+r)^{2n}} \frac{1}{v'w'} \int_{|v'-w'|}^{|v'+w'|} (R-rx^2)^{-\frac{1}{2}}\{[(R-rx^2)^{\frac{1}{2}}+x]^{2n+2}-[(R-rx^2)^{\frac{1}{2}}-x]^{2n+2}\}dx \qquad (1)$$

where

$$R = (1+r)E=(1+r)(rv'^2+w'^2). \qquad (2)$$

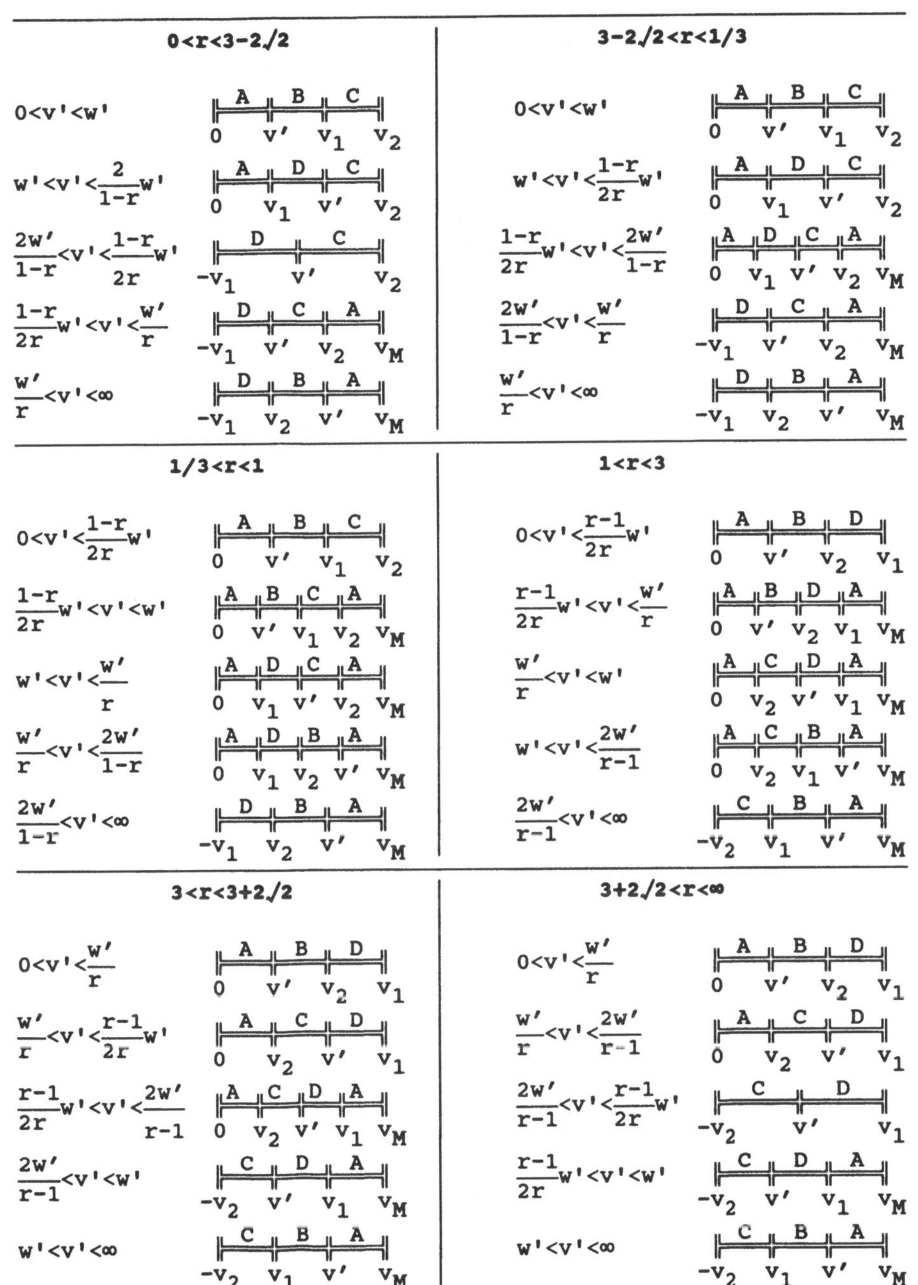

0 < r < 3 - 2√2

0 < v' < w'   A B C   0 v' v₁ v₂
w' < v' < (2/(1-r))w'   A D C   0 v₁ v' v₂
(2w'/(1-r)) < v' < ((1-r)/(2r))w'   D C   -v₁ v' v₂
((1-r)/(2r))w' < v' < (w'/r)   D C A   -v₁ v' v₂ v_M
(w'/r) < v' < ∞   D B A   -v₁ v₂ v' v_M

3 - 2√2 < r < 1/3

0 < v' < w'   A B C   0 v' v₁ v₂
w' < v' < ((1-r)/(2r))w'   A D C   0 v₁ v' v₂
((1-r)/(2r))w' < v' < (2w'/(1-r))   A D C A   0 v₁ v' v₂ v_M
(2w'/(1-r)) < v' < (w'/r)   D C A   -v₁ v' v₂ v_M
(w'/r) < v' < ∞   D B A   -v₁ v₂ v' v_M

1/3 < r < 1

0 < v' < ((1-r)/(2r))w'   A B C   0 v' v₁ v₂
((1-r)/(2r))w' < v' < w'   A B C A   0 v' v₁ v₂ v_M
w' < v' < (w'/r)   A D C A   0 v₁ v' v₂ v_M
(w'/r) < v' < (2w'/(1-r))   A D B A   0 v₁ v₂ v' v_M
(2w'/(1-r)) < v' < ∞   D B A   -v₁ v₂ v' v_M

1 < r < 3

0 < v' < ((r-1)/(2r))w'   A B D   0 v' v₂ v₁
((r-1)/(2r))w' < v' < (w'/r)   A B D A   0 v' v₂ v₁ v_M
(w'/r) < v' < w'   A C D A   0 v₂ v' v₁ v_M
w' < v' < (2w'/(r-1))   A C B A   0 v₂ v₁ v' v_M
(2w'/(r-1)) < v' < ∞   C B A   -v₂ v₁ v' v_M

3 < r < 3 + 2√2

0 < v' < (w'/r)   A B D   0 v' v₂ v₁
(w'/r) < v' < ((r-1)/(2r))w'   A C D   0 v₂ v' v₁
((r-1)/(2r))w' < v' < (2w'/(r-1))   A C D A   0 v₂ v' v₁ v_M
(2w'/(r-1)) < v' < w'   C D A   -v₂ v' v₁ v_M
w' < v' < ∞   C B A   -v₂ v₁ v' v_M

3 + 2√2 < r < ∞

0 < v' < (w'/r)   A B D   0 v' v₂ v₁
(w'/r) < v' < (2w'/(r-1))   A C D   0 v₂ v' v₁
(2w'/(r-1)) < v' < ((r-1)/(2r))w'   C D   -v₂ v' v₁
((r-1)/(2r))w' < v' < w'   C D A   -v₂ v' v₁ v_M
w' < v' < ∞   C B A   -v₂ v₁ v' v_M

Table 1

It easily seen from Eq.(3.1) that I_n is even with respect to both v' and w', and is homogeneous of order 2n, namely $I_n(tv',tw')=t^{2n}I_n(v',w')$. Moreover, one can write, by resorting to a binomial expansion,

$$I_n(v',w') = \frac{2^{n-1} n!}{(2n+2)!} \frac{1}{(1+r)^{2n}} \frac{w'^{2n+1}}{v'} \sum_{k=0}^{n} \binom{2n+2}{2k+1} J_{nk}(v'/w'), \tag{3}$$

where

$$J_{nk}(u) = \int_{|u-1|}^{|u+1|} y^{2k+1}[(1+r)(ru^2+1)-ry^2]^{n-k} \, dy \tag{4}$$

is seen by inspection to be an odd polynomial of degree 2n+1 in u. Therefore I_n must be an even polynomial of degree 2n with respect to both v' and w', and take thus necessarily the same convenient form

$$I_n(v',w') = \sum_{l=0}^{n} \mu_{nl}(r) \frac{v'^{2l}}{(2l+1)!!} \frac{w'^{2n-2l}}{(2n-2l+1)!!} \tag{5}$$

which is in order for the classical case r=1, where $\mu_{nl}=1/(n+1)$[16]. For the evaluation of the coefficients $\mu_{nl}(r)$ with $r\neq1$, one could further cast the integrals (3.4) as incomplete beta functions[10]

$$J_{nk}(u) = \frac{(1+r)^{n+1}(ru^2+1)^{n+1}}{2r^{k+1}} \cdot$$

$$\cdot \left\{ B\left[k+1,n-k+1;\frac{r(u+1)^2}{(1+r)(ru^2+1)}\right] - B\left[k+1,n-k+1;\frac{r(u-1)^2}{(1+r)(ru^2+1)}\right] \right\}. \tag{6}$$

This leads, after some manipulations, to the representation

$$J_{nk}(u) = \frac{1}{2} \frac{k!(n-k)!}{(n+1)!} \cdot$$

$$\cdot \frac{1}{r^{k+1}} \sum_{s=k+1}^{n+1} \binom{n+1}{s} \left[r^s(u+1)^{2s}(ru-1)^{2n+2-2s} - r^s(u-1)^{2s}(ru+1)^{2n+2-2s} \right] \tag{7}$$

where the same sum, from s=0 to s=n+1, would vanish, and where further binomial expansions would provide explicitly J_{nk} in powers of u, allowing finally the determination of the sought coefficients μ_{nl} via Eqs.(3.3) and (3.5). Such a straightforward procedure is however very ackward in practice, bacause of the several coupled, though finite, sums involved.

A slightly simpler form for the coefficients μ_{nl} is achieved by using the definition (1.9) directly, in connection with Table 1 and Eqs.(2.11). Although the task seems almost hopeless, it can actually be verified that the moment I_n is the same in all 30 cases (details are omitted), and that, integrating by parts, all inverse sine functions cancel out. Furthermore, the remaining irrational integrals

$$\int v^{2k}[E-rv^2]^{-\frac{1}{2}} \, dv \tag{8}$$

can be handled by recursion[10], and one ends up, after quite some algebra, with

$$I_n(v',w') = \frac{2^{n-2}}{(n+1)(n+1)!} \frac{(1+r)^2}{4r} \cdot$$
$$\cdot \sum_{k=0}^{n} \frac{(n-k)!(n-k)!}{(2n+1-2k)!(4r)^k} \frac{2v'^{2n+1-2k}w'+v_1^{2n+1-2k}w_1+v_2^{2n+1-2k}w_2}{v'w'} E^k \tag{9}$$

where the v_i's and w_i's are given by Eq.(2.10). Two binomial expansions allow now to single out the required powers of v' and w' and to achieve, after some algebra, the new form

$$I_n(v',w') = \frac{2^{n-1}}{(n+1)(n+1)!} \frac{1}{4r} \cdot$$
$$\cdot \sum_{l=0}^{n} v'^{2n-2l}w'^{2l} \sum_{k=0}^{n} \frac{(n-k)!(n-k)!}{2^{2k}(2n+1-2k)!} \frac{1}{(1+r)^{2n-2k}} \sum_{s=s_1}^{s_2} \binom{k}{s} \frac{1}{r^s} G_{n-k,l-s}(r), \tag{10}$$

where

$$s_1 = \max(0,l-n+k) \qquad\qquad s_2 = \min(k,l) \tag{11}$$

and

$$G_{kl}(r) = 2^{2l} \cdot$$
$$\cdot \left[\delta_{l0}(1+r)^{2k+2} + 4r(1-r)^{2k-2l} \binom{2k+1}{2l+1} - (1-r)^{2k-2l+2} \binom{2k+1}{2l} \right], \tag{12}$$

δ being the Kronecker symbol. Comparison to Eq.(3.5) and an inversion in the summation order leads finally to the result

$$\mu_{nl}(r) = \frac{1}{n+1} \left[\frac{2}{1+r} \right]^{2n} \frac{(2l+1)! \; (2n-2l+1)!}{2^{2n+3} r l! (n-l)! (n+1)!}$$

$$\cdot \sum_{k=0}^{l} \sum_{s=l-k}^{n-k} \frac{(n-s)!(n-s)!}{(2n-2s+1)!} \frac{1}{r^{s-l+k}} \left[\frac{1+r}{2} \right]^{2s} \left(\begin{array}{c} s \\ s-l+k \end{array} \right) G_{n-s,n-k-s}(r), \qquad (13)$$

$n=0,1,2,...,$ $l=0,1,...,n$, which generalize the analogous coefficients introduced already in Ref.18, and determined there only up to n=2. For special values of l, much simpler expressions can be obtained. In particular

$$\mu_{no}(r) = \frac{1}{n+1} \frac{1}{(1+r)^{2n}} \frac{(2n+1)!}{8rn!(n+1)!} H_n(r) \qquad (14a)$$

where

$$H_n(r) = \sum_{s=0}^{n} \frac{(n-s)!(n-s)!}{(2n-2s+1)!} \frac{1}{r^s} \left[\frac{1+r}{2} \right]^{2s} G_{n-s,n-s}(r), \qquad (14b)$$

obeys the recursion formula

$$H_{n+1}(r) = \frac{(n+1)!(n+1)!}{(2n+3)!} G_{n+1,n+1}(r) + \frac{1}{r} \left[\frac{1+r}{2} \right]^2 H_n(r), \qquad (15a)$$

from which, on using Eq.(3.12), one gets by induction

$$H_n(r) = 8r \frac{n! \; (n+1)!}{(2n+1)!} 2^{2n}, \qquad (15b)$$

and thus

$$\mu_{no}(r) = \frac{1}{n+1} \left[\frac{2}{1+r} \right]^{2n} . \qquad (16)$$

Analogously there results

$$\mu_{nn}(\overset{\circ}{r}) = \frac{1}{n+1} \frac{1}{(1+r)^{2n}} \frac{(2n+1)!}{8rn!(n+1)!} K_n(r), \qquad (17a)$$

where

$$K_n(r) = \sum_{k=0}^{n} \frac{k!k!}{(2k+1)!} \left[\frac{1+r}{2} \right]^{2(n-k)} G_{ko}(r) \qquad (17b)$$

satisfies now the recurrence relation

$$K_{n+1}(r) = \frac{(n+1)!(n+1)!}{(2n+3)!} G_{n+1,o}(r) + \left[\frac{1+r}{2} \right]^2 K_n(r). \qquad (18a)$$

Since $K_0(r)=8r$, it is not difficult to show by induction, bearing Eq.(3.12) in mind, that

$$K_n(r) = 2 \frac{n! \ (n+1)!}{(2n+1)!} \left[(1+r)^{2n+2} - (1-r)^{2n+2} \right] , \qquad (18b)$$

so that, setting $\alpha=[(1-r)/(1+r)]^2$, we obtain finally

$$\mu_{nn}(r) = \frac{1}{n+1} \frac{1-\alpha^{n+1}}{1-\alpha} , \qquad (19)$$

reproducing the eigenvalues of the linear collision term for a Maxwellian distribution of the field particles[18].

4. Moment equations and polynomial reconstruction

We are now able to take moments of Eq.(1.1), according to Eq.(1.5), and get a set of selfcontained ordinary differential equations for the moments M_n^i. Using Eq.(3.5) yields in fact at once

$$\frac{dM_n^i}{dt} + M_n^i(t) \sum_{j=1}^{N} C_{ij} M_o^j(t) = \sum_{j=1}^{N} C_{ij}^S \sum_{l=0}^{n} \mu_{nl}^{ij}(r_{ij}) M_l^i(t) M_{n-l}^j(t) + Q_n^i(t), \qquad (1)$$

$i=1,...,N$, $n=0,1,...$, with initial conditions $M_n^i(0)$, where $Q_n^i(t)$ and $M_n^i(0)$ are the corresponding moments of external sources and initial distribution functions. A nice feature of Eq.(4.1), which remains valid in the present physical context as it was in the classical case of Ref.12, is that it can be solved in cascade up to the desired order, and the equations are linear at any step, except for n=0 (continuity equations).

Once the moments M_n^i have been determined, the distribution functions may be obtained either by a procedure based on a numerical Fourier transform inversion[13], or by a generalization of the Laguerre polynomial reconstruction described in Ref.9 for the standard Boltzmann equation. It can be shown in fact[3] that the Fourier transforms

$$\tilde{f}_i(k,t) = \int f_i(v,t) \, \exp(-i\hat{\underline{k}}\cdot\underline{v}) \, d\underline{v} \tag{2}$$

obey the set of transformed functional equations

$$\frac{\partial \tilde{f}_i}{\partial t} + \tilde{f}_i(k,t) \sum_{j=1}^{N} C_{ij}\rho_j(t) =$$

$$= \tilde{Q}_i(k,t) + \tfrac{1}{2} \sum_{j=1}^{N} C_{ij}^S \int_{-1}^{1} \tilde{f}_i\left[k \, \frac{(r_{ij}^2+2r_{ij}\mu+1)^{\frac{1}{2}}}{r_{ij}+1} , t \right] \tilde{f}_j\left[k \, \frac{(2-2\mu)^{\frac{1}{2}}}{r_{ij}+1} , t \right] d\mu \tag{3}$$

The structure of Eq.(4.3) is such that the ansatz

$$\tilde{f}_i(k,t) = \exp(-k^2/4\beta_i) \, \tilde{\Phi}_i(k,t) \tag{4}$$

leaves it essentially invariant, provided the parameters β_i are related by

$$\beta_i/\beta_j = r_{ij} \; . \tag{5}$$

In fact, the factor $\exp(-k^2/4\beta_i)$ cancels out from the i-th equation, and leaves for $\tilde{\Phi}_i$ essentially the same equation as for $\tilde{f}_i$, only with external source and initial condition obviously multiplied by $\exp(k^2/4\beta_i)$. Such a feature is the counterpart of the invariance under Bobylev transformation of the classical Boltzmann equation[9], and amounts to the physical fact that, in absence of removal and sources, the system is in equilibrium if all distribution functions are Maxwellians at the same temperature. The inverse Fourier transform of $\exp(-k^2/4\beta_i)$ is in fact the normalized Maxwellian (1.8), and Eq.(5) corresponds to the requirement that $\beta_i/m_i=(2KT_i)^{-1}$ be independent of i. From a mathematical point of view, the inverse $\Phi_i(v,t)$ of $\tilde{\Phi}_i$ exists, at least in the distributional sense, if f_i is smooth enough[9]. In this case, one has

$$f_i(v,t) = \int M(|\underline{v}-\underline{w}|,\beta_i) \, \Phi_i(w,t) \, d\underline{w} \; , \tag{6}$$

and it is possible to define moments of Φ_i, labeled $\hat{M}_n^{\,i}(t)$ in the sequel, according to Eq.(1.5). These moments satisfy, on the basis of the previous discussion, the same set of equations (4.1) as for $M_n^{\,i}$, only with inhomogeneous and initial terms correspondingly modified in $\hat{Q}_n^{\,i}(t)$ and $\hat{M}_n^{\,i}(0)$.

The interesting point is that the ordinary moments $\hat{M}_n{}^i$ of Φ_i are simply proportional to the Laguerre moments $\xi_n{}^i$ of f_i, defined in Eq.(1.6). The use of Laguerre polynomials in this context is not surprising, if one bears in mind their success in treating one species classical model-Boltzmann equations[9], and the fact that they appear explicitly in the eigenfunctions of the linear collision operator, for Maxwellian distribution of the field particles[18]. In any event, since Eq.(4.6) can be rewritten as

$$f_i(v,t) = \left(\frac{\beta_i}{\pi} \right)^{3/2} \exp(-\beta_i v^2) \int \exp(-\beta_i w^2) \frac{\sinh(2\beta_i vw)}{2\beta_i vw} \Phi_i(w,t)\ d\underline{w}\ , \qquad (7)$$

the defition (1.6) yields directly

$$\xi_n{}^i(t) = \frac{2}{\sqrt{\pi}} \frac{2^n\ n!}{(2n+1)!} \cdot$$

$$\cdot \sum_{k=0}^{\infty} \frac{2^{2k}\beta_i{}^k}{(2k+1)!} \int w^{2k}\ \exp(-\beta_i w^2)\ \Phi_i(w,t)\ d\underline{w} \int_0^{\infty} x^{k+\frac{1}{2}}\exp(-x)L_n^{(\frac{1}{2})}(x)dx, \qquad (8)$$

and, after evaluating the integral[10] and summing terms, one ends up with

$$\xi_n{}^i(t) = (-1)^n(2\beta_i)^n\ \hat{M}_n{}^i(t). \qquad (9)$$

Therefore, the equations for the Laguerre coefficients read as

$$\frac{d\xi_o{}^i}{dt} + \xi_o{}^i(t) \sum_{j=1}^{N} C_{ij}^R \xi_o{}^j(t) = S_o{}^i(t) \qquad (10a)$$

for n=0 (with $\xi_0{}^i = M_0{}^i = \rho_i$), and

$$\frac{d\xi_n{}^i}{dt} + \xi_n{}^i(t) \sum_{j=1}^{N} \left[C_{ij} - \lambda_{nn}^{ij}(r_{ij})C_{ij}^S - \delta_{ij}\lambda_{no}^{ij}(r_{ij})C_{ij}^S \right] \xi_o{}^i(t) -$$

$$- \xi_o{}^i(t) \sum_{j\neq i} \lambda_{no}^{ij}(r_{ij})C_{ij}^S \xi_n{}^j(t) - \sum_{j=1}^{N} C_{ij}^S \sum_{l=1}^{n-1} \lambda_{nl}^{ij}(r_{ij})\xi_l{}^i(t)\xi_{n-l}{}^j(t) + S_n{}^i(t) \qquad (10b)$$

for $n\geq 1$, where $S_n{}^i$ denotes the n-th Laguerre moment of Q_i, and

$$\lambda_{nl}^{ij}(r_{ij}) = r_{ij}^{n-l}\ \mu_{nl}^{ij}(r_{ij}). \qquad (11)$$

Eq.(4.10) retains all nice features of Eq.(4.1), and may be regarded as a generalization of the set of equations established in Ref.18 for a single nonscattering species in a background medium. The main advantage now is the possibility of analytically expressing all distribution functions by means of Eq.(1.7). Notice that the distribution $f_i = \rho_i M(v, \beta_i)$ corresponds to the option $\xi_n{}^i = \rho_i \delta_{no}$, namely all Laguerre moments, but the first, are vanishing.

A numerical investigation of Eq.(4.10) will be object of future work. It is worth remarking here that several competing effects will affect time evolution of the distribution functions. Consider for instance the very simple case of a single species in a host medium (labeled below by a hat, in order to avoid unnecessary indeces) at a given Maxwellian distribution $M(v, \hat{\beta})$. Setting $r = m/\hat{m}$ and $\beta = r\hat{\beta}$, the Laguerre moments for the background collapse simply to $\hat{\xi}_n = \hat{\rho} \delta_{no}$, and the test particle distribution function follows from the usual Laguerre expansion (1.7), with Laguerre coefficients that are solution to

$$\dot{\xi}_n(t) + \hat{C}\hat{\rho}\xi_n(t) + C\xi_o(t)\xi_n(t) = \hat{C}_S\hat{\rho}\mu_{nn}(r)\xi_n(t) + C_S \frac{1}{n+1} \sum_{l=0}^{n} \xi_l(t)\xi_{n-l}(t) + S_n(t) \quad (12)$$

In absence of source, removal and background, the distribution function f would relax to a Maxwellian at a temperature corresponding to the initial average kinetic energy per particle. In absence of other effects, scattering by field particles destroys energy conservation and tends to establish a Maxwellian at the field particle temperature. Simultaneously, the source Q modifies continuously the test particle shape function $f(v,t)/\rho(t)$, as well as their number density. It is clear that the asymptotic shape function does not need to be a Maxwellian. Evidence of this fact may be found even in the very restricted class of known analytical solutions. Assume for a moment an initial distribution

$$f(v,0) = \rho_o \left[\frac{m(1+\beta_o)}{2\pi KT} \right]^{3/2} \left\{ 1 + \beta_o \left[\frac{m(1+\beta_o)}{2KT} v^2 - \frac{3}{2} \right] \right\} \exp\left[-\frac{m(1+\beta_o)}{2KT} v^2 \right] \quad (13)$$

with $0 < \beta_o \leq 2/3$, in order to ensure positivity. Then, for $Q=0$ and $\hat{C}_S=0$, the present problem can be solved analytically[6], and the solution is given again by Eq.(4.13), with ρ_o replaced by $\rho(t)$, and β_o replaced by $\beta(t) = \beta_o\Phi(t)/\{1+\beta_o[1-\Phi(t)]\}$. The monotonically decreasing and positive functions ρ and Φ are given by

$$\rho(t) = \frac{\hat{\rho}\hat{C}_R\rho_o}{(\hat{\rho}\hat{C}_R + \rho_o C_R)\exp(\hat{\rho}\hat{C}_R t) - \rho_o C_R} \quad , \quad (14a)$$

and

$$\Phi(t) = \left[\frac{\hat{\rho}\hat{C}_R + \rho_o C_R - \rho_o C_R \exp(-\hat{\rho}\hat{C}_R t)}{\hat{\rho}\hat{C}_R} \right]^{-C_S/6C_R} . \quad (14b)$$

Thus ρ vanishes asymptotically, but Φ tends to a positive limit for t→∞, so that also β has a positive asymptotic value. Therefore, the shape function does not relax to a Maxwellian, even though driving effects of source and scattering by field particles are missing, and energy is conserved. Notice that relaxation to a Maxwellian actually occurs in the limiting case $\hat{C}_R$→0, C_R→0 of the famous BKW mode[2,12], as well as for $\hat{C}_R$→0 only[15].

Acknowledgement

Work supported by the National Group for Mathematical Physics of C.N.R., in the frame of the activities PS-AITM.

References

[1]. Amaldi,E.: The production and slowing down of neutrons. Handbuch der Physik, Vol. 38/2, 1-659. Berlin: Springer 1959.

[2]. Bobylev,A.V.:Exact solutions of the Boltzmann equation. Dokl. Acad. Nauk SSSR $\underline{225}$ (1975) 1296-1299.

[3]. Bobylev,A.V.: On the structure of spatially homogeneous normal solutions of the nonlinear Boltzmann equation for a mixture of gases. Sov. Phys. Dokl. $\underline{25}$ (1980) 30-32.

[4]. Boffi,V.C.; Molinari,V.G.: Nonlinear transport problems by factorization of the scattering probability. Nuovo Cimento $\underline{65B}$ (1981) 29-44.

[5]. Boffi,V.C.; Franceschini,V.; Spiga,G.: Dynamics of a gas mixture in an extended kinetic theory. Phys. Fluids $\underline{28}$ (1985) 3232-3236.

[6]. Boffi,V.C.; Spiga,G.: Exact time dependent solutions to the nonlinear Boltzmann equation. Proceedings of the 15th Intern. Symp. on Rarefied Gas Dynamics, Vol.1, 55-63. Stuttgart: Teubner 1986.

[7]. Boffi,V.C.; Spiga,G.: Calculation of the number densities in an extended kinetic theory of gas mixtures. Trans. Th. Stat. Phys. $\underline{16}$ (1987) 175-188.

[8]. Dukek,G.; Rupp,D.: Boltzmann equation for nonlinear particle transport in a host medium: moment methods and Lie group techniques. Proceedings of the 15 Intern. Symp. on Rarefied Gas Dynamics, Vol.1, 64-74. Stuttgart: Teubner 1986.

[9]. Ernst,M.H.: Nonlinear model-Boltzmann equations and exact solutions. Phys. Rep. $\underline{78}$ (1981) 1-71.

[10]. Gradshteyn,I.S.; Rizhik,I.M.: Tables of integrals, series and products. New York: Academic Press 1965.

[11]. Holloway,J.P.; Dorning,J.J.: The dynamics of coupled nonlinear model Boltzmann equations. J. Stat. Phys., submitted for publication.

[12]. Krook,M.; Wu,T.T.: Formation of Maxwellian tails. Phys. Rev. Lett. 36 (1976) 1107-1109.

[13]. Rupp,D.; Nonnenmacher,T.F.: Solutions to the nonlinear Boltzmann equation for particle transport in a host medium. Phys. Fluids 29 (1986) 2746-2747.

[14]. Spiga,G.: Nonlinear problems in particle transport theory. Application of Mathematics in Technology, 430-447. Stuttgart: Teubner 1984.

[15]. Spiga,G.: A generalized BKW solution of the nonlinear Boltzmann equation with removal. Phys. Fluids 27 (1984) 2599-2600.

[16]. Spiga,G.; Nonnenmacher,T.F.; Boffi,V.C.: Moment equations for the diffusion of the particles of a mixture via the scattering kernel formulation of the nonlinear Boltzmann equation. Physica 131A (1985) 431-448.

[17]. Spiga,G.: Dynamical systems in nonlinear transport theory. Dept. Math. Univ. Bari, Rep.7/87, 1-29. Bari: 1987.

[18]. Spiga,G.; Premuda,F.: Nonlinear Boltzmann equation for particle transport in a background gas with Maxwellian distribution. Proceedings of the 15th Intern. Symp. on Rarefied Gas Dynamics, Vol.1, 100-109. Stuttgart: Teubner 1986.

[19]. Zanette,D.H.: Two velocity gas diffusion with removal and regeneration process. Physica 148A (1988) 288-297.

[20]. Zanette,D.H.; Barrachina,R.O.: Nonlinear particle diffusion in a time dependent host medium. Phys.Fluids, in press.

+ *Istituto Ricerche Matematica Applicata del C.N.R., Dipartimento di Matematica, via G.Fortunato, 70125 Bari, Italy.*

x *Laboratorio di Ingegneria Nucleare, Università di Bologna, via dei Colli 16, 40136 Bologna, Italy.*

* *Dipartimento di Matematica, Università di Bari, via G.Fortunato, 70125 Bari, Italy.*

A HORIZONTALLY TWODIMENSIONAL CLIMATE MODEL

Gerhard Dziuk

1. Climate Models

In the last time there has been an accelerated growth of interest
in the earth's climate, specially in its sensitivity with respect
to CO_2-concentration of the atmosphere. Thus mathematical climate
modeling has become more and more important.

The physical system contains atmosphere, hydrosphere, cryosphere
land surface and last but not least man. The sun is the external
energy source. The complete threedimensional system normally
exceeds todays mathematical and numerical capability.

The basic system of equations for climate models is obtained by
combining the Navier-Stokes equations for compressible flow in
a rotating coordinate frame with the first law of thermodynamics.
The approximate equations for large-scale flow read as follows:

$$u_t + u \cdot \nabla u - a_0 \nabla \cdot (r_0 \nabla u) + \nabla p + R \times u = \gamma\, T\, n$$

$$(1)$$

$$\nabla \cdot (r_0 u) = 0$$

$$(2) \qquad T_t + u \cdot \nabla T - a_0 \nabla \cdot (r_0 \nabla T) + \gamma\, u \cdot n = Q(\cdot ,T)$$

in $(t,x) \in I \times G$, where I is some time interval and G is the
spherical shell with radii $0 < R_1 < R_2$, $\quad G = \left\{ x \in \mathbb{R}^3 \mid R_1 < |x| < R_2 \right\}$

with boundary conditions

$$u = 0 \text{ on } \partial G, \quad T = 0 \text{ on } |x| = R_2, \quad \partial T/\partial n = 0 \text{ on } |x| = R_1$$

and appropriate initial conditions. Here $n(x) = x/|x|$ is the radial unit vector, $u = (u_1, u_2, u_3)$ is the velocity, T is the absolute temperature, r_0 is a given function of $|x|$ alone and R represents the angular rotation rate about the x_3-axis. $a_0, \gamma, |R|$ are constants and Q is the thermodynamical source term involving nonlinear feedback-mechanisms such as the ice-albedo coupling or the greenhouse effect.

A derivation of these equations can be found in the paper of J. Dutton [3]. See also the work of B. Saltzman [8].

The system (1), (2) gives reason for a crude classification of statistical-dynamical climate models. Thermodynamical or Energy-Balance Models are deduced from (2), Momentum Models from (1). In both cases parameterizations model the effects of the other equation.

Our aim is to treat a stationary version of the vertically (radially) averaged system (1), (2). The nonlinearity Q in the energy equation is of Budyko-Sellers-Ghil type. It has been used by M. Ghil in [6] in a onedimensional EBM and there creates three stationary solutions which can be viewed as "climatic ground states". The physical constants and functions are taken from [3] and [6]. We present a horizontally twodimensional "global circulation model" and solve it numerically using special finite element methods for the Beltrami Operator on the sphere developed in [4] and for tangential Stokes Flow [5]. We shall compare the result with reality. The result shows that temperature distribution models the vertical mean temperature on earth fairly well.

2. The Equations

We look for solutions (u, p, T) on $S = \{x \in \mathbb{R}^3 \mid |x| = 1\}$ of the system

$$-\nu \underline{\Delta} u - 2u + u \cdot \underline{\nabla} u + \underline{\nabla} p + R \times u - \gamma \underline{\nabla} T \times n = 0$$

$$\underline{\nabla} \cdot u = 0$$

(3)

$$u \cdot n = 0$$

$$-\nu \underline{\Delta} T + u \cdot \underline{\nabla} T - Q(\cdot, T) = 0$$

on S, where $\underline{\nabla}$ is the tangential gradient and $\underline{\Delta}$ represents the Laplace-Beltrami Operator on S, $R = |R| (0,0,1)$, and ν, γ, $|R|$ are given positive constants. According to J. Dutton we take the Reynolds number to be about 1000. Notice that we are looking for a velocity vector u which is tangential to the sphere everywhere.

Now we have to explain what the nonlinearity Q looks like.
For $x=(x_1, x_2, x_3)$ on S,

$$Q(x,T) = q(x_3) (1 - a(x,T)) - s_0 \, \max\{T,0\}^4 \, ,$$

where q models the distribution of incoming solar radiation, which is just north-south dependent, s_0 is the Stefan-Boltzman constant, $a(x,T) = \alpha(T) w(x)$ with the albedo α parametrized by the function

$$\alpha(T) = \begin{cases} 0.85 & (T < 210) \\ \text{linear} & (210 \leq T \leq 272) \\ 0.25 & (T > 272). \end{cases}$$

$w(x)$ is taken to be a smoothened version of the function which is equal to 1 for x on land and equal to 0.7 for x above water.

We do not want to prove existence and regularity for (3) in this paper but we should mention that it is not too difficult to prove such a result by a Galerkin method since our problem is two-dimensional and the nonlinearity in the temperature equation ensures boundedness and positivity of T.

We want to solve (3) numerically. So now it is important to provide a good Finite Element Method. This will be done seperately for Poisson equations and Stokes equations on S and then put together to solve (3).

3. A FEM for the Poisson equation on the sphere

In [4] the author constructed a simple FEM for Beltrami's operator on arbitrary surfaces. This method we shall use to solve the temperature equation in (3).

Let the polyhedron S_h be an approximation of S. S_h is made up out of triangles of regular type and size proportional to h with vertices on the smooth surface S. The suitable Hilbert space for the solution of the continuous problem

$$(4) \qquad - \underline{\Delta} w = f \quad \text{on S,} \quad \oint_S w = 0$$

with $f \in L^2(S)$, $\oint_S f = 0$, is

$$X = \left\{ w \in H^1(S) \mid \oint_S w = 0 \right\}.$$

To approximate the solution of (4) we use conforming linear Finite Elements on the polyhedron S_h. Let

$$X_h = \left\{ w_h \in C^0(S_h) \mid w_h = \text{a linear polynomial on each triangle,} \oint_{S_h} w_h = 0 \right\}.$$

X_h is contained in $H^1(S_h)$, not in $H^1(S)$. Note that $H^2(S_h)$ is not defined. The correspondig discrete problem now reads as follows: find $w_h \in X_h$ such that for every $\varphi_h \in X_h$

$$(4_h) \qquad \int_{S_h} \underline{\nabla}_h w_h \, \underline{\nabla}_h \varphi_h = \int_{S_h} f_h \varphi_h.$$

Here $f_h = f - \oint f$ for example, and $\underline{\nabla}_h$ represents the tangential gradient on the S polyhedron. From [4] we have the following convergence theorem.

__Theorem:__ Let $w \in H^2(S)$ be the solution of (4) and let w_h be the solution of (4_h). For $W_h(x/|x|) = w_h(x)$ $(x \in S_h)$

$$\| w_h - W_h \|_{L^2(S)} + h \, \| \underline{\nabla} (w_h - W_h) \|_{L^2(S)} \; \leq \; c \, h^2 \, \| f \|_{L^2(S)}.$$

Let us remark that the only difference to a FEM in plane domains is that the vertices of our triangulation are threedimensional vectors instead of twodimensional ones.

4. A FEM for Tangential Stokes Flow on the Sphere

At this point we need some information about the correct Stokes problem on surfaces resp. on the sphere. Replacing the Laplace operator by Beltrami's operator, the gradient by the tangential gradient and chosing the velocity vector to be tangential does not model Stokes flow on the sphere. One has to think about the correct deformation tensor. For a correct formulation on surfaces and existence and regularity results see A. Takeshita [9], A. Avez and Y. Bamberger [1, 2]. Because of physical interpretation the deformation tensor has to be tangential. In the case of the sphere this leads to the Stokes problem

$$(5) \qquad -\underline{\Delta} v - 2\,v + \underline{\nabla} q \; = g, \quad \underline{\nabla} \cdot v = 0, \; v \cdot n = 0$$

on S. This problem is unique up to rigid body motions B of the sphere for given $g \cdot n = 0$, $g \in L^2(S)$, $\oint_S g = 0$. The continuous solution is constructed in

$$V = \left\{ v \in H^1(S)^3 \; \middle| \; \underline{\nabla} \cdot v = 0, \; v \cdot n = 0 \right\} / \, B.$$

In order to solve the corresponding discrete problem on the polyhedron S_h we use exactly divergence free nonconforming linear elements, tangentially continuous in the midside nodes of the triangulation of S. Let n_h be the normal vector to S_h. Then

$$V_h = \{ v_h \in L^\infty(S_h)^3 \mid v_h = \text{(linear polynomial)}^3, \underline{\nabla}_h \cdot v_h = 0,$$
$$v_h \cdot n_h = 0 \text{ on each triangle, } v_h \text{ is tangentially continuous}$$
$$\text{at each midside node} \} / B_h$$

Here "tangentially continuous" means that

$$v_h\big|_{\Delta_h}(b) \cdot t = - v_h\big|_{\Delta'_h}(b) \cdot t' ,$$
$$v_h\big|_{\Delta_h}(b) \cdot \nu = - v_h\big|_{\Delta'_h}(b) \cdot \nu' ,$$

where b is the midside node on the edge joining the two triangles Δ_h and Δ'_h. $t = - t'$ is the tangential vector of the edge with different orientation on each triangle, ν and ν' are outer normal vectors to Δ_h and Δ'_h at b. Note that $\nu \neq \nu'$.

A simple basis for this space is given in [5, 10]. For plane problems this element is well known. For references the reader is referred to F. Thomasset [10]. The discrete analogon for (5) now reads as follows:

Find $v_h \in V_h$ such that for every $\varphi_h \in V_h$

$$(5_h) \qquad \sum_{\Delta_h} \int_{\Delta_h} \underline{\nabla}_h v_h \underline{\nabla}_h \varphi_h - 2 v_h \varphi_h = \int_{S_h} g \; \varphi_h .$$

In [5] we prove

<u>Theorem</u>: Let $v \in H^2(S)^3 \cap V$ be the solution of (5) and let $v_h \in V_h$ be the discrete solution of (5_h). For

$$V_h(x/|x|) = v_h(x) - v_h(x) \, n(x/|x|) \, n(x/|x|) \quad (x \in S_h)$$

we have the estimate

$$\left(\sum_\Delta \| \underline{\nabla}(v_h - V_h) \|^2_{L^2(\Delta)} \right)^{1/2} \leq c \; h \; \| g \|_{L^2(S)}$$

where Δ is the projection of $\Delta_h \subset S_h$ onto S.

5. Discretization of the Model

Now we are able to formulate the discrete horizontally twodimensional
climate model. Since Reynolds number is fairly small we apply Newton's
method to (3). But we shall skip these details and just formulate
the discrete version of (3) using the FEM schemes developed in
parts 3 and 4 of this paper. Let X_h be as in part 3 but without the
condition that the mean value is zero. Then we want to find functions

$$(u_h, T_h) \in V_h \times X_h$$

such that for every $v_h \in V_h$ and every $t_h \in X_h$

$$\sum_{\Delta_h} \int_{\Delta_h} (\nu \underline{\nabla}_h u_h \ \nabla_h v_h - 2\ u_h\ v_h + u_h \cdot \underline{\nabla}_h u_h\ v_h + R \times u_h\ v_h - \gamma \underline{\nabla}_h T_h \times n_h\ v_h) = 0 \tag{3_h}$$

$$\int_{S_h} (\nu\ \underline{\nabla}_h T_h \underline{\nabla}_h t_h + u_h \cdot \underline{\nabla}_h T_h\ t_h - Q(\cdot, T_h)\ t_h\ ) = 0.$$

6. Numerical Result

Of course in general there is no uniqueness for (3) and (3_h).
As mentioned above the nonlinearity Q in the energy equation
is expected to generate three stationary solutions when flow
is parametrized as a diffusion coefficient in this equation.

We used Newton to solve (3_h). As initial value for the iteration
we took the warmest solution of the temperature equation with
$u = 0$ and $\underline{\nabla} \cdot (k(x_3) \underline{\nabla} T)$ instead of $\underline{\Delta}\ T$, where k is taken from the
work of Ghil [6]. Fig. 1 shows the temperature achieved by this
iteration. Fig. 2 shows reality.

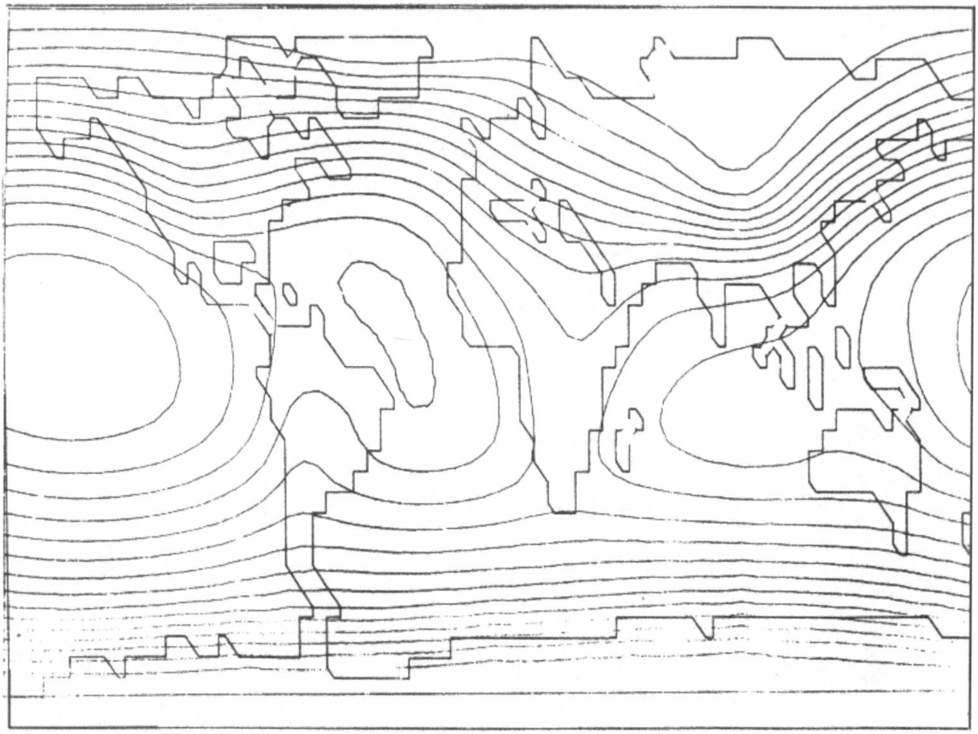

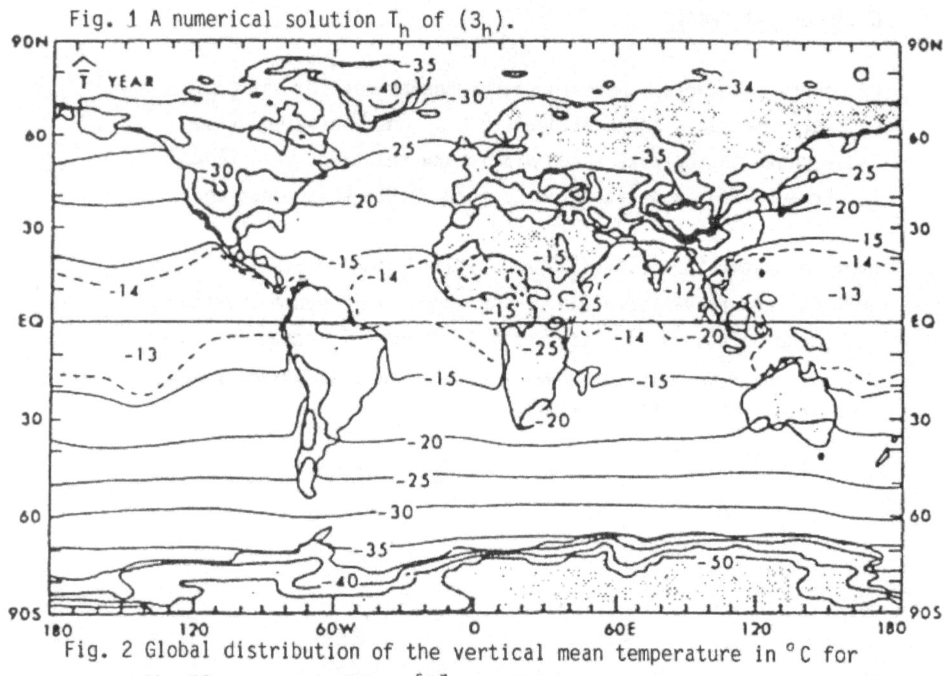

Fig. 1 A numerical solution T_h of (3_h).

Fig. 2 Global distribution of the vertical mean temperature in $^\circ$C for
the 10-year period from [7], p. 394

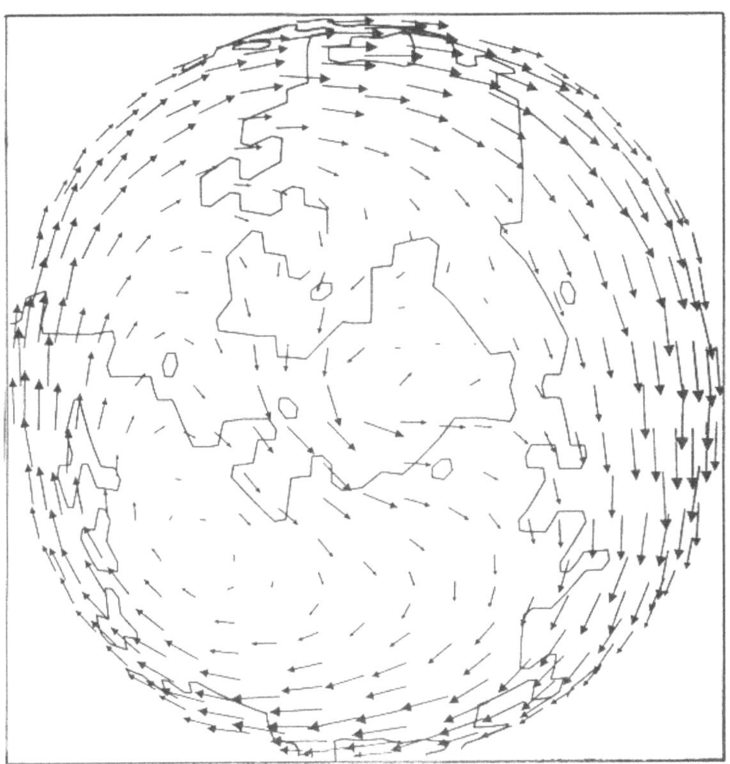

Fig. 3 Calculated velocity u_h seen from the northpole

References

[1] Avez, A.; Bamberger, Y.: Mouvements spheriques des fluides visquex incompressibles. C. R. Acad. Sc., Paris, t. 282, serie A (1976) 341 - 344

[2] Avez, A.; Bamberger, Y.: Mouvements spheriques des fluides visqueux incompressibles. Journal de Mecanique 17, 1 (1978) 107 - 145

[3] Dutton, J. A.: Fundamental theorems of climate theory - some proved, some conjectured. SIAM Rev. 24, Nr. 1 (1982) 1 - 32

[4] Dziuk, G.: Finite Elements for the Beltrami operator on arbitrary surfaces. SFB 256 Bonn Preprint Nr. 10 (1987)

[5] Dziuk, G.: A Finite Element Method for tangential Stokes flow on surfaces. To appear

[6] Ghil, M.: Climate stability for a Sellers-type model. J. Atmos.
Sci. 33 (1976) 3 - 200

[7] Oort, A. H.; Peixoto, J. P.: Global angular momentum and energy
balance requirements from observations. In "Theory of Climate"
edited by B. Saltzman, Adv. Geophys. 25 (1983) 355 - 490

[8] Saltzman, B.: A survey of statistical-dynamical models of the
terrestrial climate. Adv. Geophys. 20 (1978) 183 - 303

[9] Takeshita, A.: Existence and non-existence of solutions to the
stationary Navier-Stokes equations on compact Riemannian
manifolds. Nagoya University Preprint series Nr. 12 (1983)

[10] Thomasset, F.: Implementation of Finite Element Methods for
Navier-Stokes equations. New York, Heidelberg, Berlin:
Springer 1981

Gerhard Dziuk
Institut für Angewandte Mathematik
Wegelerstraße 6
D-5300 Bonn
FRG

ON THE COUPLING OF HYPERBOLIC AND PARABOLIC SYSTEMS: ANALITICAL AND NUMERICAL APPROACH

Fabio Gastaldi and Alfio Quarteroni***

* Dipartimento di Matematica dell'Università – Pavia – Italy
** Dipartimento di Matematica, Università Cattolica – Brescia – Italy
and Istituto di Analisi Numerica del C.N.R. – Pavia – Italy

ABSTRACT

We deal with the coupling of hyperbolic and parabolic systems in a domain Ω divided into two disjoint subdomains Ω^+ and Ω^-. Our main concern is to find out the proper interface conditions to be fulfilled at the surface separating the two domains. Next, we will use them in the numerical approximation of the problem. The justification of the interface conditions is based on a singular perturbation analysis, that is, the hyperbolic system is rendered parabolic by adding a small artificial "viscosity". As this goes to zero, the coupled parabolic–parabolic problem degenerates into the original one, yielding some conditions at the interface. These we take as interface conditions for the hyperbolic–parabolic problem. Actually, we discuss two alternative sets of interface conditions according to whether the regularization procedure is variational or nonvariational. We show how these conditions can be used in the frame of a numerical approximation to the given problem. Furthermore, we discuss a method of resolution which alternates the resolution of the hyperbolic problem within Ω^- and of the parabolic one within Ω^+. The spectral collocation method is proposed, as an example of space discretization (different methods could be used as well); both explicit and implicit time–advancing schemes are considered. The present study is a preliminary step toward the analysis of the coupling between Euler and Navier–Stokes equations for compressible flows.

1. INTRODUCTION

In this work we deal with (initial) boundary value problems for partial differential equations (or systems) which change their character within the domain under consideration. Precisely, we consider problems which are of hyperbolic type in a subdomain Ω^- of the whole domain Ω and of parabolic type in the complement Ω^+.

This interest is motivated by various applications. Among others, we emphasize the case of fluid dynamical problems for viscous, compressible flows in presence of a body, governed by the system of Navier–Stokes equations (see, e.g., [CM], [S]). A convenient numerical approach to the solution to these problems relies upon the splitting of the physical region where flow occurs in two computational domains, one (say, Ω^+) close to the body, where viscous terms are to be taken into account and another far away (say, Ω^-), where viscous terms may be neglected.

This leads precisely to a coupled problem involving Euler equations (hyperbolic) in the far region and the complete Navier–Stokes equations in the near region.

From a computational point of view, this splitting procedure carries obvious advantages. In particular, we mention the possibility of using different solvers for the two subproblems. Of course, a crucial point in this framework is how to relate the two problems to each other at the interface separating the two subregions. This feature must be investigated for the differential problem, first: suitable conditions at the interface will be derived. Whatever numerical scheme is used, it must take these conditions into account.

Identifying such conditions is generally understood whenever the two differential problems in the subregions are of the same kind. For instance, for the interaction of two second order elliptic problems, the interface conditions consist in requiring the continuity of the unknowns and of the flux (these conditions are transferred to any numerical scheme easily). Analogously, when coupling two Navier–Stokes problems, we must impose the continuity of the velocity and of the normal stress at the interface. Eventually, when coupling two hyperbolic systems of first order, we request the continuity of the unknown at the interface (unless the interface is a discontinuity line, in which case Rankine–Ugoniot equations ought to be fulfilled).

When coupling Euler and Navier–Stokes equations, the proper interface conditions are not obvious, in advance. A possible way of deducing them is to see the coupled problem as a limit of two coupled Navier–Stokes problems with vanishing viscous terms in Ω^-.

This approach can be adopted also in a simplified version of the problem, namely considering a coupling between hyperbolic and parabolic linear systems in one space variable (as well as their stationary counterpart). In this framework, the hyperbolic–parabolic problem is seen as limit of a coupling between two parabolic problems endowed with the usual transmission conditions at the interface. As we shall see, this procedure yields certain interface conditions for the limit problem. Although these conditions are somewhat physically meaningful, there is a sensible loss of continuity in passing from the approaching problems to the limit one. In what follows we shall detail also a different type of limit procedure, which maintains a higher order of regularity for the limit problem.

To be more precise, let us state the problem we will discuss in this paper:

In the interval (a,c), find $\mathbf{w}$ such that (b is any point internal to (a,c))

$$\mathbf{w}_t^- + A\,\mathbf{w}_x^- + B\,\mathbf{w}^- = \mathbf{F} \quad \text{for } x \in \Omega^- = (a,b),\, t > 0, \quad (1.1)$$

$$\mathbf{w}_t^+ - (\nu\,\mathbf{w}_x^+)_x + A\,\mathbf{w}^+{}_x + B\,\mathbf{w}^+ = \mathbf{F} \quad \text{for } x \in \Omega^+ = (b,c),\, t > 0, \quad (1.2)$$

with an initial condition and proper boundary conditions at $x = a$ and $x = c$. Here A and B are two constant 3×3 matrices, while $\mathbf{F}$ is a given vector function with three components ($\mathbf{w}$ is an unknown three dimensional vector); $\nu = \nu(x,t) \geq \nu_0 > 0$ is a given viscosity. We assume that A has three real, nonvanishing eigenvalues (p_0 of them are positive and $3-p_0$ are negative): in particular, this implies that the system (1.1) is hyperbolic.

For the above problem we are going to specify the interface conditions obtainable by the arguments previously mentioned. By the first approach, which we will refer to as *variational*, we find the following interface conditions at $x = b$, for all $t > 0$:

$$T_n\,\mathbf{w}^+ = T_n\,\mathbf{w}^-, \qquad (1.3)$$

$$-\,\nu\mathbf{w}_x^+ + A\,\mathbf{w}^+ = A\,\mathbf{w}^-: \qquad (1.4)$$

the rectangular matrix T_n has $3 - p_0$ rows given by the left eigenvectors corresponding to the negative eigenvalues of A (see section 3.1).

With the second approach, which we will refer to as *nonvariational*, the interface conditions at $x = b$ and for all $t > 0$ are:

$$\mathbf{w}^+ = \mathbf{w}^-, \qquad (1.5)$$

$$T_n\,\mathbf{w}_x^+ = T_n\,\mathbf{w}_x^-. \qquad (1.6)$$

In particular, note that (1.5) does imply continuity of all unknowns at the interface, while (1.4) gives continuity of the "flux" at the interface, allowing a discontinuity on the unknowns (actually, a mild discontinuity, as the jump has the same order of the viscosity coefficient ν at the interface).

The above results are presented in section 4, as a consequence of a procedure of "increasing difficulty" carried out throughout sections 2 and 3. Precisely, in section 3 we deal with the steady counterpart of (1.1), (1.2) and in section 2 we detail the coupling between two time independent equations, one of first order and the other of second order (the proofs of the abstract results are given in the Appendix). Although the problems of sections 2 and 3 might be regarded as autonomous problems, actually they are treated as intermediate steps toward the analysis of the main problem (1.1), (1.2). For each and every problem, we present the numerical approximation based on the spectral collocation method and show how the interface conditions are used in this frame. This could be done for numerical methods based on different approaches, as well. Here we just remark that, in the numerical scheme, we must supplement the above interface conditions suitable compatibility relations at the interface. These arise from the hyperbolic nature of the problem in Ω^-: a thorough discussion is made in sections 2.2, 3.2, 4.2.

We end this introduction by noticing that (1.1), (1.2) present some similarities with the coupling between Euler and Navier–Stokes equations we mentioned at the beginning as a driving motivation for our work. The relevant difference lies in that the viscous terms in Navier–Stokes equations do not enjoy the particular diagonal structure as in the right hand side of (1.2). Since our analysis relies heavily upon this feature, there is no immediate application of our results to the coupling between Euler and Navier–Stokes equations. Nevertheless, it seems that several elements of our approach can be useful in that problem, too. From this point of view, the present work is an intermediate step toward our goal.

2. HYPERBOLIC-ELLIPTIC INTERACTION: THE SCALAR CASE

In this section we consider a one dimensional, linear, scalar problem. The two subsections are devoted to the analysis of the continuous problem (with special concern to different elliptic regularizations) and to its numerical approximation, respectively.

2.1. The differential problem

We begin by stating the boundary value problem, as follows. Let

(i) a, b, c be real numbers, with $a < b < c$;

(ii) α, β, ν be functions defined in $[a,c]$, with $\alpha \neq 0$;

(iii) f be a function defined in $[a,c]$.

Then, consider the problem

$$(\mathbf{P})\colon \text{find } u \text{ defined in } [a,b], \; v \text{ defined in } [b,c] \text{ such that}$$

$$\alpha \, u_x + \beta \, u = f \qquad in \; (a,b); \tag{2.1}$$

$$- (\nu v_x)_x + \alpha \, v_x + \beta \, v = f \qquad in \; (b,c); \tag{2.2}$$

$$v(c) = 0; \tag{2.3}$$

$$u(a) = 0, \qquad if \; \alpha > 0 \; in \; [a,c]. \tag{2.4}$$

Clearly, the formulation of problem $(\mathbf{P})$ is incomplete: it needs one coupling condition between u and v at the interface b, when $\alpha > 0$ in $[a,c]$, while two coupling conditions are required if $\alpha < 0$ in $[a,c]$ (in this case, (2.4) does not hold). Moreover, we may allow (2.3) to be substituted by $v_x(c) = 0$, if $\alpha > 0$ in $[a,c]$.

Remark 2.1 Problem $(\mathbf{P})$ may be regarded as a stationary problem (in this case β might vanish identically) or else as a time discretization of an evolution advection diffusion problem (hyperbolic in (a,b) and parabolic in (b,c)) by an implicit method (in this case, β behaves essentially like the reciprocal of the time discretization step). For this reason, we will always refer to problem $(\mathbf{P})$ as to a "hyperbolic–elliptic" problem, even if $(\mathbf{P})$ is a purely steady problem. By the way, we just note that the characteristic lines of the evolution hyperbolic problem enter the domain $(a,b) \times (0,+\infty)$ across $\{a\} \times (0,+\infty)$, when $\alpha > 0$ and across $\{b\} \times (0,+\infty)$, when $\alpha < 0$. This is the reason why we choose to impose condition (2.4) among others, which are equally admissible for the time–independent problem. When $\alpha < 0$, the same argument suggests not to impose any boundary condition at $x = a$ (though admissible for the very equation (2.1)); on the contrary, we are led to consider a condition on u at $x = b$. In the frame of the global problem (2.1), (2.2), this condition reads as an interface condition.

Two different types of elliptic regularizations are possible for problem **(P)**, both acceptable for some reason. We will see that the two ways are essentially different as for the behavior at the interface.

The case $\alpha > 0$.

Given $\epsilon > 0$, consider the problem

$(\mathbf{P}_\epsilon)$: to find u_ϵ defined in $[a,b]$, v_ϵ defined in $[b,c]$ such that

$$- \epsilon\, u_{\epsilon,xx} + \alpha\, u_{\epsilon,x} + \beta\, u_\epsilon = f \qquad in \ (a,b); \tag{2.5}$$

$$- (\nu v_{\epsilon,x})_x + \alpha\, v_{\epsilon,x} + \beta\, v_\epsilon = f \qquad in \ (b,c); \tag{2.6}$$

$$u_\epsilon(a) = 0; \tag{2.7}$$

$$v_\epsilon(c) = 0; \tag{2.8}$$

$$\left. \begin{array}{l} (i) \quad u_\epsilon = v_\epsilon \\ (ii) \quad \epsilon\, u_{\epsilon,x} = \nu\, v_{\epsilon,x} \end{array} \right\} \quad at \ x = b. \tag{2.9}$$

$(\mathbf{P}_\epsilon)$ is equivalent to a *variational* problem on the whole of (a,c); condition (2.9) expresses that u_ϵ and v_ϵ join continuously at b and that the flux across b is continuous, too.

About the existence of solutions to problem $(\mathbf{P}_\epsilon)$ and their behavior as $\epsilon \to 0$, the following result holds (see Appendix, where the appropriate choices of functional spaces are made and the regularity assumptions on the data are specified).

Proposition 2.1 *Assume the coerciveness condition*

$$2\beta - \alpha_x \geqslant 0 \qquad in \ [a,c]. \tag{2.10}$$

Then, problem $(\mathbf{P}_\epsilon)$ has a unique solution. Furthermore, as $\epsilon \to 0$, u_ϵ and v_ϵ converge to a pair of functions u, v which satisfy (2.1), (2.2), (2.3), (2.4) and the interface condition

$$\alpha\, u = -\nu\, v_x + \alpha\, v \qquad at \ x = b. \tag{2.11}$$

$\square$

Remark 2.2 (2.11) means that the flux across b is conserved, as $\epsilon \to 0$. On the contrary, analytical solution of $(\mathbf{P}_\epsilon)$ shows that u and v *do not join continuously* at b, in general. Actually, the closed form of the solution (as well as numerical experiments, see subsection 2.2.3) shows that the jump between u and v at b has the same order as v, when $v \to 0$.

A second approach is to consider the following problem

> $(\mathbf{Q}_\epsilon)$: to find u_ϵ defined in $[a,b]$, v_ϵ defined in $[b,c]$ such that (2.5), (2.6), (2.7), (2.8) and (2.9i) hold, along with the condition

$$u_{\epsilon,x} = v_{\epsilon,x} \qquad at \quad x = b. \tag{2.12}$$

$(\mathbf{Q}_\epsilon)$ is equivalent to a *nonvariational* elliptic problem on the whole of (a,c): now we are looking for a pair of functions u_ϵ, v_ϵ which have a $\mathbf{C}^1$ junction at b.

Proposition 2.2 *Assume the coerciveness condition (2.10). Let u_ϵ, v_ϵ solve problem $(\mathbf{Q}_\epsilon)$. As $\epsilon \to 0$, u_ϵ and v_ϵ converge to a pair of functions u, v which satisfy (2.1), (2.2), (2.3), (2.4) and the continuity condition*

$$u(b) = v(b) \tag{2.13}$$

at the interface.

$\qquad\qquad\qquad\qquad\qquad\qquad\qquad\qquad\qquad\qquad\qquad\qquad\qquad\quad\square$

We remark that (2.12) is not preserved, in general, as $\epsilon \to 0$: this can be checked on the closed form of the solutions to problem $(\mathbf{Q}_\epsilon)$, in some particular cases. Moreover, this feature is clearly shown by the numerical results presented in subsection 2.2.3. Thus, we are approaching a solution to problem $(\mathbf{P})$ which is continuous but not $\mathbf{C}^1$ at b.

The case $\alpha < 0$.

In this case, one can consider the same problems $(\mathbf{P}_\epsilon)$ and $(\mathbf{Q}_\epsilon)$ as before. However, for a reason which will be clear in section 3, we prefer to perform a slight change in the two problems, namely replacing the Dirichlet condition (2.7) with a Neumann one. Note that the original problem $(\mathbf{P})$ has no condition at all for $x = a$. Thus, we are dealing with a new couple of

130

problems, which we denote by $(\mathbf{P}_\epsilon)_N$ and $(\mathbf{Q}_\epsilon)_N$, respectively. For clarity, we state them in detail.

> $(\mathbf{P}_\epsilon)_N$: to find u_ϵ defined in $[a,b]$, v_ϵ defined in $[b,c]$ such that (2.5), (2.6), (2.8) and (2.9) hold, along with the condition
>
> $$u_{\epsilon,x}(a) = 0. \tag{2.14}$$
>
> $(\mathbf{Q}_\epsilon)_N$: to find u_ϵ defined in $[a,b]$, v_ϵ defined in $[b,c]$ such that (2.5), (2.6), (2.8), (2.9i), (2.12) and (2.14) hold.

The difference with respect to the case $\alpha > 0$ lies in the asymptotic behavior and, more precisely, in the interface conditions (remind that the limit problem (**P**) needs *two* conditions at b, in this case). The abstract analysis shown in the Appendix yields the following results (again, we do not specify the regularity on the data and on the unknowns here).

Proposition 2.3 *Assume the coerciveness condition (2.10). Then, problem $(\mathbf{P}_\epsilon)_N$ has a unique solution. Furthermore, as $\epsilon \to 0$, u_ϵ and v_ϵ converge to a pair of functions u, v which satisfy (2.1), (2.2), (2.3) and the following interface conditions:*

$$u(b) = v(b), \tag{2.15}$$
$$v_x(b) = 0. \tag{2.16}$$

□

Proposition 2.4 *Assume the coerciveness condition (2.10); moreover, suppose that $\beta \geqslant \beta_0 > 0$ in $[a,b]$. Let u_ϵ, v_ϵ solve problem $(\mathbf{Q}_\epsilon)_N$. As $\epsilon \to 0$, u_ϵ and v_ϵ converge to a pair of functions u, v which satisfy (2.1), (2.2), (2.3) and the following interface conditions:*

$$u(b) = v(b), \tag{2.17}$$
$$u_x(b) = v_x(b). \tag{2.18}$$

□

We point out that the condition at a for both $(\mathbf{P}_\epsilon)_N$ and $(\mathbf{Q}_\epsilon)_N$ is lost in the limit, as it is natural for this kind of problems.

Remark 2.3 By means of both approaches, the limit functions u and v enjoy a continuous junction at b. But the derivatives behave in a very different way (see (2.16) and (2.18)). Indeed, the limit of the solution to $(\mathbf{P}_\epsilon)_N$ shows an angle at b, in general, while the limit of the solution to $(\mathbf{Q}_\epsilon)_N$ is $\mathbf{C}^1$ at b. Thus, as in the previous case, the nonvariational approach is able to preserve an order of regularity higher by one, with respect to the variational one.

Remark 2.4 The two regularized problems with the original Dirichlet condition (2.7) have the same type of asymptotic behavior as the problems with the Neumann condition (2.14). The difference lies in that in the Dirichlet case the value $u_\epsilon(a)$ *does not* converge to the corresponding value $u(a)$, which is true for the Neumann case of problems $(\mathbf{P}_\epsilon)_N$ and $(\mathbf{Q}_\epsilon)_N$.

Remark 2.5 A comment is needed about (2.18). This condition calls into play the first derivative of the solution to (2.1) at b: but (2.1) is a first order equation, hence (2.18) involves a boundary operator *of the same order* as the interior equation. Thus, the left hand side of (2.18) must be compatible with the collocation of the equation (2.1) at b. Precisely, whenever the data are smooth, we expect equation (2.1) to hold at b, hence (2.17) and (2.18) imply

$$\alpha \, v_x + \beta \, v = f \qquad at \ x = b. \tag{2.19}$$

2.2. The numerical approximation

Set $\Omega^- = \,]a,b[$, $\Omega^+ = \,]b,c[$. On the reference interval $[-1,1]$, let us consider the Chebyshev collocation points

$$x_j^* = -\cos\frac{\pi \, j}{N}, \qquad j = 0,\cdots,N, \tag{2.20}$$

whose images in the interval $\overline{\Omega}^\mp$ are denoted by $\{x_j^\pm\}$. Note that $x_0^- = a$, $x_N^- = x_0^+ = b$, $x_N^+ = c$.

As an initial step, we consider two separate boundary value problems: a first order problem in Ω^- and a second order elliptic problem in Ω^+. Next, we introduce their numerical approximations based on the spectral collocation method. This presentation has the aim of providing the reader a guideline to the numerical approach of the coupled problem (**P**).

2.2.1. The split model problem

The two separate differential problems in Ω^- and Ω^+ are the following (we keep the same terminology as in section 2.1).

"Hyperbolic" problem in Ω^-:

$$\left|\begin{array}{rl} \alpha\, u_x + \beta\, u = f & in \ \ \Omega^-, \\ u(a) = u_a & if \ \alpha > 0, \\ u(b) = u_b & if \ \alpha < 0, \end{array}\right. \qquad (2.21)$$

where u_a and u_b are given. The motivation for the different choice of boundary conditions is given in Remark 2.1.

Elliptic problem in Ω^+:

$$\left|\begin{array}{rl} -(\nu v_x)_x + \alpha\, v_x + \beta\, v = f & in \ \ \Omega^+, \\ B_b\, v = v_b & at \ x = b, \\ B_c\, v = v_c & at \ x = c, \end{array}\right. \qquad (2.22)$$

where v_b and v_c are given and $B_b\, v$ and $B_c\, v$ are suitable combinations of v and v_x leading to a well posed problem.

The *spectral collocation approximation* to (2.21) is as follows (see, e.g., [CHQZ], Ch. 10 and 11). We look for $u_N \in \mathbf{P}_N$ (the space of algebraic polynomials of degree $\leqslant N$) such that

$$\alpha\, u_{N,x} + \beta\, u_N = f \qquad\qquad at \ x_j^-, \quad j = 1, \cdots, N-1, \quad (2.23)$$

supplemented by the two boundary equations:

$$if \ \alpha > 0 \ \left|\begin{array}{ll} (i) & u_N = u_a \quad at \ x_0^-, \\ (ii) \ \ \alpha\, u_{N,x} + \beta\, u_N = f & at \ x_N^-; \end{array}\right. \qquad (2.24)$$

$$if \ \alpha < 0 \ \left|\begin{array}{ll} (i) \ \ \alpha\, u_{N,x} + \beta\, u_N = f & at \ x_0^-, \\ (ii) & u_N = u_b \quad at \ x_N^-. \end{array}\right. \qquad (2.25)$$

The numerical approximation to (2.22), based on the spectral collocation method, is as follows. We look for $v_N \in \mathbf{P}_N$ satisfying

$$- \left[I_N \left(\nu v_{N,x} \right) \right]_x + \alpha \, v_{N,x} + \beta \, v_N = f \qquad at \ \ x_j^+ , \ j = 1, \cdots, N-1, \qquad (2.26)$$

$$B_b \, v_N = v_b \qquad at \ \ x_0^+ , \qquad (2.27)$$

$$B_c \, v_N = v_c \qquad at \ \ x_N^+ , \qquad (2.28)$$

where I_N is the interpolation operator at the points x_j^+.

2.2.2. The original coupled problem

Now we are in a position to describe the numerical approximation to the original coupled problem $(\mathbf{P})$, taking (2.23)–(2.28) into account.

1. At the *interior points* of Ω^- and Ω^+ , we impose the set of equations (2.23) and (2.26), respectively.

2. At $x = a$, we impose either (2.24i) (with $u_a = 0$) or (2.25i), according to the sign of α.

3. At $x = c$, we always enforce $v_N = 0$ (which corresponds to (2.28) with $v_c = 0$ and $B_c \, v_N = v_N$).

4. At $x = b$, we need two equations, in order to close the algebraic system. These depend both on the sign of α and on the interface conditions provided by either elliptic regularization (see section 2.1). In particular:

   (a) if $\alpha > 0$, we impose (2.24ii), along with either

$$- \nu v_{N,x} + \alpha \, v_N = \alpha \, u_N \qquad (\textit{variational approach}) \qquad (2.29)$$

   or

$$v_N = u_N \qquad (\textit{nonvariational approach}); \qquad (2.30)$$

   (b) if $\alpha < 0$, we impose the condition

$$u_N = v_N \qquad (2.31)$$

   (i.e. (2.25ii), with $u_b = v_N \left(x_0^+ \right)$); the remaining equation is given by either

$$v_{N,x} = 0 \qquad (\textit{variational approach}) \qquad (2.32)$$

   or

$$v_{N,x} = u_{N,x} \qquad (\textit{nonvariational approach}). \qquad (2.33)$$

We note that (2.29), (2.30), (2.32) and (2.33) are but special versions of (2.27), with suitable choices of B_b and v_b . These are specified in table 1, which summarizes the equations to be fulfilled by the numerical solution at each collocation point (including boundary and interface).

Collocation points	$\alpha > 0$		$\alpha < 0$	
	variational	nonvariational	variational	nonvariational
$x = a$	(2.24i)	(2.24i)	(2.23)	(2.23)
$x_j^{(1)}$, $1 \leqslant j \leqslant N-1$	(2.23)	(2.23)	(2.23)	(2.23)
$x = b$	(2.27)$^{(1)}$ (L) (2.23) (C)	(2.27)$^{(2)}$ (L) (2.23) (C)	(2.25ii) (L) (2.27)$^{(3)}$ (L)	(2.25ii) (L) (2.27)$^{(4)}$ (L)
$x_j^{(2)}$, $1 \leqslant j \leqslant N-1$	(2.26)	(2.26)	(2.26)	(2.26)
$x = c$	(2.28)	(2.28)	(2.28)	(2.28)

Table 1. *Numerical approximation to problem* **P** *by spectral collocation method: (L) = limit condition given by the asymptotic analysis; (C) = compatibility condition (the transport equation must be collocated at the outflow boundary of Ω_1);*

(1) with $B_- v_N = -\gamma\, v_{N,x} + \alpha v_N$ and $v_b = \alpha u_N (x_N^{(1)})$;

(2) with $B_- v_N = v_N$ and $v_b = u_N (x_N^{(1)})$;

(3) with $B_- v_N = v_{N,x}$ and $v_b = 0$;

(4) with $B_- v_N = v_{N,x}$ and $v_b = u_{N,x} (x_N^{(1)})$.

2.2.3. Some numerical results

Now we present several numerical experiments which support the theoretical results obtained in the previous subsections. We deal with the elliptic regularizations of problem (**P**), taking $(a,c) = (-1,1)$, with $b = 0$. In all cases ($\alpha > 0$ or $\alpha < 0$, variational or nonvariational approach), the equations are

$$- \epsilon\, u_{\epsilon,xx} + \alpha\, u_{\epsilon,x} + \beta\, u_\epsilon = f \qquad in \ (-1,0); \qquad (2.34)$$

$$- (\nu v_{\epsilon,x})_x + \alpha\, v_{\epsilon,x} + \beta\, v_\epsilon = f \qquad in \ (0,1). \qquad (2.35)$$

The interface conditions change according to the regularization chosen:

$$(variational) \quad \begin{cases} (i) \quad u_\epsilon = v_\epsilon, \\ (ii) \quad \epsilon\, u_{\epsilon,x} = \nu\, v_{\epsilon,x} \end{cases} \quad at \ x = 0 \qquad (2.36)$$

or

$$(nonvariational) \quad \begin{cases} (i) \quad u_\epsilon = v_\epsilon, \\ (ii) \quad u_{\epsilon,x} = v_{\epsilon,x} \end{cases} \quad at \ x = 0. \qquad (2.37)$$

The boundary conditions will be distinguished later.

These problems are solved by the Chebyshev collocation method described in advance for fully elliptic problems of the form (2.22).

(To be more precise, we have implemented the collocation method in a domain decomposition framework, in order to achieve the highest precision. To this end, three subdomains are used; within each of them, we take 50 points; the middle subdomain includes the interface point $x = 0$. At each interface between subdomains the $\mathbf{C}^1$ continuity is enforced directly (see [FQZ]).)

The data we have used are the following

$$a = -1, \ b = 0, \ c = 1, \ f \equiv 1, \ \nu \equiv 1, \ \beta \equiv 1 + x^2. \qquad (2.38)$$

A homogeneous Dirichlet condition is enforced at $x = 1$. About the point $x = -1$, we consider the case of a homogeneous Dirichlet condition, to begin with.

In Figure 2.1 we graph the results obtained for the variational approach, with $\epsilon = 0.005$ and $\epsilon = 0.1$, when $\alpha = 1$. In agreement with our theoretical results (see Remark 2.2), the solution exhibits a discontinuity as $\epsilon \to 0$ at the interface point $x = 0$. The discontinuity is revealed by the presence of oscillations near the interface, due to the Gibbs phenomenon. However, the jump is of the same order as the viscosity coefficient ν, as shown in Figure 2.2.

Figure 2.3 displays the results obtained for the nonvariational case, using the same data as in Figure 2.1. Note that, as $\epsilon \to 0$, the solution is continuous (though not C^1) at the interface point, as predicted by (2.17).

The comparison between variational and nonvariational approaches is clearer in Figure 2.4, where we take $\epsilon = 0.005$.

In Figures 2.5 and 2.6 we present the results obtained using the two approaches, with the same data as before, but with $\alpha = -1$. As predicted by the theory (see Propositions 2.2 and 2.3), as $\epsilon \to 0$ the nonvariational solution remains C^1, while the variational one is just C^0.

Finally, Figure 2.7 reports the results obtained with $\alpha = -1$, $\epsilon = 0.005$ and a homogeneous Neumann condition at $x = -1$ (rather than the Dirichlet one), using the variational and nonvariational approaches.

Figure captions

Figure 2.1   *Results for the variational approach, with $\alpha = 1$: $\epsilon = 0.1$ (dashed line), $\epsilon = 0.005$ (solid line).*

Figure 2.2   *Results for the variational approach, with $\alpha = 1$, $\epsilon = 0.005$: $v = 1$ (solid line), $v = 0.1$ (dash-dot line), $v = 0.01$ (dash-dash line)*

Figure 2.3   *Results for the nonvariational approach, with $\alpha = 1$: $\epsilon = 0.1$ (dashed line), $\epsilon = 0.005$ (solid line).*

Figure 2.4   *Comparison between the two approaches, with $\alpha = 1$, $\epsilon = 0.005$: nonvariational approach (dashed line), variational approach (solid line).*

Figure 2.5   *Results for the variational approach, with $\alpha = -1$: $\epsilon = 0.1$ (dashed line), $\epsilon = 0.005$ (solid line).*

Figure 2.6   *Results for the nonvariational approach, with $\alpha = -1$: $\epsilon = 0.1$ (dashed line), $\epsilon = 0.005$ (solid line).*

Figure 2.7   *Comparison between the two approaches, with $\alpha = -1$, $\epsilon = 0.005$, with homogeneous Neumann boundary condition at $x = -1$ and homogeneous Dirichlet boundary condition at $x = 1$: nonvariational approach (dashed line), variational approach (solid line).*

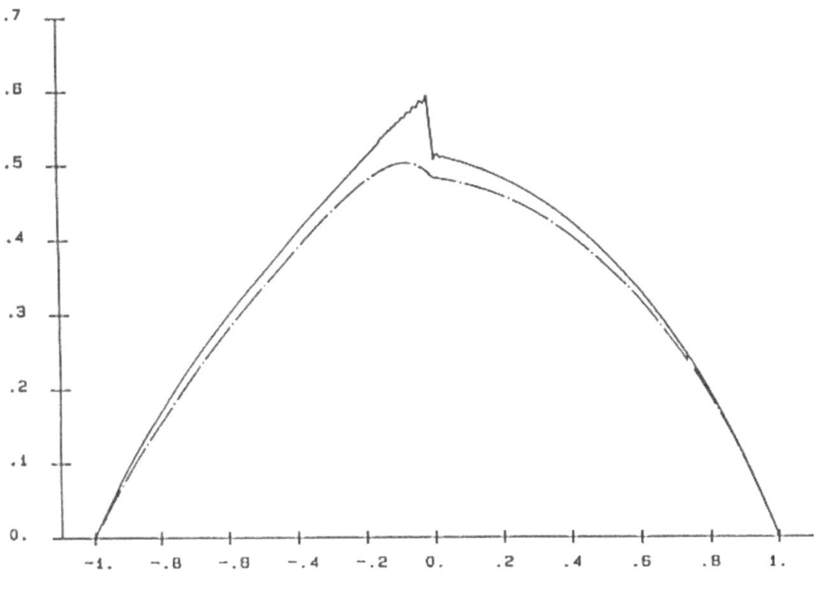

Fig. 2.1

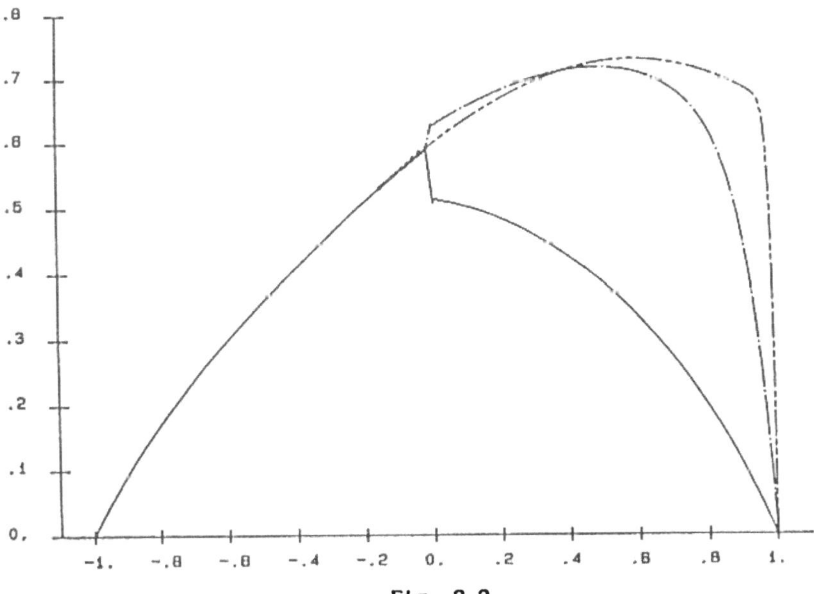

Fig. 2.2

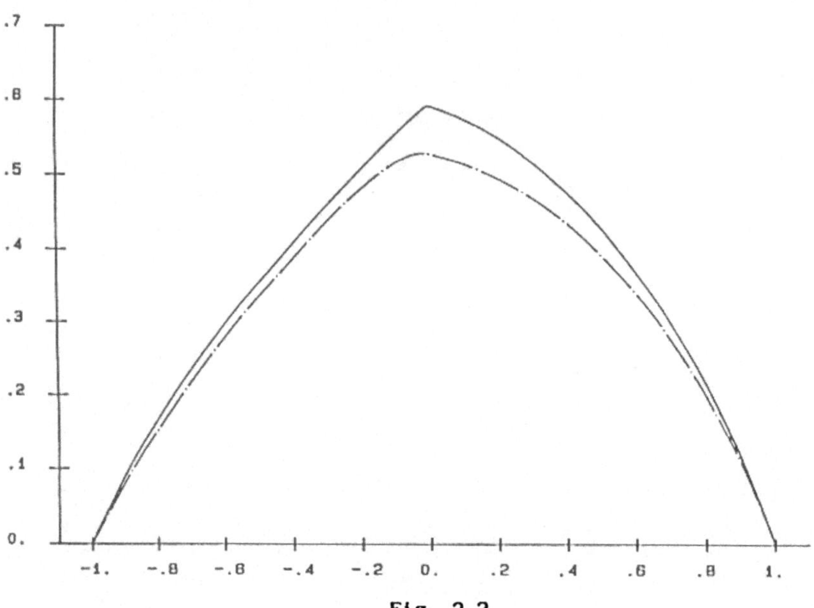

Fig. 2.3

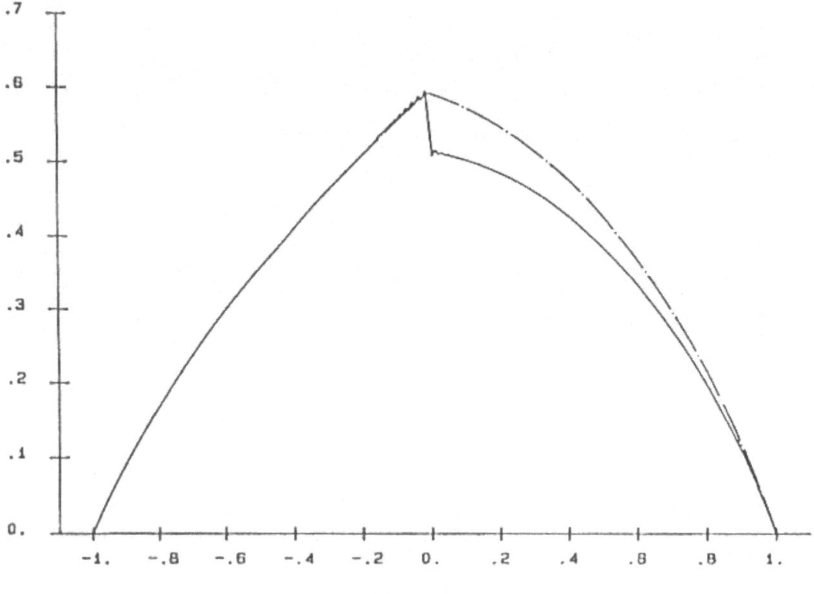

Fig. 2.4

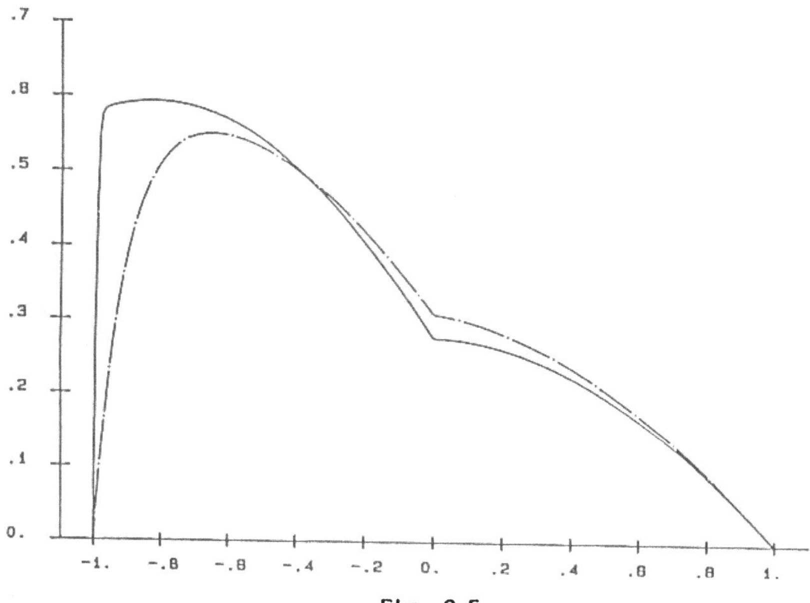

Fig. 2.5

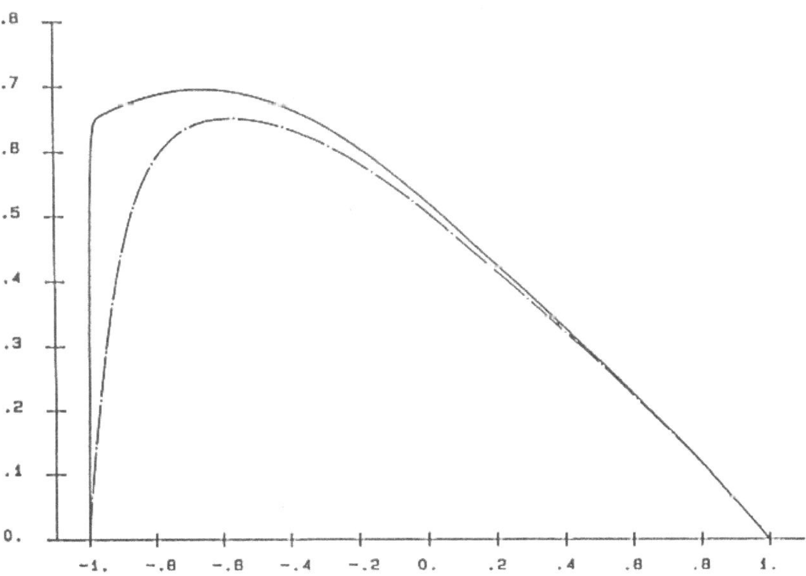

Fig. 2.6

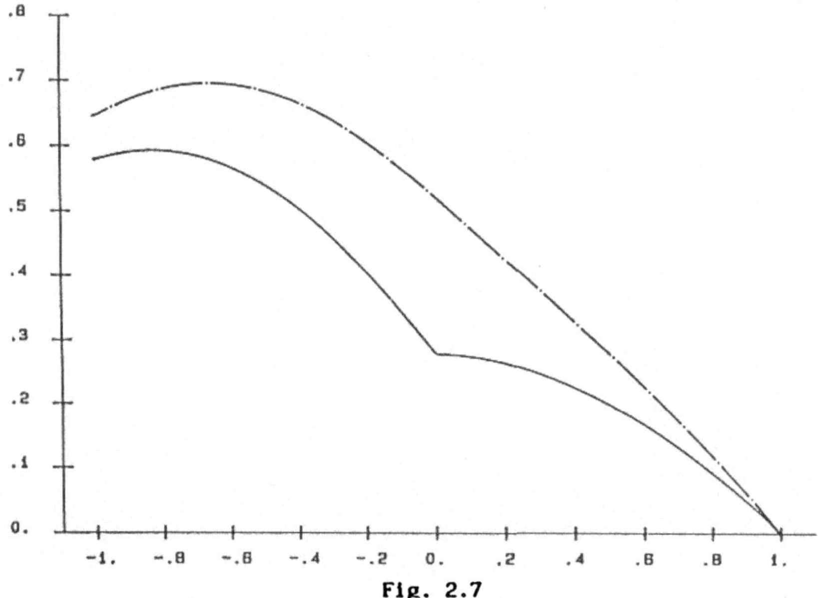

Fig. 2.7

3. HYPERBOLIC-ELLIPTIC INTERACTION: THE (TIME INDEPENDENT) VECTOR CASE

In this section we consider a boundary value problem for a system of three linear equations. Precisely, we deal with the stationary problem associated to (1.1), (1.2).

3.1. The differential problem

With the notations of the introduction, we seek for a pair of three dimensional vector functions $\mathbf{w}^-$ and $\mathbf{w}^+$ such that

$$A\,\mathbf{w}_x^- + B\,\mathbf{w}^- = \mathbf{F} \qquad in \;\; \Omega^-, \qquad (3.1)$$

$$-(\nu\mathbf{w}_x^+)_x + A\,\mathbf{w}_x^+ + B\,\mathbf{w}^+ = \mathbf{F} \qquad in \;\; \Omega^+. \qquad (3.2)$$

About boundary conditions, we must distinguish between the points a and c.

At $x = a$, we prescribe exactly p_0 conditions on $\mathbf{w}^-$, where p_0 is the number of positive eigenvalues of the matrix A. These conditions take the form

$$G^-\mathbf{w}^- = \mathbf{q}^- \qquad at \;\; x = a, \qquad (3.3)$$

where $\mathbf{q}^-$ is a given vector with p_0 components, while G^- is a $p_0 \times 3$ matrix with rank p_0. The choice of G^- is subject to some restrictions that will be specified later.

At $x = c$, the boundary conditions can be written in the general form

$$G^+\mathbf{w}^+ + H^+\mathbf{w}_x^+ = \mathbf{q}^+ \qquad at \;\; x = c, \qquad (3.4)$$

where G^+ and H^+ are 3×3 matrices and $\mathbf{q}^+$ is a given vector with three components. (3.4) must provide 3 independent equations, which are admissible for the elliptic system (3.2). In general, (3.4) yields a coupling between the three components of $\mathbf{w}^+$ and their derivatives. However, in some special circumstances, (3.4) might lead to three equations, each of them containing only one component and/or its derivative.

Problem (3.1)–(3.4) needs $(6 - p_0)$ further conditions at the interface point $x = b$. Essentially, three of them are requested by the elliptic system (3.2), while $(3 - p_0)$ (the number of negative eigenvalues of A) pertain to the hyperbolic system (3.1).

To write down the interface conditions, let us introduce the matrix T which diagonalizes A and denote by

$$\Lambda = T \; A \; T^{-1} \tag{3.5}$$

the (diagonal) eigenvalue matrix. We write Λ as

$$\Lambda = \begin{pmatrix} \Lambda_p & 0 \\ 0 & \Lambda_n \end{pmatrix}, \tag{3.6}$$

where Λ_p is the diagonal matrix of the p_0 positive eigenvalues of A, while Λ_n is made by the remaining $3 - p_0$ negative eigenvalues. Correspondingly, we write T as

$$T = \begin{pmatrix} T_p \\ T_n \end{pmatrix}, \tag{3.7}$$

where T_p is the submatrix of the first p_0 rows of T and T_n is the rest (note that the rows of T are made by the left eigenvectors of A).

The interface conditions we consider here are of two types: either

$$\left\{ \begin{array}{llr} (i) & -\nu \mathbf{w}_x^+ + A\,\mathbf{w}^+ = A\,\mathbf{w}^- & (3\ cond.s) \\ (ii) & T_n \mathbf{w}^+ = T_n \mathbf{w}^- & (3 - p_0\ cond.s) \end{array} \right. \tag{3.8}$$

(variational approach) or

$$\left\{ \begin{array}{llr} (i) & \mathbf{w}^+ = \mathbf{w}^- & (3\ cond.s) \\ (ii) & T_n \mathbf{w}_x^+ = T_n \mathbf{w}_x^- & (3 - p_0\ cond.s) \end{array} \right. \tag{3.9}$$

(nonvariational approach).

In both cases, we impose as many conditions as requested.

The rest of this subsection is devoted to a mathematical justification of (3.8) and (3.9) by means of the asymptotic procedure on elliptic regularizations, in analogy with the scalar case.

Precisely, for a given $\epsilon > 0$, we consider the regularized problem

$$- \epsilon \mathbf{w}_{\epsilon,xx}^- + A\,\mathbf{w}_{\epsilon,x}^- + B\,\mathbf{w}_\epsilon^- = \mathbf{F} \qquad in \ \ \Omega^-, \tag{3.10}$$

$$- (\nu\,\mathbf{w}_{\epsilon,x}^+)_x + A\,\mathbf{w}_{\epsilon,x}^+ + B\,\mathbf{w}_\epsilon^+ = \mathbf{F} \qquad in \ \ \Omega^+, \tag{3.11}$$

with boundary conditions

$$\left\{ \begin{array}{ll} G^- \mathbf{w}_\epsilon^- = \mathbf{q}^- & \\ T_n \mathbf{w}_{\epsilon,x}^- = 0 & at \ x = a, \end{array} \right. \tag{3.12}$$

$$G^+ \mathbf{w}_\epsilon^+ + H^+ \mathbf{w}_{\epsilon,x}^+ = \mathbf{q}^+ \quad at \ x = c \tag{3.13}$$

and interface conditions at the point $x = b$

$$(variational\ approach) \quad \left| \begin{array}{l} \mathbf{w}_\epsilon^- = \mathbf{w}_\epsilon^+ \\ \epsilon\, \mathbf{w}_{\epsilon,x}^- = \nu\, \mathbf{w}_{\epsilon,x}^+ \end{array} \right. \tag{3.14}$$

or

$$(nonvariational\ approach) \quad \left| \begin{array}{l} \mathbf{w}_\epsilon^- = \mathbf{w}_\epsilon^+ \\ \mathbf{w}_{\epsilon,x}^- = \mathbf{w}_{\epsilon,x}^+ \,. \end{array} \right. \tag{3.15}$$

In (3.12), the original boundary condition has been added a homogeneous Neumann condition on $T_n \mathbf{w}_\epsilon^-$: this is not the only possibility, but it is optimal, in some sense (see Remark 3.1).

In order to exploit the results of section 2.1, it is natural to diagonalize the system (3.10), (3.11). This is done by introducing the characteristic variables associated to the system, namely

$$\mathbf{z}_\epsilon^\pm = T\, \mathbf{w}_\epsilon^\pm :$$

denote by $(\mathbf{z}_\epsilon^\pm)_p$ the first p_0 components of $\mathbf{z}_\epsilon^\pm$ and by $(\mathbf{z}_\epsilon^\pm)_n$ the remaining $3 - p_0$ components. Thus, (3.10)–(3.15) imply that $\mathbf{z}_\epsilon^\pm$ satisfy the equations:

$$- \epsilon\, \mathbf{z}_{\epsilon,xx}^- + \Lambda \mathbf{z}_{\epsilon,x}^- + B_T\, \mathbf{z}_\epsilon^- = \mathbf{F}_T \quad in \ \Omega^-, \tag{3.16}$$

$$- (\nu\, \mathbf{z}_{\epsilon,x}^+)_x + \Lambda \mathbf{z}_{\epsilon,x}^+ + B_T\, \mathbf{z}_\epsilon^+ = \mathbf{F}_T \quad in \ \Omega^+ \tag{3.17}$$

(where $B_T = T\, B\, T^{-1}$ and $\mathbf{F}_T = T\, \mathbf{F}$), boundary conditions

$$\left| \begin{array}{ll} G_T^-\, \mathbf{z}_\epsilon^- = \mathbf{q}^- & \\ (\mathbf{z}_{\epsilon,x}^-)_n = 0 & at \ x = a\,, \end{array} \right. \tag{3.18}$$

$$G_T^+ \mathbf{z}_\epsilon^+ + H_T^+ \mathbf{z}_{\epsilon,x}^+ = \mathbf{q}^+ \quad at \ x = c \tag{3.19}$$

(where $G_T^\pm = G^\pm\, T^{-1}$ and $H_T^+ = H^+\, T^{-1}$) and interface conditions at the point $x = b$

$$(variational\ approach) \quad \left| \begin{array}{l} \mathbf{z}_\epsilon^- = \mathbf{z}_\epsilon^+ \\ \epsilon\, \mathbf{z}_{\epsilon,x}^- = \nu\, \mathbf{z}_{\epsilon,x}^+ \end{array} \right. \tag{3.20}$$

or

$$(nonvariational\ approach) \quad \left| \begin{array}{l} \mathbf{z}_\epsilon^- = \mathbf{z}_\epsilon^+ \\ \mathbf{z}_{\epsilon,x}^- = \mathbf{z}_{\epsilon,x}^+ \,. \end{array} \right. \tag{3.21}$$

In (3.18), the matrix G_T^- must satisfy the following assumpti‹

$$\text{the submatrix given by the first}$$
$$p_0 \text{ columns of } G_T^- \text{ is nonsingular.}$$

(3.22) poses restrictions on the choice of G^- in (3.3), dependi›

We are now in a position to use Propositions 2.1–2.4, wl‹ following convergence results (we suppose existence of soluti‹ larized problems).

Proposition 3.1 *(Variational approach) As $\epsilon \to 0$, the sol· (3.16)-(3.20) converges to a pair of functions $\mathbf{z}^-$, $\mathbf{z}^+$ wl›*

$$\Lambda \mathbf{z}_x^- + B_T \mathbf{z}^- = \mathbf{F}_T \qquad in \ \Omega^-,$$

$$- (\nu \, \mathbf{z}_x^+)_x + \Lambda \mathbf{z}_x^+ + B_T \mathbf{z}^+ = \mathbf{F}_T \qquad in \ \Omega^+,$$

$$G_T^- \mathbf{z}^- = \mathbf{q}^- \qquad at \ x = a,$$

$$G_T^+ \mathbf{z}^+ + H_T^+ \mathbf{z}_x^+ = \mathbf{q}^+ \qquad at \ x = c,$$

$$- \nu \, \mathbf{z}_{p,x}^+ + \Lambda_p \, \mathbf{z}_p^+ = \Lambda_p \, \mathbf{z}_p^- \qquad at \ x = b,$$

$$\mathbf{z}_n^+ = \mathbf{z}_n^- \qquad at \ x = b,$$

$$\mathbf{z}_{n,x}^+ = 0 \qquad at \ x = b$$

(in (3.27)-(3.29), $\mathbf{z}_p^\pm$ denotes the first p_0 components denotes the remaining $3 - p_0$ components).

Proposition 3.2 *(Nonvariational approach) As $\epsilon \to 0$, ‹/ $\mathbf{z}_\epsilon^+$ to (3.16)-(3.19) and (3.21) converges to a pair of f‹ which satisfy (3.23)-(3.26) and*

$$\mathbf{z}^+ = \mathbf{z}^- \qquad at \ x = b,$$

$$\mathbf{z}_{n,x}^+ = \mathbf{z}_{n,x}^- \qquad at \ x = b.$$

By re–transforming these results in terms of the physical v‹ find (3.1)–(3.4) and either

$$\left\{ \begin{array}{llr} (i) & - \nu T_p \, \mathbf{w}_x^+ + \Lambda_p \, T_p \mathbf{w}^+ = \ \Lambda_p \, T_p \, \mathbf{w}^- & (p\cdot \\ (ii) & T_n \mathbf{w}^+ = \ T_n \, \mathbf{w}^- & (3 - p \\ (iii) & T_n \mathbf{w}_x^+ = \ 0 & (3 - p\cdot \end{array} \right.$$

or (3.9), according to the regularization chosen. It remains to show that (3.32) is equivalent to (3.8). Actually, taking (3.32ii) into account, (3.32iii) can be written as

$$- \nu T_n \mathbf{w}_x^+ + \Lambda_n T_n \mathbf{w}^+ = \Lambda_n T_n \mathbf{w}^-.$$

Together with (3.32i), this last condition gives

$$- \nu T \mathbf{w}_x^+ + \Lambda T \mathbf{w}^+ = \Lambda T \mathbf{w}^-.$$

Multiplying by T^{-1} and recalling that ν is a scalar function, we get

$$- \nu \mathbf{w}_x^+ + A \mathbf{w}^+ = A \mathbf{w}^-,$$

whence (3.8) follows.

Remark 3.1 A Dirichlet condition on $T_n \mathbf{w}_\epsilon^-$ in (3.12) (i.e. a Dirichlet condition on $(\mathbf{z}_\epsilon^-)_n$ in (3.18)) is as good as the Neumann condition we considered in (3.12), provided (3.25) involves the p_0 characteristic variables corresponding to positive eigenvalues only. This means that the last $3 - p_0$ columns of G_T^- ought to vanish identically. Essentially, the reason of this drawback is that the Dirichlet condition cannot guarantee the convergence of $(\mathbf{z}_\epsilon^-)_n (a)$ to $\mathbf{z}_n^- (a)$. Thus, the strategy of reducing the analysis of the system to that of the scalar case cannot deal with a condition of type (3.3) involving the value of $\mathbf{z}_n^- (a)$. However, a more sophisticated vector approach could be performed, capable of overcoming this difficulty (see [L] and the references quoted there).

3.2. The numerical approximation

We adopt the notations of section 2.2 for the collocation points.

The spectral collocation approximation to problem (3.1)–(3.4) reads as follows. We look for $\mathbf{w}_N^- \in (\mathbf{P}_N)^3$ and $\mathbf{w}_N^+ \in (\mathbf{P}_N)^3$ satisfying:

$$A \mathbf{w}_{N,x}^- + B \mathbf{w}_N^- = \mathbf{F} \quad at \ x_j^-, \ j = 1, \cdots, N-1, \quad (3.33)$$

$$- [I_N (\nu \mathbf{w}_{N,x}^+)]_x + A \mathbf{w}_{N,x}^+ + B \mathbf{w}_N^+ = \mathbf{F} \quad at \ x_j^+, \ j = 1, \cdots, N-1. \quad (3.34)$$

The conditions at $x = a$ are of two types:

(i) p_0 prescribed boundary conditions (see (3.3)):

$$G^- \mathbf{w}_N^- = \mathbf{q}^- \qquad at \ x_0^-, \qquad (3.35)$$

(ii) $(3 - p_0)$ compatibility conditions:

$$T_n \, [A \, \mathbf{w}_{\bar{N}}\,,_x + B \, \mathbf{w}_{\bar{N}} - \mathbf{F}] = 0 \qquad at \ \ x_0^- \ . \qquad (3.36)$$

Note that (3.36) are nothing but the collocation at x_0^- of the equations on the characteristic variables corresponding to negative eigenvalues: they generalize to the vector case the compatibility condition (2.25i) for the scalar case, yielding a stable and consistent scheme (see, e.g., [CQ]).

At the right boundary point c, we enforce the prescribed boundary conditions (3.4) on the discrete solution, namely

$$G^+ \mathbf{w}_{\bar{N}}^+ + H^+ \mathbf{w}_{\bar{N}}^+\,,_x = \mathbf{q}^+ \qquad at \ \ x_N^+ \ . \qquad (3.37)$$

Now, we come to the conditions at the interface point b. As usual, we distinguish between the variational and the nonvariational approaches which have been used. The results of the analysis presented in section 3.1 (see Propositions 3.1 and 3.2) suggest the proper continuity conditions to be enforced at the interface point.

(a) Variational approach.

(i) p_0 compatibility conditions on the equations corresponding to the positive characteristic variables:

$$T_p \, [A \, \mathbf{w}_{\bar{N}}\,,_x + B \, \mathbf{w}_{\bar{N}} - \mathbf{F}] = 0 \quad at \ \ b \ (= x_N^-); \qquad (3.38)$$

(ii) $(3 - p_0)$ conditions of continuity on the negative characteristic variables:

$$T_n \, \mathbf{w}_{\bar{N}}^- = T_n \, \mathbf{w}_{\bar{N}}^+ \qquad at \ \ b \ (= x_N^-), \qquad (3.39)$$

obtainable from (3.8ii);

(iii) 3 conditions of continuity of the "flux" on the physical variables:

$$- \nu \mathbf{w}_{\bar{N}}^+\,,_x + A \, \mathbf{w}_{\bar{N}}^+ = A \, \mathbf{w}_{\bar{N}}^- \quad at \ \ b \ (= x_0^+). \qquad (3.40)$$

obtainable from (3.8i).

Remark 3.2 Notice that the hyperbolic system (3.33) has been supplemented three conditions at the interface point b $(= x_N^-)$ in (i) and (ii) (see (3.38) and (3.39)). Similarly, the elliptic system (3.34) has been given three Newton–like conditions at the interface point b $(= x_0^+)$ in (iii) (see (3.40)).

147

(b) Nonvariational approach.

(i) p_0 compatibility conditions on the equations corresponding to the positive characteristic variables, given by (3.38);

(ii) $(3 - p_0)$ conditions of continuity on the negative characteristic variables, given by (3.39);

(iii) p_0 conditions of continuity on the positive characteristic variables:

$$T_p \mathbf{w}_N^+ = T_p \mathbf{w}_N^- \qquad at \ b \ (= x_0^+): \qquad (3.41)$$

both (ii) and (iii) are obtainable from (3.9i);

(iv) $(3 - p_0)$ conditions of continuity of first derivatives on the negative characteristic variables:

$$T_n \mathbf{w}_{N,x}^+ = T_n \mathbf{w}_{N,x}^- \qquad at \ b \ (= x_0^+), \qquad (3.42)$$

obtainable from (3.9ii).

The same kind of considerations as in Remark 3.2 can be made in this case, too.

We note that (ii) and (iii) amount to require that $\mathbf{w}_N^+ = \mathbf{w}_N^-$ at b.

Remark 3.3 An efficient (and quite natural) method to solve problems of the form (3.33)–(3.37), supplemented with the interface conditions (3.38)–(3.40) (or (3.38), (3.39), (3.41) and (3.42)), relies upon an iterative procedure alternating the solution of a hyperbolic problem in Ω^- and of an elliptic one in Ω^+.

At each step, the iterative method entails within Ω^- the solution of the hyperbolic problem (3.33) with the boundary conditions (3.35) and (3.36) at the left hand boundary x_0^- , and (3.38), (3.39) at the right hand boundary x_N^- . Next, in Ω^+ we solve the elliptic problem (3.34) with the boundary condition (3.37) at the right hand boundary x_N^+ and the conditions (3.40) (or (3.41), (3.42)) at the left hand boundary x_0^+ . Finally, a relaxation procedure on the interface variables is generally needed, in order to ensure the convergence of the above process.

The details and the convergence analysis will be presented in a forth-coming paper.

148

4. HYPERBOLIC-PARABOLIC SYSTEMS FOR TIME DEPENDENT PROBLEMS

In this section we consider the problem (1.1), (1.2) presented in the introduction, endowed with its boundary, initial and interface conditions.

4.1. The differential problem

With a, b, c chosen in the usual way, we look for a three dimensional vector valued function $\mathbf{w}^{\pm}$ defined for $x \in \Omega^{\pm}$, $t > 0$, satisfying

$$\mathbf{w}_t^- + A\,\mathbf{w}_x^- + B\,\mathbf{w}^- = \mathbf{F} \qquad \text{for } x \in \Omega^- = (a,b), \, t > 0, \quad (4.1)$$

$$\mathbf{w}_t^+ - (\nu\mathbf{w}_x^+)_x + A\,\mathbf{w}_x^+ + B\,\mathbf{w}^+ = \mathbf{F} \qquad \text{for } x \in \Omega^+ = (b,c), \, t > 0, \quad (4.2)$$

where A, B, $\mathbf{F}$ and ν are given as like as in the introduction.

The system (4.1), (4.2) must be given an initial condition

$$\mathbf{w}^{\pm}(x,0) = \mathbf{w}_0^{\pm}(x), \qquad x \in \Omega^{\pm} \qquad (4.3)$$

and boundary conditions, which we take again of the form (3.3) and (3.4), namely

$$G^-\mathbf{w}^- = \mathbf{q}^- \qquad at \; x = a, \, t > 0, \qquad (4.4)$$

$$G^+\mathbf{w}^+ + H^+\mathbf{w}_x^+ = \mathbf{q}^+ \qquad at \; x = c, \, t > 0, \qquad (4.5)$$

where G^-, G^+, H^+, $\mathbf{q}^-$ and $\mathbf{q}^+$ may depend on t.

Analogously, at the interface line $\{b\} \times (0,+\infty)$ we impose conditions which are the natural extension of (3.8) and (3.9) to the evolution case: either

$$(variational \atop approach) \left\{ \begin{array}{ll} (i) & T_n\mathbf{w}^+ = T_n\mathbf{w}^- \\ (ii) & -\nu\mathbf{w}_x^+ + A\mathbf{w}^+ = A\mathbf{w}^- \end{array} \right. \qquad (4.6)$$

or

$$(nonvariational \atop approach) \left\{ \begin{array}{ll} (i) & \mathbf{w}^+ = \mathbf{w}^- \\ (ii) & T_n\mathbf{w}_x^+ = T_n\mathbf{w}_x^- \end{array} \right. \qquad (4.7)$$

for $x = b$ and for $t > 0$.

The interface conditions (4.6) or (4.7) might be derived directly by means of regularized parabolic problems, in analogy to the procedure presented in section 3.1.

On the other hand, several heuristic justifications of these conditions may be given. For instance, one may take the Laplace transform of (4.1), (4.2), at least formally: the new unknowns satisfy a problem similar to (3.1)-(3.4). This means that the interface conditions for the new unknowns are precisely (3.8) or (3.9): by anti–transforming these conditions one gets exactly (4.6) or (4.7).

Furthermore, problem (3.1)-(3.4) can be viewed as a (possible) steady state for the time–dependent problem (4.1)-(4.5), or else as the time–discretization (at any time level) of problem (4.1)-(4.5), using an implicit time–stepping scheme. In both cases, in section 3.1 we have seen that the interface conditions (3.8) or (3.9) are appropriate for problem (3.1)-(3.4). Thus, (4.6) or (4.7) turn out to be appropriate for problem (4.1)-(4.5).

4.2. The numerical approximation

First, we consider a *semidiscrete (continuous in time)* approximation of problem (4.1)-(4.5), endowed either with (4.6) or with (4.7). Keeping the same notations of the preceding sections 2.2, 3.2, we apply the spectral collocation method in space, that is, we look for two mappings

$$t \; \rightarrow \; \mathbf{w}_N^{\mp}(t) \in (\mathbf{P}_N)^3$$

satisfying, for all $t > 0$ and all $j = 1, \cdots, N-1$,

$$\mathbf{w}_{N,t}^- + A\,\mathbf{w}_{N,x}^- + B\,\mathbf{w}_N^- = \mathbf{F} \quad at \; x_j^-, \qquad (4.8)$$

$$\mathbf{w}_{N,t}^+ - [I_N(\nu\mathbf{w}_{N,x}^+)]_x + A\,\mathbf{w}_{N,x}^+ + B\,\mathbf{w}_N^+ = \mathbf{F} \quad at \; x_j^+, \qquad (4.9)$$

At the left boundary we impose the conditions

$$G^-\mathbf{w}_N^- = \mathbf{q}^-, \quad T_n\,[\mathbf{w}_{N,t}^- + A\,\mathbf{w}_{N,x}^- + B\,\mathbf{w}_N^- - \mathbf{F}] = 0 \qquad (4.10)$$

for $x = x_0^-$ and $t > 0$, while at the right boundary the conditions are

$$G^+\mathbf{w}_N^+ + II^+\mathbf{w}_{N,x}^+ = \mathbf{q}^+ \qquad (4.11)$$

for $x = x_N^+$ and $t > 0$. Eventually, the two alternative sets of interface conditions to be requested for $x = x_N^- = x_0^+$ and $t > 0$ are the following:

(a) Variational approach,

$$T_p\,[\mathbf{w}_{N,t}^- + A\,\mathbf{w}_{N,x}^- + B\,\mathbf{w}_N^- - \mathbf{F}] = 0, \qquad (4.12)$$

$$T_n\,\mathbf{w}_N^- = T_n\,\mathbf{w}_N^+, \qquad (4.13)$$

$$-\nu\mathbf{w}_{N,x}^+ + A\,\mathbf{w}_N^+ = A\,\mathbf{w}_N^-. \qquad (4.14)$$

150

(b) Nonvariational approach.

$$T_p \left[\mathbf{w}_{\bar{N}\,,t} + A\,\mathbf{w}_{\bar{N}\,,x} + B\,\mathbf{w}_{\bar{N}} - \mathbf{F}\right] = 0, \tag{4.15}$$

$$\mathbf{w}_{\bar{N}} = \mathbf{w}_{\bar{N}}^{+}, \tag{4.16}$$

$$T_n\,\mathbf{w}_{\bar{N}\,,x}^{+} = T_n\,\mathbf{w}_{\bar{N}\,,x}^{-}. \tag{4.17}$$

A *fully discrete approximation* to problem (4.1)–(4.5), endowed either with (4.6) or with (4.7) can be achieved by applying a time–stepping procedure to (4.8), (4.9). Whatever scheme (either implicit or explicit) one uses to advance from a known time level t^{k} to a new one t^{k+1}, the interface conditions, as well as the boundary conditions , need to be imposed at the new time t^{k+1}.

If an *explicit scheme* is used in this regard, at the time t^{k+1} the unknown vectors $\{\mathbf{w}_{\bar{N}}(x_j^{-})\}$ and $\{\mathbf{w}_{\bar{N}}^{+}(x_j^{+})\}$, $j = 1,\cdots,N-1$, can be computed independently of the boundary and interface values. Once these internal values are available, the boundary equations (4.10) and (4.11), together with the interface conditions (4.12)–(4.14) (or (4.15)–(4.17)), can be solved to provide the remaining values at boundary and interface points. Actually, we note that the presence of derivatives in space among boundary and interface conditions relates boundary and interface values to each other. We also note that the differential equations between brackets in (4.10) and in (4.12) (or (4.15)) ought to be advanced by the same explicit scheme which was used for the equations at the internal points.

When an *implicit time marching scheme* is used, the internal unknowns are not decoupled from the remaining ones any more. As an example, we detail the case of the simplest implicit scheme, namely the first order forward Euler scheme.

Denoting by Δt the time step, by $t^{k} = k\,\Delta t$ the k-th time level and by $(\mathbf{w}_{\bar{N}}^{\pm})^{k}$ the spectral solutions at the time t^{k}, the corresponding problem reads:

$$(\mathbf{w}_{\bar{N}}^{-})^{k+1} + \Delta t\,[A\,\mathbf{w}_{\bar{N}\,,x} + B\,\mathbf{w}_{\bar{N}} - \mathbf{F}]^{k+1} - (\mathbf{w}_{\bar{N}}^{-})^{k} = 0, \tag{4.18}$$

$$(\mathbf{w}_{\bar{N}}^{+})^{k+1} + \Delta t\,\{-[I_N(\nu\mathbf{w}_{\bar{N}\,,x}^{+})]_x +$$
$$+ A\,\mathbf{w}_{\bar{N}\,,x}^{+} + B\,\mathbf{w}_{\bar{N}}^{+} - \mathbf{F}\}^{k+1} - (\mathbf{w}_{\bar{N}}^{+})^{k} = 0. \tag{4.19}$$

The boundary equations (4.10) and (4.11) are discretized as follows:

$$\left\{ \begin{array}{l} [\,G^{-}\mathbf{w}_{\bar{N}}^{-} - \mathbf{q}^{-}\,]^{k+1} = 0, \\[2mm] T_n\,\{(\mathbf{w}_{\bar{N}}^{-})^{k+1} + \Delta t\,[A\,\mathbf{w}_{\bar{N}\,,x} + B\,\mathbf{w}_{\bar{N}} - \mathbf{F}]^{k+1} - (\mathbf{w}_{\bar{N}}^{-})^{k}\} = 0 \end{array} \right. \tag{4.20}$$

at x_0^- ;

$$[\, G^+ \mathbf{w}_N^+ + H^+ \mathbf{w}_{N,x}^+ - \mathbf{q}^+ \,]^{k+1} = 0 \qquad (4.21)$$

at x_N^+. Analogously, the interface conditions (4.12)–(4.14) give:

$$T_p \{\, (\mathbf{w}_N^-)^{k+1} + \Delta t \,[A\,\mathbf{w}_{N,x}^- + B\,\mathbf{w}_N^- - \mathbf{F}]^{k+1} - (\mathbf{w}_N^-)^k \,\} = 0, \quad (4.22)$$

$$T_n \,[(\mathbf{w}_N^-)^{k+1} - (\mathbf{w}_N^+)^{k+1}] = 0, \qquad (4.23)$$

$$-\,\nu^{k+1}(\mathbf{w}_{N,x}^+)^{k+1} + A\,(\mathbf{w}_N^+)^{k+1} = A\,(\mathbf{w}_N^-)^{k+1}, \qquad (4.24)$$

at x_0^+ . The alternative interface equations (4.15)–(4.17) read:

$$T_p \{\, (\mathbf{w}_N^-)^{k+1} + \Delta t \,[A\,\mathbf{w}_{N,x}^- + B\,\mathbf{w}_N^- - \mathbf{F}]^{k+1} - (\mathbf{w}_N^-)^k \,\} = 0, \quad (4.25)$$

$$(\mathbf{w}_N^-)^{k+1} = (\mathbf{w}_N^+)^{k+1}, \qquad (4.26)$$

$$T_n \,[(\mathbf{w}_{N,x}^-)^{k+1} - (\mathbf{w}_{N,x}^+)^{k+1}] = 0. \qquad (4.27)$$

We notice that the structure of the system would be the same when using other implicit time–marching schemes (such as, for instance, the second order Beam & Warming scheme).

Remark 4.1 We note that (4.18)–(4.21) with the interface conditions (4.22)–(4.24) (or (4.25)–(4.27)) have the same shape as the time independent problem (3.33)–(3.42) considered in the previous section. Clearly, in (3.33)–(3.42) we must replace $\mathbf{w}$ by $\mathbf{w}_N^{k+1}$, B by $B + (\Delta t)^{-1}I$ and $\mathbf{F}$ by $\mathbf{F}^{k+1} + (\Delta t)^{-1}\mathbf{w}_N^k$, respectively. Therefore, the same iterative procedure can be used in order to decouple the hyperbolic problem in Ω^- and the elliptic one in Ω^+.

APPENDIX: abstract analysis of the regularizing problems presented in section 2

In this Appendix, we detail the existence and asymptotic convergence results stated in Propositions 2.1–2.4 for problems $(\mathbf{P}_\epsilon)$, $(\mathbf{Q}_\epsilon)$, $(\mathbf{P}_\epsilon)_N$ and $(\mathbf{Q}_\epsilon)_N$.

As a standard notation, whenever Ω is an open interval and k is a positive integer we introduce the Sobolev space (see [A])

$$\mathbf{H}^k(\Omega) = \{\, v \in \mathbf{L}^2(\Omega)\colon D^m v \in \mathbf{L}^2(\Omega),\ m = 1, \cdots, k \,\}. \qquad (A.1)$$

$\mathbf{H}^k(\Omega)$ is a Hilbert space with norm

$$\| v \|_{\mathbf{H}^k(\Omega)} = \left[\, \| v \|_{\mathbf{L}^2(\Omega)}^2 + \sum_{m=1}^{k} \| D^m v \|_{\mathbf{L}^2(\Omega)}^2 \,\right]^{1/2}.$$

Since Ω is one dimensional, we have that

$$\mathbf{H}^{k}(\Omega) \subset \mathbf{C}^{k-1}(\overline{\Omega}), \tag{A.2}$$

for all positive integer k, the embedding being compact. In particular, $\mathbf{H}^{1}(\Omega)$ is made by functions continuous up to the boundary. Therefore, the following (usual) notation is meaningful:

$$\mathbf{H}_0^{1}(\Omega) = \{\, v \in \mathbf{H}^{1}(\Omega) : v = 0 \ \textit{at the endpoints of} \ \Omega\, \}. \tag{A.3}$$

Unless otherwise stated, we will make the following assumptions on the data of problem $(\mathbf{P})$:

$$v \in \mathbf{L}^{\infty}(b,c), \quad \alpha \in \mathbf{H}^{1}(a,c), \quad \beta \in \mathbf{L}^{2}(a,c), \quad f \in \mathbf{L}^{2}(a,c). \tag{A.4}$$

Problem $(\mathbf{P}_\epsilon)$.

Recall that $\alpha > 0$ in this case.

Under the assumption (A.4) (actually, under *milder* assumptions), $(\mathbf{P}_\epsilon)$ can be written in a rigorous variational form:

find $w_\epsilon \in \mathbf{W}$ such that, for all $\phi \in \mathbf{W}$,

$$\int_a^c a_\epsilon w_{\epsilon,x}\,\phi_x\,dx + \int_a^c \alpha w_{\epsilon,x}\,\phi\,dx + \int_a^c \beta w_\epsilon \phi\,dx = \int_a^c f\,\phi\,dx, \tag{A.5}$$

where

$$\mathbf{W} = \mathbf{H}_0^{1}(a,c), \qquad a_\epsilon = \begin{cases} \epsilon \ \textit{in} \ (a,b) \\ v \ \textit{in} \ (b,c). \end{cases} \tag{A.6}$$

If w_ϵ solves (A.5), then the functions

$$u_\epsilon = w_{\epsilon|(a,b)}, \qquad v_\epsilon = w_{\epsilon|(b,c)} \tag{A.7}$$

solve (2.5)–(2.9): this is easily checked by means of suitable choices of ϕ in (A.5). In particular, (A.5) entails the equation (in the distribution sense)

$$-(a_\epsilon w_{\epsilon,x})_x + \alpha w_{\epsilon,x} + \beta w_\epsilon = f \qquad \textit{in} \ (a,c), \tag{A.8}$$

whence

$$a_\epsilon w_{\epsilon,x} \in \mathbf{H}^{1}(a,c). \tag{A.9}$$

By (A.2) it follows that both w_ϵ and $a_\epsilon w_{\epsilon,x}$ are continuous in $[a,c]$, hence (2.7), (2.8) and (2.9) have the classical meaning.

In order to achieve an existence result for $(\mathbf{P}_\epsilon)$, from now on we make the following requests:

$$v \geqslant v_0 \quad in \ (b,c),\tag{A.10}$$

for a suitable strictly positive constant v_0 and

$$2\beta - \alpha_x \geqslant 0 \quad in \ (a,c).\tag{A.11}$$

Lemma A.1 *Under the assumptions (A.4), (A.10), (A.11), $(\mathbf{P}_\epsilon)$ has a unique solution.*

Proof. It is possible to apply Lax–Milgram lemma, because (A.10), (A.11) and Poincaré inequality imply that problem $(\mathbf{P}_\epsilon)$ is coercive. It goes without saying that coerciveness fails as $\epsilon \to 0$.

$\square$

Now, let us discuss the asymptotic behavior of w_ϵ as $\epsilon \to 0$. We recall the notations (A.7) and the assumptions (A.4), (A.10) and (A.11), which still hold.

Lemma A.2 *There is a constant $C > 0$ such that*

$$\| w_\epsilon \|_{L^2(a,c)} \leqslant C,\tag{A.12}$$

$$\| v_{\epsilon,x} \|_{L^2(b,c)} \leqslant C,\tag{A.13}$$

$$\| \sqrt{\epsilon}\, u_{\epsilon,x} \|_{L^2(a,b)} \leqslant C.\tag{A.14}$$

Proof. Plug the function ϕ in (A.5), with $\phi = e^{-x}\, w_\epsilon$ in (a,b), $\phi = e^{-b}\, w_\epsilon$ in (b,c), then integrate by parts. The assumptions and Poincaré inequality give the results.

$\square$

Lemma A.3 *The L^2 norm of $u_{\epsilon,x}$ is bounded in a right neighborhood of the left boundary $x = a$.*

Proof. Let ψ be a smooth function in (a,c), vanishing outside a right neighborhood of a. Take the L^2 scalar product of (A.8) by $\psi w_{\epsilon,x}$: the assertion follows by (A.12) and (A.14).

$\square$

Now, let us introduce the function

$$\Phi_\epsilon = a_\epsilon w_{\epsilon,x} - \alpha w_\epsilon : \tag{A.15}$$

We already know that $\Phi_\epsilon \in \mathbf{H}^1(a,c)$ (see (A.9)).

Lemma A.4 *The* $\mathbf{H}^1$ *norm of* Φ_ϵ *is bounded in* (a,c).

Proof. Lemma A.2 gives the boundedness of Φ_ϵ in $\mathbf{L}^2(a,c)$; (A.8) and (A.12) give the boundedness of $\Phi_{\epsilon,x}$ in $\mathbf{L}^2(a,c)$.

$\square$

Now, we are in a position to give the following result, which completes and refines the statement of Proposition 2.1.

Proposition A.1 *Assume (A.4), (A.10), (A.11). There are* $u \in \mathbf{L}^2(a,b)$ *and* $v \in \mathbf{L}^2(b,c)$ *which satisfy*

$$\alpha\, u_x + \beta\, u = f \qquad in\ \mathbf{L}^2(a,b); \tag{A.16}$$

$$-(\nu v_x)_x + \alpha\, v_x + \beta\, v = f \qquad in\ \mathbf{L}^2(b,c); \tag{A.17}$$

$$u(a) = 0; \tag{A.18}$$

$$v(c) = 0; \tag{A.19}$$

$$\alpha u = -\nu v_x + \alpha v \qquad at\ x = b. \tag{A.20}$$

Proof. As a consequence of Lemmas A.2–A.4, of Banach–Alaoglu–Bourbaki theorem and of (A.2), we can find $u \in \mathbf{L}^2(a,b)$, $v \in \mathbf{L}^2(b,c)$ and $\Phi \in \mathbf{H}^1(a,c)$ such that (upon extracting a subfamily)

    (i) $u_\epsilon \to u$ weakly in $\mathbf{L}^2(a,b)$;

    (ii) $v_\epsilon \to v$ weakly in $\mathbf{H}^1(b,c)$;

    (iii) $\Phi_\epsilon \to \Phi$ weakly in $\mathbf{H}^1(a,c)$;

    (iv) $u_\epsilon(a) \to u(a)$;

    (v) $v_\epsilon(c) \to v(c)$;

    (vi) $v_\epsilon(b) \to v(b)$;

    (vii) $\epsilon u_{\epsilon,x} \to 0$ strongly in $\mathbf{L}^2(a,b)$.

Note that the value $u(a)$ is well defined, because of Lemma A.3. (i)–(iii) and (vii) permit to pass to the limit in (A.5): this gives (A.16) and (A.17). (A.18) and (A.19) follow by (iv) and (v), respectively, since $u_\epsilon(a) = v_\epsilon(c) = 0$.

Finally, (i)–(iii) and (vii) entail that $\Phi = -\alpha u$ in (a,b) and $\Phi = \nu v_x - \alpha v$ in (b,c), whence (A.20) follows, by (vi).

$\square$

Remark A.1 Analogous results could be proved when replacing the homogeneous Dirichlet condition at c by a Neumann condition or by a Newton–type condition.

$$Problem\ (\mathbf{P}_\epsilon)_N \ .$$

Recall that $\alpha < 0$ in this case.

For this problem, the variational formulation is still (A.5), just changing the function space: now we take

$$\mathbf{W} = \{\, v \in \mathbf{H}^1(a,c): v(c) = 0 \,\}. \tag{A.21}$$

The existence holds the same way as in the previous case and the asymptotic analysis is analogous. For completeness, we detail the main steps. Again, we assume (A.4), (A.10), (A.11) and use the notations (A.7).

Lemma A.5 *There is a constant* $C > 0$ *such that*

$$\| w_\epsilon \|_{\mathbf{L}^2(a,c)} \leqslant C, \tag{A.22}$$

$$\| v_{\epsilon,x} \|_{\mathbf{L}^2(b,c)} \leqslant C, \tag{A.23}$$

$$\| u_{\epsilon,x} \|_{\mathbf{L}^2(a,b)} \leqslant C, \tag{A.24}$$

$$\sqrt{\epsilon}\,| u_{\epsilon,x}(b)| \leqslant C, \tag{A.25}$$

$$\| (\nu v_{\epsilon,x})_x \|_{\mathbf{L}^2(b,c)} \leqslant C. \tag{A.26}$$

Proof. (A.22) and (A.23) follow by plugging the function ϕ in (A.5), with $\phi = e^x\, w_\epsilon$ in (a,b), $\phi = e^b\, w_\epsilon$ in (b,c), then integrating by parts. The assumptions and Poincaré inequality give the results. Moreover, the same computation shows that

$$| u_\epsilon(a)| \leqslant C. \tag{A.27}$$

To prove (A.24) and (A.25), let χ be the characteristic function of the interval $[a,b]$. First, take the $\mathbf{L}^2$ scalar product of (A.8) by χ, then integrate by parts in the integrals containing derivatives of u_ϵ. By (A.22), (A.23) and (A.27) it follows that

$$\epsilon \,|\, u_{\epsilon,x}(b)\,|\; \leqslant\; C. \qquad\qquad (A.28)$$

Now, take the $\mathbf{L}^2$ scalar product of (A.8) by χ, as before. This time, integrate by parts only in the integral containing the second derivative of u_ϵ: by (A.22) and (A.28) we get the uniform boundedness of the $\mathbf{L}^1(a,b)$ norm of $u_{\epsilon,x}$. Therefore, a well known theorem gives the uniform boundedness of the $\mathbf{L}^\infty(a,b)$ norm of u_ϵ. Keeping this estimate in mind, take the $\mathbf{L}^2$ scalar product of (A.8) by $\chi w_{\epsilon,x}$, then integrate by parts. (A.24) and (A.25) follow easily.

Finally, (A.26) follows by (A.8), (A.22) and (A.23).

$$\square$$

Thus, we are in a position to prove the main result, which was summarized in Proposition 2.2.

Proposition A.2 *Assume (A.4), (A.10), (A.11). Moreover, assume that v is continuous at $x = b$. Then, there are $u \in \mathbf{H}^1(a,b)$ and $v \in \mathbf{H}^1(b,c)$ which satisfy (A.16), (A.17), (A.19) and the interface conditions*

$$u(b) = v(b), \qquad\qquad (A.29)$$

$$v_x(b) = 0. \qquad\qquad (A.30)$$

Proof. As a consequence of the previous Lemma and of (A.2), we can show the existence of $u \in \mathbf{H}^1(a,b)$ and $v \in \mathbf{H}^1(b,c)$ such that (upon extracting a subfamily)

      (i) $u_\epsilon \to u$ weakly in $\mathbf{H}^1(a,b)$;

      (ii) $v_\epsilon \to v$ weakly in $\mathbf{H}^1(b,c)$;

      (iii) $\nu v_{\epsilon,x} \to \nu v_x$ weakly in $\mathbf{H}^1(a,c)$;

      (iv) $v_\epsilon(c) \to v(c)$;

      (v) $u_\epsilon(b) \to u(b)$ and $v_\epsilon(b) \to v(b)$;

      (vi) $(\nu v_{\epsilon,x})(b) \to (\nu v_x)(b)$;

      (vii) $\epsilon u_{\epsilon,x}(b) \to 0$.

(i)–(iii) permit to pass to the limit in (A.5). The conditions at $x = c$ and $x = b$ follow by (iv)–(vii), noting that $v(b) > 0$ (see (A.10)).

$$\square$$

Remark A.2 If we take a homogeneous Dirichlet condition at $x = a$ instead of the Neumann one, then $(\mathbf{P}_\epsilon)_N$ coincides with $(\mathbf{P}_\epsilon)$; so does its variational formulation. But now we are assuming $\alpha < 0$, hence the asymptotic behavior is different from that of the case $\alpha > 0$. It is easy to see that the final Proposition A.2 still holds, with u found in $\mathbf{L}^2(a,b)$: actually, the convergence of u_ϵ to u is only $\mathbf{L}^2(a,b)$ (weak), whence we cannot have a convergence of $u_\epsilon(a)$ to $u(a)$, in general. Actually, Figure 2.5 shows a numerical evidence of a boundary layer for u_ϵ at $x = a$, although the limit function u is obviously continuous in $[a,b]$ (see (A.16) and (A.2)). This feature makes $(\mathbf{P}_\epsilon)_N$ preferable, especially in view of the applications to systems (sections 3 and 4).

$$Problems \ (\mathbf{Q}_\epsilon) \ and \ (\mathbf{Q}_\epsilon)_N \ .$$

Now, the two problems do not admit a "natural" global variational formulation and the question of existence and the asymptotic behavior are somewhat more complicate. Nevertheless if we assume that (A.4) holds and that

$$v \ is \ continuous \ at \ x = b , \tag{A.31}$$

then the equations and the boundary and interface conditions defining $(\mathbf{Q}_\epsilon)$ and $(\mathbf{Q}_\epsilon)_N$ make sense, provided the solutions are sought for in $\mathbf{H}^1(a,b)$ and $\mathbf{H}^1(b,c)$, respectively.

We begin with problem $(\mathbf{Q}_\epsilon)$, recalling that $\alpha > 0$.

Lemma A.6 *Assume (A.4), (A.10), (A.11), (A.31): if ϵ is small, then $(\mathbf{Q}_\epsilon)$ has a unique solution.*

Proof. The proof is carried out by means of a fixed point procedure. Let λ be a real number and solve the two separate boundary value problems:

$$- \epsilon u_{\lambda,xx} + \alpha \, u_{\lambda,x} + \beta \, u_\lambda = f \qquad in \ (a,b); \tag{A.32}$$

$$u_\lambda(a) = 0; \tag{A.33}$$

$$u_\lambda(b) = \lambda; \tag{A.34}$$

$$- (v v_{\lambda,x})_x + \alpha \, v_{\lambda,x} + \beta \, v_\lambda = f \qquad in \ (b,c); \tag{A.35}$$

158

$$v_\lambda(b) = \lambda; \tag{A.36}$$

$$v_\lambda(c) = 0. \tag{A.37}$$

Both problems (A.32)–(A.34) and (A.35)–(A.37) have a unique solution, by the assumptions (A.4), (A.10), (A.11). Let

$$\Phi(\lambda) \equiv v_{\lambda,x}(b) - u_{\lambda,x}(b): \tag{A.38}$$

clearly, solving (Q_ϵ) is equivalent to finding a zero of Φ. Now, Φ is a linear affine function; moreover, we claim that it is stricly decreasing, at least if ϵ is small enough (this gives existence and uniqueness of the zero of Φ, at once). For, take $\lambda_1 < \lambda_2$ and denote by u_i, v_i the solutions to (A.32)–(A.37) corresponding to λ_i, $i = 1,2$. Let

$$\lambda = \lambda_2 - \lambda_1 > 0, \quad u = u_2 - u_1, \quad v = v_2 - v_1. \tag{A.39}$$

By difference, we see that λ, u, v solve

$$-\epsilon\, u_{xx} + \alpha\, u_x + \beta\, u = 0 \quad in\ (a,b); \tag{A.40}$$

$$u(a) = 0; \tag{A.41}$$

$$u(b) = \lambda; \tag{A.42}$$

$$-(v v_x)_x + \alpha\, v_x + \beta\, v = 0 \quad in\ (b,c); \tag{A.43}$$

$$v(b) = \lambda; \tag{A.44}$$

$$v(c) = 0. \tag{A.45}$$

Now, take the $\mathbf{L}^2$ scalar product of (A.40) and of (A.43) by $\dfrac{u}{\epsilon}$ and by $\dfrac{v}{v(b)}$, respectively, then integrate by parts and add term by term. Taking (A.41), (A.42), (A.44), (A.45) and (A.11) into account, it follows

$$\lambda(u_x(b) - v_x(b)) \geqslant \frac{\lambda^2}{2}\,\alpha(b)\left|\frac{1}{\epsilon} - \frac{1}{v(b)}\right| +$$

$$+ \int_a^b u_x^{\,2}\,dx + \frac{1}{v(b)} \int_b^c v v_x^{\,2}\,dx. \tag{A.46}$$

Thus, $u_x(b) - v_x(b)$ is nonnegative, provided ϵ is small enough, whence

$$\Phi(\lambda_1) \geqslant \Phi(\lambda_2).$$

Moreover, this inequality must be *strict*, otherwise (A.46) would imply $\lambda = 0$, which is impossible. Therefore, Φ is strictly decreasing and the existence of a unique solution to (Q_ϵ) holds.

$$\square$$

From now on, u_ϵ and v_ϵ denote the pair of functions which solve (Q_ϵ). Their asymptotic behavior is being investigated now, under the assumptions (A.4), (A.10), (A.11), (A.31).

Lemma A.7 *There is a constant $C > 0$ such that*

$$\| u_\epsilon \|_{\mathbf{L}^2(a,b)} \leqslant C, \tag{A.47}$$

$$\| \sqrt{\epsilon} u_{\epsilon,x} \|_{\mathbf{L}^2(a,b)} \leqslant C, \tag{A.48}$$

$$| u_\epsilon(b) | \leqslant C. \tag{A.49}$$

Proof. (i) Take the $\mathbf{L}^2(a,b)$ scalar product of (2.5) by $v(b)e^{-x} u_\epsilon$, then integrate by parts.

(ii) Take the $\mathbf{L}^2(b,c)$ scalar product of (2.6) by $\epsilon e^{-b} v_\epsilon$, then integrate by parts.

(iii) Add the two equations provided by (i) and (ii), term by term: the conclusion follows by Poincaré inequality.

$\square$

Lemma A.8 *There is a constant $C > 0$ such that*

$$\| v_\epsilon \|_{\mathbf{H}^1(b,c)} \leqslant C, \tag{A.50}$$

$$| v_{\epsilon,x}(b) | \leqslant C, \tag{A.51}$$

$$\| u_\epsilon \|_{\mathbf{H}^1(a,b)} \leqslant C. \tag{A.52}$$

Proof. Let $\zeta_\epsilon \in \mathbf{H}^1(b,c)$ be the solution of

$$-(v\zeta_{\epsilon,x})_x = 0 \ \ in \ (b,c), \ \ \zeta_\epsilon(b) = v_\epsilon(b), \ \ \zeta_\epsilon(c) = 0.$$

By (A.49), the $\mathbf{H}^1(b,c)$ norm of ζ_ϵ is bounded, as well as the value of $\zeta_{\epsilon,x}(b)$. Moreover, the function $d_\epsilon \equiv v_\epsilon - \zeta_\epsilon$ belongs to $\mathbf{H}_0^1(b,c)$ and satisfies

$$-(v d_{\epsilon,x})_x + \alpha d_{\epsilon,x} + \beta d_\epsilon = g_\epsilon, \tag{A.53}$$

where $g_\epsilon = f - \alpha\zeta_{\epsilon,x} - \beta\zeta_\epsilon$ is bounded in $\mathbf{L}^2(b,c)$. Multiplying (A.53) in $\mathbf{L}^2(b,c)$ by d_ϵ, it follows that the $\mathbf{H}^1(b,c)$ norm of d_ϵ is bounded, whence (A.50).

Next, we multiply (A.53) by $\psi v d_{\epsilon,x}$, where ψ is a smooth function vanishing outside a right neighborhood of b: (A.51) follows easily.

Finally, (A.52) can be proved by taking the $\mathbf{L}^2(a,b)$ scalar product of (2.5) by $u_{\epsilon,x}$ and using (A.47), (A.51).

$\square$

From Lemma A.8 we get the following proposition (see Proposition 2.3).

Proposition A.3 *Assume (A.4), (A.10), (A.11), (A.31). There are $u \in \mathbf{H}^1(a,b)$ and $v \in \mathbf{H}^1(b,c)$ which satisfy (A.16), (A.17), (A.18), (A.19) and (A.29).*

Proof. Let u_ϵ, v_ϵ solve $(\mathbf{Q}_\epsilon)$. By Lemma A.8, there are $u \in \mathbf{H}^1(a,b)$ and $v \in \mathbf{H}^1(b,c)$ such that (upon extracting a subfamily)

 (i) $u_\epsilon \rightharpoonup u$ weakly in $\mathbf{H}^1(a,b)$;

 (ii) $v_\epsilon \rightharpoonup v$ weakly in $\mathbf{H}^1(b,c)$;

 (iii) $u_\epsilon(a) \to u(a)$;

 (iv) $v_\epsilon(c) \to v(c)$;

 (v) $u_\epsilon(b) \to u(b)$ and $v_\epsilon(b) \to v(b)$.

All of these properties permit to pass to the limit in the regularized problem $(\mathbf{Q}_\epsilon)$. Thus, the proof follows easily.

$\square$

Now we come to problem $(\mathbf{Q}_\epsilon)_N$: recall that $\alpha < 0$.

This case looks somewhat trickier than the previous one and the natural choices for test functions do not seem to be appropriate, in proving both the existence and the a priori estimates. Even more, it can be shown that problem $(\mathbf{Q}_\epsilon)_N$ may fail to have a solution under the assumptions (A.4), (A.10), (A.11), (A.31) (which were sufficient for existence in the previous case).

This trouble seems to be motivated by the lack of a maximum principle under the sole coerciveness condition (A.11) on β. For this reason, we discuss problem $(\mathbf{Q}_\epsilon)_N$ under the further hypothesis:

$$\beta(x) \geqslant 0 \ \ for \ x \ a.e. \ in \ (a,b). \tag{A.54}$$

We just note that such an assumption is not strongly restrictive if the problem we are dealing with is regarded as a time discretization of an evolution problem by an implicit method (see section 4.2).

Now we are able to answer the question of existence of solutions to $(\mathbf{Q}_\epsilon)_N$.

Lemma A.9 *Assume (A.4), (A.10), (A.11), (A.31), (A.54): then* $(\mathbf{Q_\epsilon})_N$ *has a unique solution.*

Proof. The procedure is analogous to the one applied in Lemma A.6. Keeping the same notations, for given λ we construct the function Φ as in (A.38), where v_λ solves (A.35), (A.36), (A.37) and u_λ solves (A.32), (A.34) and the boundary condition at a :

$$u_{\lambda,x}(a) = 0. \tag{A.55}$$

As in the previous case, we show that Φ is strictly decreasing. For, take $\lambda_1 < \lambda_2$, denote by u_i , v_i ($i = 1,2$) the corresponding solutions and recall the notation (A.39). Obviously, λ, u, v solve (A.40), (A.42)–(A.45) and

$$u_x(a) = 0. \tag{A.56}$$

Now, take the $\mathbf{L}^2$ scalar product of (A.40) by $v(b)\phi u_\epsilon$, where

$$\phi(x) = \frac{1}{\epsilon} \exp\left\{\frac{1}{\epsilon} \int_x^b \alpha(t)\,dt\right\}: \tag{A.57}$$

note that $\epsilon\phi_x + \alpha\phi = 0$ in (a,b) and $\phi(b) = \frac{1}{\epsilon}$. Then, integrate by parts. Next, take the $\mathbf{L}^2$ scalar product of (A.43) by v, then integrate by parts. Adding term by term the two results and taking (A.11), (A.42), (A.44), (A.45), (A.54), (A.56) into account, it follows

$$\lambda(u_x(b) - v_x(b)) \geqslant \epsilon\int_a^b \phi u_x{}^2 dx + \frac{1}{v(b)} \int_b^c v v_x{}^2 dx. \tag{A.58}$$

Thus, it follows $u_x(b) - v_x(b) \geqslant 0$, whence Φ is decreasing (actually, *strictly decreasing*, otherwise (A.58) would imply $u_x = v_x \equiv 0$, which is impossible). Therefore, the existence of a unique solution to $(\mathbf{Q_\epsilon})_N$ holds. $\square$

From now on, u_ϵ and v_ϵ denote the pair of functions which solve $(\mathbf{Q_\epsilon})_N$. Their asymptotic behavior is being investigated now, under the assumptions (A.4), (A.10), (A.11), (A.31), (A.54). For technical reasons, we will confine the situation a bit more, making the further hypothesis:

$$f \in \mathbf{L}^\infty(a,b), \quad \beta(x) \geqslant \beta_0 > 0 \ \textit{for } x \ \textit{a.e. in } (a,b), \tag{A.59}$$

for some β_0 . This allows us to get low order estimates on u_ϵ and v_ϵ . Later on, we will make further assumptions in order to find higher order estimates.

Lemma A.10 *There is a constant $C > 0$ such that*

162

$$\| v_\epsilon \|_{\mathbf{H}^1(b,c)} \leqslant C, \tag{A.60}$$

$$| v_{\epsilon,x}(b) | \leqslant C, \tag{A.61}$$

$$\| (\nu v_{\epsilon,x})_x \|_{\mathbf{L}^2(b,c)} \leqslant C. \tag{A.62}$$

Proof. (i) Take the $\mathbf{L}^2(a,b)$ scalar product of (2.5) by $\nu(b)\phi u_\epsilon$, where ϕ is defined in (A.57). Then, integrate by parts.

(ii) Take the $\mathbf{L}^2(b,c)$ scalar product of (2.6) by v_ϵ, then integrate by parts.

(iii) Add the two equations provided by (i) and (ii), term by term. Recalling (A.10), (A.11), (A.59), we find that

$$\epsilon\, \nu(b)\int_a^b \phi u_{\epsilon,x}{}^2 dx + \nu_0\int_b^c v_{\epsilon,x}{}^2 dx + \nu(b)\,\beta_0\int_a^b \phi u_\epsilon{}^2 dx - \frac{1}{2}\,\alpha(b)v_\epsilon{}^2(b) \leqslant$$

$$\leqslant \nu(b)\int_a^b f\, \phi u_\epsilon dx + \int_b^c f\, v_\epsilon dx\ . \tag{A.63}$$

Now, by (A.59) we have

$$\int_a^b f\, \phi u_\epsilon dx \leqslant \| f \|_{\mathbf{L}^\infty(a,b)}\int_a^b \phi \mid u_\epsilon \mid dx,$$

so that Poincaré inequality in (A.63) gives

$$\epsilon \int_a^b \phi u_{\epsilon,x}{}^2 dx + k_1\int_b^c v_{\epsilon,x}{}^2 dx + \beta_0\int_a^b \phi u_\epsilon{}^2 dx + k_2 v_\epsilon{}^2(b) \leqslant$$

$$\leqslant \| f \|_{\mathbf{L}^\infty(a,b)}\int_a^b \phi \mid u_\epsilon \mid dx + k_3\ , \tag{A.64}$$

where k_i are positive constants, $i = 1,2,3$. In particular, it follows that

$$\beta_0\int_a^b \phi u_\epsilon{}^2 dx \leqslant \| f \|_{\mathbf{L}^\infty(a,b)}\int_a^b \phi \mid u_\epsilon \mid dx + k_3$$

and an elementary computation shows that the integral $\int_a^b \phi \mid u_\epsilon \mid dx$ is bounded. Thus, (A.64) and Poincaré inequality imply (A.60) and the boundedness of $v_\epsilon(b)$.

To show (A.61), take the $\mathbf{L}^2(b,c)$ scalar product of (2.6) by $\nu\psi v_{\epsilon,x}$, where ψ is a nonnegative, smooth function, vanishing near c, with $\psi(b) = 1$. After integration by parts, (A.61) follows by (A.60).

Finally, (A.62) follows by (A.60) and by the very equation (2.6).

□

Lemma A.11 *There is a constant $C > 0$ such that*

$$\| u_\epsilon \|_{\mathbf{H}^1(a,b)} \leqslant C. \tag{A.65}$$

Proof. Take the $\mathbf{L}^2(a,b)$ scalar product of (2.5) by $e^z u_\epsilon$, then integrate by parts. By Lemma (A.10), it follows that the $\mathbf{L}^2(a,b)$ norm of u_ϵ is bounded, as well as the value $u_\epsilon(a)$.

Next, take the $\mathbf{L}^2(a,b)$ scalar product of (2.5) by $u_{\epsilon,x}$, then integrate by parts. The conclusion follows by Lemma (A.10) and by the first part of this proof.

$\square$

Now, we are in a position to prove part of the results stated in Proposition 2.4.

Proposition A.4 *Assume (A.4), (A.10), (A.11), (A.31), (A.59). There are $u \in \mathbf{H}^1(a,b)$ and $v \in \mathbf{H}^1(b,c)$ which satisfy (A.16), (A.17), (A.19), (A.29).*

Proof. Let u_ϵ, v_ϵ solve $(\mathbf{Q_\epsilon})_N$. By Lemmas A.10, A.11, there are $u \in \mathbf{H}^1(a,b)$ and $v \in \mathbf{H}^1(b,c)$ such that (upon extracting a subfamily)

    (i) $u_\epsilon \to u$ weakly in $\mathbf{H}^1(a,b)$;

    (ii) $v_\epsilon \to v$ weakly in $\mathbf{H}^1(b,c)$;

    (iii) $v_\epsilon(c) \to v(c)$;

    (iv) $u_\epsilon(b) \to u(b)$ and $v_\epsilon(b) \to v(b)$.

All of these properties permit to pass to the limit in the regularized problem $(\mathbf{Q_\epsilon})_N$. Thus, the proof follows easily.

$\square$

To complete Proposition 2.4, it remains to show that the derivatives of the limit functions u, v of the preceding Proposition join continuously. To this end, we assume that the data α, β, f are more regular than it was until now, precisely:

$$\alpha \text{ is Lipschitz continuous in } [a,b],$$
$$\beta \in \mathbf{H}^1(a,b), f \in \mathbf{H}^1(a,b). \tag{A.66}$$

Lemma A.12 *Assume (A.4), (A.10), (A.11), (A.31), (A.59), (A.66). The $\mathbf{L}^2$ norm of $u_{\epsilon,xx}$ is bounded in a left neighborhood of the interface*

164

point $x = b$.

Proof. Take the $L^2(a,b)$ scalar product of (2.5) by $u_{\epsilon,xx}$ (which lies in $L^2(a,b)$ because of the equation itself). Next, integrate by parts in all terms except in the first. Recalling (A.61), (A.65) and (A.66), we get

$$\| \sqrt{\epsilon} u_{\epsilon,xx} \|_{L^2(a,b)} \leqslant C, \tag{A.67}$$

for some $C > 0$. Finally, take the derivative of (2.5) and multiply it in $L^2(a,b)$ by $\phi u_{\epsilon,xx}$, where ϕ is smooth, nonnegative, with $\phi(a)=0$. By (A.65), (A.67) and recalling that $\alpha < 0$, the assertion follows.

$\square$

Proposition A.5 *Assume (A.4), (A.10), (A.11), (A.31), (A.59), (A.66). The functions* u, v *considered in Proposition A.4 satisfy*

$$u_x = v_x \quad at \quad x = b. \tag{A.68}$$

Proof. Since the property holds for u_ϵ and v_ϵ (see (2.12)), it is enough to prove that:

(i) $u_{\epsilon,x}(b) \to u_x(b)$;

(ii) $v_{\epsilon,x}(b) \to v_x(b)$.

(i) follows by Lemma A.12 and by (A.2); (ii) follows by (A.2) and by (A.62), recalling (A.10) and (A.31).

$\square$

Thus, the proof of Propositions 2.1–2.4 is complete.

REFERENCES

[A]   Adams, R. *Sobolev Spaces*. Academic Press, New York (1975).

[CQ]   Canuto, C. and Quarteroni, A. *The boundary treatment for spectral approximations to hyperbolic systems*, J. Comput. Phys., **71** (1987), 100–110.

[CHQZ] Canuto, C., Hussaini, M.Y., Quarteroni, A. and Zang, T.A. *Spectral Methods in Fluid Dynamics*. Springer–Verlag, New York Heidelberg Berlin (1988).

[FQZ]   Funaro, D., Quarteroni, A. and Zanolli, P. *An iterative procedure with interface relaxation for domain decomposition methods*, SIAM J. Numer. Anal., to appear.

[CM]   Chorin, A.J. and Marsden, J.E. *A Mathematical Introduction to Fluid Mechanics*. Springer–Verlag, New York Heidelberg Berlin (1979).

[L]   Lions, J.L. *Perturbations Singulières dans les Problèmes aux Limites et en Controle Optimal*. Springer–Verlag, Berlin Heidelberg New York (1973).

[S]   Saad, M.A. *Compressible Fluid Flow*. Prentice Hall, Englewood Cliffs (1985).

NEW CONTRIBUTIONS TO NONLINEAR STABILITY OF THE MAGNETIC BENARD PROBLEM

Galdi G.P. & M. Padula, Ferrara

One of the most significant subject in hydromagnetics is the so-called Magnetic Benard problem, which studies the effect of an externally impressed magnetic field on the onset of thermal instability in electrically conducting fluids. This problem has attrcted the attention of many authors, see THOMPSON (1951), CHANDRASEKHAR (1981), concerning linear theory, and the more recent nonlinear contribution of BUSSE (1975), PROCTOR & GALLOWAY (1979), RUDRAIAH (1981), WEISS (1981) based upon formal expansion procedure, see also PROCTOR & WEISS (1982). More recently, GALDI (1985) presented a rigorous approach to the problem, employing a generalized nonlinear energy method. Further progress along the ideas advocated by GALDI has been made by RIONERO (1988), RIONERO & MULONE (1988). For the analysis to be carried here, it turns useful to have two Reynolds number-like parameters, say R^2 the Rayleigh number and Q^2 the Chandrasekhar number. Precisely, the results of all the above authors essentially predict that the critical value R_{crit} of R is an increasing function of Q; moreover, R_{crit} depends also on the ratio $p = P_m/P_r$ - P_r, P_m Prandtl and magnetic Prandtl numbers, respectively- in such a way that R_{crit} reaches a maximum independent of p for $p \leq 1$, while R_{crit} diminishes as soon as p increases away from one. As we are going to show, the method here employed predicts that R_{crit} depends crucially also on P_m whenever $P_m \neq 1$. In particular, in this range of parameters we prove that R_{crit} becomes an in creasing function of $1/P_m$ and so P_m <u>acts as another stabilizing parameter.</u> A similar but only qualitative result can be found in RUDRAIAH (1981), RUDRAI-AH, KUMUDINI & UNNO (1981).

To this end, we shall employ a new stability theory, developed by the authors, see GALDI & PADULA, forthcoming. Precisely, our theory starts from a criticism of the usual nonlinear energy method, cf. SERRIN (1959), JOSEPH (1976), to explain why frequently results obtained by that method can be very conservative. Actually, GALDI & STRAUGHAN (1985) pointed out the connection between the stability results of the classical energy theory and the symmetry (more generally simmetrization) of the linear part L. In fact, they prove that in bounded domains the energy stability criteria, sufficient for nonlinear stability, become also necessary for linear stability whenever L is symmetric. This is achieved by adding to the usual energy E_0 a fictitious one E_1 which dominates the nonlinearities. By adopting the symmetry as a reading key of the equations we can furnish, for a sufficiently large class of systems, a method of construction of the "right energy" functional with respect to which one should study stability. The interest of the new "energy" functional relies upon the fact that it now depends on the "basic flow".

Let us consider a horizontal layer of electrically conducting viscous fluid heated from below, upon which a uniform magnetic field $H=H\hat{z}$ orthogonal to the layer is impressed. Denote by S_0 the state in which steady adverse temperature gradient is mantained and there is no kinetic motion. As is well known, CHANDRASEKHAR (1981), in the Boussinesq approximation, the pertu rbations $\mathbf{u}$, p, θ, $\mathbf{h}$ to S_0 of the kinetic field, the pressure, the temperature and the magnetic field, respectively, obey the equations (in dimensionless form)

$$\mathbf{u}_{,t} + \mathbf{u}.\nabla\mathbf{u} - P_m\mathbf{h}.\nabla\mathbf{h} = -\nabla p + R\theta\,\mathbf{k} + \text{curlcurl}\mathbf{u} + Q\mathbf{h}_{,z}$$

$$\nabla.\mathbf{u} = 0$$

(1.1)
$$P_r(\theta_{,t} + \mathbf{u}.\nabla\theta) = Rw + \Delta\theta$$

$$P_m(\mathbf{h}_{,t} + \mathbf{u}.\nabla\mathbf{h} - \mathbf{h}.\nabla\mathbf{u}) = \text{curlcurl}\mathbf{h} + Q\mathbf{u}_{,z}$$

$$\nabla.\mathbf{h} = 0$$

where $R^2 = g\alpha\beta d^4/\nu k$, $Q^2 = \mu_m H^2 d^2/4\pi\rho\nu\eta_m$, $P_r = \nu/k$, $P_m = \nu/\eta_m$ are the Rayleigh, Chandrasekhar, Prandtl and magnetic Prandtl numbers, respectively, μ_m the magnetic permeability, k the thermometric conductivity, ν the kinematic viscosity, η_m the resistivity, β the adverse temperature gradient, g the gravity acceleration, α the coefficient of volume expansion, d the depth of the layer, ρ the (constant) density. Moreover, $._{,t}$ means the partial derivative with respect to t. To (1.1) we append the boundary conditions ($\mathbf{k}$ is the vertical ascending direction)

$$(\mathbf{u}\times\mathbf{k})_{,z} = \mathbf{u}.\mathbf{k} = 0 \quad\text{at } z = 0,1 \text{ (stress free boundary)}$$
(1.2)
$$\theta = \mathbf{h}\times\mathbf{k} = 0 \quad\text{at } z = 0,1$$

see also PECKOVER & WEISS (1972). To exclude rigid motions and constant vertical magnetic fields we impose in the "periodicity cell" $C = \{x,y,z \;\epsilon$

$$[0, \pi'a_1) \times [0, \pi/a_2) \times [0,1) \ \}$$

$$(1.3) \qquad\qquad \int_C \mathbf{u} \times \mathbf{k}\ dx = \int_C \mathbf{h}.\mathbf{k}\ dx$$

Let us notice that conditions $(1.1)_5$ and $(1.3)_2$ are automatically fulfilled at any time $t>0$ once they are satisfied at $t=0$.

We now put system $(1.1)-(1.3)$ in the abstract form

$$Bu_{,t} = Lu + Nu \ ,$$

where u is the seven-component vector $(\mathbf{u}, \theta\ , \mathbf{h})$, $u \in C^8(C)$,

$$Bu = \begin{pmatrix} \mathbf{u} \\ P_r\, \theta \\ P_m \mathbf{h} \end{pmatrix}$$

$$Nu = \begin{pmatrix} -\pi(\mathbf{u}.\nabla\mathbf{u} - P_m\mathbf{h}.\nabla\mathbf{h}) \\ -P_r\,\mathbf{u}.\nabla\ \theta \\ -P_m(\mathbf{u}.\nabla\mathbf{h} - \mathbf{h}.\nabla\mathbf{u}) \end{pmatrix}$$

and L is a linear operator. Moreover, we set $H = J(C) \times L^2(C) \times J^\#(C)$ with $J(C)$, $J^\#(C)$ subspaces of $[L^2(C)]^3$ consisting of solenoidal vector functions $\mathbf{u}$, $\mathbf{h}$, with $\mathbf{u}.\mathbf{k} = \mathbf{h} \times \mathbf{k} = 0$, at $z = 0,1$. π is the orthogonal projection of $[L^2(C)]^3$ onto J.

In order to apply our theory we shall discriminate in L the symmetric and skew-symmetric part, precisely we set

$$L = A + RS + QM \ ,$$

with

$$Au = \begin{pmatrix} -\mathrm{curl\,curl\,}\mathbf{u} \\ \Delta\ \theta \\ -\mathrm{curl\,curl\,}\mathbf{h} \end{pmatrix}$$

$$Su = \begin{vmatrix} \Pi(\theta \, \mathbf{k}) \\ w \\ 0 \end{vmatrix}$$

$$Mu = \begin{vmatrix} \mathbf{h}_{,z} \\ 0 \\ \mathbf{u}_{,z} \end{vmatrix}$$

As in the standard energy stability analysis of Bénard problem, it is prominent the role of the "energy parameter" σ_E, given now by

$$(1.4) \qquad \frac{1}{R_E} = -\max_{\psi \, \phi' \, \chi} \frac{2 \int_C \phi \, \psi . \, \mathbf{k} \, dx}{\int_C [\, |\nabla\psi|^2 + |\nabla\phi|^2 + |\nabla\chi|^2 \,]dx}$$

where the maximum is taken over the class $Y(L)$ —$Y(.)$ denotes the domain of the associated operator . It can be shown that R_E given in (1.4) coincides with the critical Rayleigh number of the classical Bénard problem, i.e. R_E =25,641, see RIONERO (1964). We, now, introduce the coupling functional $\Gamma(u)$ which plays a central role in the theory of GALDI & PADULA, forthcaming

$$\Gamma(u) \equiv (B^{-1}L_S u, Mu) \equiv \Gamma_1(u) + \Gamma_2(u)$$

with

$$\Gamma_1(u) \equiv R \int_C \mathbf{h}_{,z} . \Pi(\theta \, \mathbf{k})dx$$

$$\Gamma_2(u) \equiv (1-P_m)\int_C \mathbf{u} . \mathbf{h}_{,z} dx$$

In order for M to have a stabilizing influence one <u>must</u> require $\Gamma(u) \neq 0$, GALDI & PADULA, forthcaming, thus at least one of the functionals $\Gamma_1(u)$ or $\Gamma_2(u)$ must not be zero. To reveal the "measure" of stabilization of the magnetic fi<u> </u>

eld, we shall include the functionals $\Gamma_1(u)$ and $\Gamma_2(u)$ in the "energy" with respect to which we prove stability. We notice, now, that the inhibiting effe ct of the magnetic field H on the onset of convection reveals itself thruogh two physical effects: the Lorentz force ($\Gamma_1(u)\neq 0$), the Joule effect ($\Gamma_2\neq 0$). In this connection, we notice that $\Gamma_1(u)$ just coincides with the coupling fu nctional introduced heuristically in GALDI (1985) sect.3, and reconsidered by RIONERO & MULONE (1988), formula (3.4). On the other hand, the coupling effect $\Gamma_2(u)$ is completely new, and its contribution may became more relev- ant the more P_m differs from one.

We are now in position to construct the right energy functional. Precisely, theory of GALDI & PADULA suggests to add the coupling functional $\lambda\Gamma_1(u)$ and $\lambda_1\Gamma_2(u)$ to the classical energy $\frac{1}{2}(Bu,u)$, where λ, λ_2 are Lagrange multipliers. Of course, in order to have a positive definite energy we are forced to add the term $\lambda_2\|Mu\|^2$, with $\|\cdot\|$ the L^2-norm. We finally set

$$E = \tfrac{1}{2}\int_C \{u^2 + P_r\theta^2 + P_m h^2 + \lambda_2(|u_{,z}|^2 + P_m|h_{,z}|^2)$$

(1.5)
$$+ 2\lambda(P_r P_m)^{\frac{1}{2}}\theta k.h_{,z} + 2\lambda_3 u.h_{,z}\}dx$$

where $\lambda_3 = \lambda_1\sqrt{P_m}(1-P_m)$. With the choice

$$\lambda_2 = \lambda Q/R\sqrt{p}, \quad \lambda = \xi Q/R\sqrt{p}, \quad \xi\epsilon(0,1),$$

(1.6)
$$\lambda_3 = \xi\nu\pi\sqrt{P_m}Q^2/pR^2, \quad |\nu|\epsilon(0,\sqrt{(1-\xi)}),$$

and the use of the Poincaré inequality it is easy to show the E is positive definite. We, next, compute the time derivative of E along (1.1). Long but straightforward calculations imply

$$\frac{dE}{dt} \le (m-1)D + N(u),$$

where

$$m = \max \frac{\int_C \{2R\theta W + \lambda[R\sqrt{p}\,wh_{,z} - \frac{(1+p)}{\sqrt{p}}\cdot\nabla h_{,z}] + \lambda_3[R\theta h_{,z} - (1+\frac{1}{Pm})\nabla u : \nabla h_{,z} - Q(\frac{|u_{,z}|^2}{Pm} - |h_{,z}|^2)]\}}{\int_C [(\lambda Q/R\sqrt{p})(|\nabla u_{,z}|^2 + |\nabla h_{,z}|^2) + |\nabla u|^2 + |\nabla\theta|^2 + |\nabla h|^2]}$$

(1.7)

$$D = \int_C [(\lambda Q/R\sqrt{p})(|\nabla u_{,z}|^2 + |\nabla h_{,z}|^2) + |\nabla u|^2 + |\nabla\theta|^2 + |\nabla h|^2]$$

here the element dx in the integrals has been omitted Moreover, it results
see GALDI & PADULA, forthcoming,

$$\|Nu\| \le (1 + P_r^2 + P_m^2)(\sup_C |u|^2 + \sup_C |h|^2)D$$

Tharefore, the dissipation D cannot dominate the terms increasing the nonlinearities. Thus, we have to consider stronger dissipative terms this is achieved employing the method
of "fictitious energy", introduced by GALDI (1985), cf. also
GALDI & STRAUGHAN (1985b), GALDI, PAYNE, PROCTOR & STRAUGHAN
(1987), GALDI, PADULA & RAJAGOPAL (1988), PADULA (1986), PA-
DULA (1988b),COSCIA & PADULA (1988), RIONERO (1988). Specifi_
cally, we consider the functional

$$E_1 = \tfrac{1}{2}\int_C (|\nabla u|^2 + P_r|\nabla\theta|^2 + P_m|\nabla h|^2)dx$$

which will play no other role than that of furnishing the
dissipation $\int_C (|\Delta u|^2 + |\Delta h|^2 + |\Delta\theta|^2)dx$. By adding E_1 to E we re_
alize that $\mathbb{E} = E + \eta E_1$ verifies

$$\frac{d}{dt}\mathbb{E} \leq (m-1+c\,\mathbb{E}^{\alpha})\,\mathbb{D}$$

with $\mathbb{D}=D+\eta\,D_1$ and c (computable) constant which is always strictly positive and is a decreasing function of Q. Therefore, the asymptotic behaviour to zero for u follows from the smallness assumption on the initial data, i.e. $\mathbb{E}^{\alpha}(0)<(m-1)/c$, whenever m is less than one. Conditions under which m<1 will then provide the stabilizing effect of the magnetic field on convection.

In order to obtain a quantitative stability criterion, we next solve the variational problem (1.7). We assume $P_m<<1$ and $P_m/P_r<<1$, as it happens, for instance, for liquid metals such as mercury at terrestrial conditions ($P_m=1.5\cdot10^{-7}$), or in the solar corona ($P_m=10^{-3}$). We begin to observe that the Euler-Lagrange equations associated to (1.7) are given by

$$m\Delta\mathbf{u}-m(\lambda Q/R\sqrt{p})\Delta\mathbf{u}_{zz}+R(\lambda+\tfrac{1}{2}\sqrt{p}h_z)\mathbf{k}+\tfrac{1}{2}\lambda_3(1+P_m^{-1})\Delta\mathbf{h}_z+(\lambda_3 Q/P_m)\mathbf{u}_{zz}-\nabla\chi=0$$

$$\nabla\cdot\mathbf{u}=\nabla\cdot\mathbf{h}=0$$

$$(1.8)\qquad m\Delta\theta+Rw+[\lambda(1+p)/2\sqrt{p}]\Delta h_z+\tfrac{1}{2}\lambda_3 Rh_z=0$$

$$m\Delta\mathbf{h}-m\frac{\lambda Q}{R\sqrt{p}}\Delta\mathbf{h}_{zz}-\tfrac{1}{2}[\frac{\lambda(1+p)}{\sqrt{p}}\Delta\theta_z+R\sqrt{p}w_z+\lambda_3 R\theta_z]\mathbf{k}$$

$$+\tfrac{1}{2}\lambda_3(1+P_m^{-1})\Delta\mathbf{u}_z-\lambda_3 Q\mathbf{h}_{zz}=0.$$

By (1.8) it is straightforward to obtain a differential equation of the form $\ell w=0$, for $w=\mathbf{u}\cdot\mathbf{k}$. Performing standard normal mode analysis with

$w=W(z)\exp\{2i(a_1 x+a_2 y)\}$, employing the appropriate boundary conditions for w we also infer that $W(z)=V\sin r_\pi z$. Substituting such w into $\ell u=0$, we obtain a forth order equation for m, and at "criticality" (m=1) we deduce with $R^2=x$

$$(1.9) \qquad x^3 + Ax^2 + Bx + C = 0$$

where A,B,C are explicit functions of $r, \pi^2 \eta = a_1^2 + a_2^2, p, P_m, Q, \lambda, \lambda_3$. We choose, now, λ and λ_3 in order that both B and C become negative, uniformly in r and η. In this way we know that there is only one positive solution to (1.9), say $R_c = R_c(r,\eta, P_m, Q)$ and set

$$R_E = R_E(P_m,Q) = \min_r \min_\eta R_c(r, \eta, P_m, Q)$$

Since it can be proved, see GALDI & PADULA, forthcoming, that $R < R_E$ implies m < 1, we state that $R_E = R_E(P_m,Q)$ is the desidered stability bound on the Rayleigh number. A numerical study of the critical curve $R_E = R_E(P_m,Q)$ reveals that, beside the magnetic field, P_m acts _independently_ as _another_ _stabili_ _zing parameter_ and a decreasing of P_m causes an enlargement of the stability region. Moreover, for $Q > Q_{crit} \equiv \pi/\sqrt{2P_m}$, the behaviour of R_E with P_m is given by the following asymptotic formula

$$(1.10) \qquad R_E^2 = \pi^3 [(1+\eta_\infty)(2+\eta_\infty)]^2 P_m^{-\frac{1}{2}} Q$$

where n_∞ is a constant lying between 1.35 and 1.4, depending on P_m^{-1}. Relation (1.10) shows that, at criticality, the Rayleigh number R_E is proportional to Q through a coefficient which behaves like $P_m^{-\frac{1}{2}}$.

From Fig.1 we see that for Q not too large the stability bounds essentially do not depend on P_m, while from Fig.2 we notice that, as Q increases, a decreasing in P_m provides an enlargement of the stability region. Moreover, for a fixed value of P_m^{-1}, a change in the slope of the stability curves should be noticed.

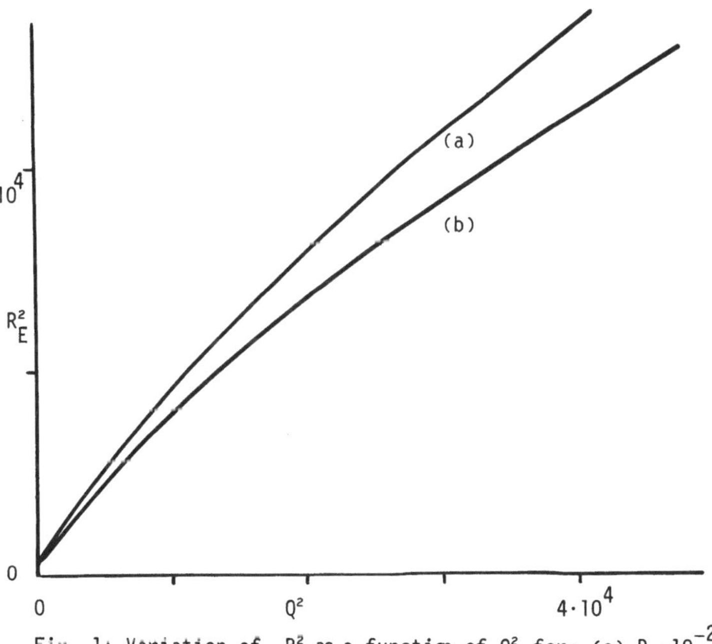

Fig. 1: Variation of R_E^2 as a function of Q^2 for: (a) $P_m=10^{-2}$; (b) $P_m=1.5 \cdot 10^{-7}$

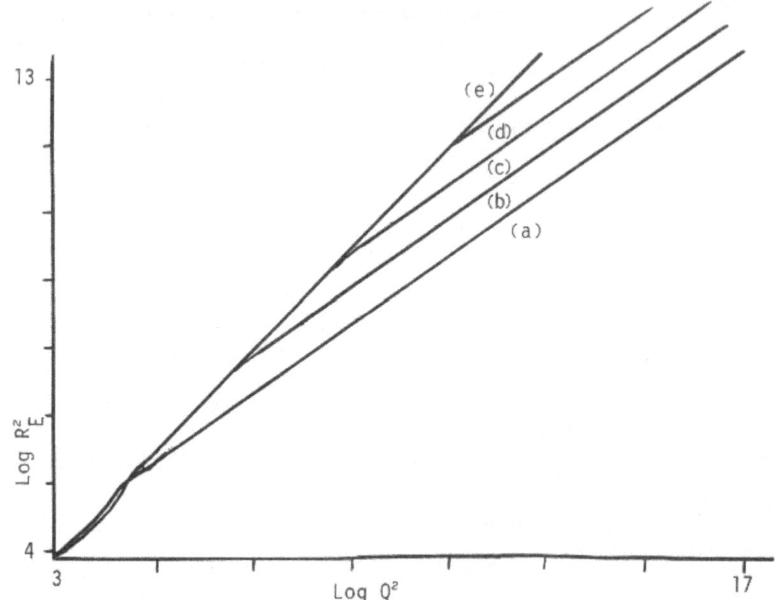

Fig.2: Variation of R_E^2 as a function of Q^2 for:(a) $P_m=10^{-2}$;(b)$P_m=10^{-3}$;(c)$P_m=10^{-4}$;(d)$P_m=10^{-5}$;(e)$P_m=1.5\cdot10^{-7}$.

Bibliografy

BUSSE, F.H., (1975), J. Fluid Mech., **71**, 193

CHANDRASEKHAR, S., (1981), Hydrodinamic and Hydromagnetic Stability, Dover Publ. Inc., N.Y.

COSCIA,V.&M.PADULA,Quantitative Estimates for the Nonlinear Stability Bounds for the Compressible Bénard Problem, Proc. 3rd Workshop on Math. Aspects of Fluids and Plasma Dynamics,SaliceTerme,Italy,forthcoming

GALDI, G.P., (1985), Arch.Ratl.Mech.Anal., **87**, 167

GALDI,G.P.&B.STRAUGHAN,(1985a), Arch.Ratl.Mech.Anal.,**89**, 211

GALDI,G.P.&B.STRAUGHAN,(1988b),Proc.Roy.Soc.London(A)**402**,257

GALDI,G.P.,PAYNE,L.H.,PROCTOR,M.R.&B.STRAUGHAN,(1987), Proc. Roy.Soc.London(A), **417**, 319

GALDI,G.,PADULA,M.&K.R.RAJAGOPAL,(1988),Arch.Ratl.Mech.Anal.
in thePress

GALDI,G.P.&M.PADULA, A new Approach to the Energy Theory in
the Stability of Fluid Motions, forthcoming

JOSEPH, D.D., (1976), Stability of Fluid Motions, Springer
Tracts in Natural Phylosophy, voll.I and II

PADULA,M.,(1986), Boll.U.M.I.,**5B**, 581

PADULA,M.,(1988a),Energy Instability Methods: an Application
to Burger Equation, Proc.Meeting ˝Energy Stability
and Convection˝,GALDI&STRAUGHAN Eds,Pitman Research
Notes in Mathematics, **168**

PADULA,M.,(1988b),Compressible convection: Nonlinear Results
Proc.Intl.Conf.on ˝Mathamatical Modelling in Scien
ce and technology˝, Madras, India

PECKOVER,R.S.&N.O.WEISS,(1972), Comp.Phys.Comm., **4**, 339

PROCTOR,M.R. & D.J.GALLOWAY, (1979),J. Fluid Mech., **90**, 273

RIONERO,S., (1988), On thc choice of the Liapunov Functional
in the Stability of Fluid Motions, Proc. Meeting˝
˝Energy Stability and Convection˝,GALDI,G.P.&B.STR
AUGHAN Eds, Pitman Research Notes in Math., **168**

RIONERO,S. & G.MULONE, (1988), Arch.Ratl.Mech.Anal., **103**,347

RUDRAIAH, N., (1981), Publ.Astron.Soc.Japan, **33**, 721

RUDRAIAH,N.,KUMUDINI,V. & W.UNNO, (1985), Publ.Astron.SOC.
Japan, **37**, 183

SERRIN, J., (1959), Arch.Ratl.Mech.Ansl., **3**, 1

THOMPSON, W.B., (1951), Phil. Mag., **42**, 1417

WEISS, N.O., (1981), J. Fluid Mech., **108**, 247

A Criticality Concept

for

Reaction-Diffusion Systems far from Thermodynamic Equilibrium

Dirk Meinköhn, Bielefeld

Summary: The singularities of a stationary reaction-diffusion system far from thermodynamic equilibrium are classified as "dissipative phase transitions", by which is implied the existence of an hierarchical order among them.

1   Introduction

The action of thermodynamic fluxes (of, e.g., energy, substance) may force an open thermodynamic system to attain stationary states far from thermodynamic equilibrium. In accordance with accepted practice, the stationary states of an open and finite system are represented as points on a state surface in an appropriate state space spanned by the state variables and the control parameters of the system. Here, the state variables (e.g., pressure, temperature) characterize the internal state of the system and are thus to be regarded as dependent variables. The control parameters, on the other hand, designate by definition the independent parameters which represent the experimenter's control of the evolution of the system (through boundary conditions, geometry etc.). Control parameters and state variables together are supposed to uniquely assign the thermodynamic state. The stationary states of a reaction-diffusion system far from thermodynamic equilibrium result as solutions of a nonlinear elliptic boundary value problem. Therefore, the relationship between control parameters (i.e., the independent parameters) and the state variables is not necessarily unique such that a projection of the state surface into the subspace of control parameters may comprise multiply covered regions. A change in the state of the system - a process - is achieved through an appropriate change in the control parameters. For an investigation of the state surface, i.e., the set of all stationary states of the system, "quasi-stationary" processes are used. This kind of process implies an extremely slow change in the control parameters. Due to nonuniqueness, a small change in the control parameters does not necessarily imply a correspondingly small displacement of the system on the state surface. As it turns out, locations on the state surface exist where the system becomes unstationary upon changes in the control parameters which are arbitrarily small. These locations mark the so-called "critical points" of the system and are found to lie on the boundary between regions of stable and unstable states (stability boundary). Upon projection of the state surface into the subspace of control parameters, the stability boundary gives rise to contours which result in consequence of an associated geometrical "visibility condition". The singularities of the visible contours are of particular interest because they represent essential physical characteristics of the system.

After having become unstationary, the system may eventually settle down to become stationary again. In the context of stationary states, such an unstationary evolution is interpreted as a jump between two states which are given as points on the state surface. This picture of jump-like transitions between certain points on the state surface bears great resemblance to phase changes for equilibrium systems (e.g., gas-liquid or liquid-liquid systems). The visible contours, in particular, which result as images of the stability boundary can be interpreted as analogues of the binodal and spinodal lines [1], [2], [3] which represent the boundaries between the regions of stable, metastable and unstable equilibrium states. Jumps between stationary states thus give indication of what have been termed "dissipative phase transitions" [2] .

2   <u>Model System</u>

The system to be modelled may be viewed as a porous fuel pellet embedded in an oxidizing atmosphere [4] , or a porous catalyst particle surrounded by gaseous reactants [5] , or a homogeneous mixture of reactants [6] . Diffusive transport supplies reactants to the system (e.g., the fuel pellet, the catalyst particle)and removes the gaseous products of the chemical reaction within. This reaction is supposed to be exothermic so as to liberate the energy which forces the system into stationary states far removed from thermodynamic equilibrium. A highly porous body possesses a large internal reaction surface and may therefore be viewed as "pseudo-homogeneous" [5] , by which assumption the chemical reaction results as homogeneous, with diffusive transport described by "effective" transport coefficients.

In contradistinction to the theory of phase transitions for equilibrium systems, a proper elliptic boundary value problem can be set up here, such that its solutions represent the stationary temperature distributions :

$$(1) \quad \Delta y + \lambda (1 - \xi y)^n \exp\left(\frac{y}{1 + \beta y}\right) = 0 \qquad \text{in } D$$

$$y = 0 \qquad \text{on } \partial D$$

Here, the following designations have been employed : D = region occupied by the system (pellet, particle), ∂D = boundary of D, y = temperature distribution, Δ = Laplacian, λ = Frank-Kamenetzki parameter, β = activation energy parameter, ξ = Lewis number, n = reaction order.

Eq.(1) is seen to be quasi-linearly elliptic, with nonlinearity entering via the function w with

$$(2) \quad w(y) = (1 - \xi y)^n \exp\left(\frac{y}{1 + \beta y}\right)$$

Here, the algebraic term represents the influence of the reactant concentrations whereas the influence of the temperature is essentially of exponential character. A stationary state thus results from the competing influences of diffusive supply of reactants and diffusive removal of heat, with a particularly high sensitivity towards temperature changes which are governed by heat production due to chemical reaction and the competing heat removal into the surrounding atmosphere (beyond ∂D).

λ , ξ, β, n represent the control parameters and y the dependent state variable. In order to include ordinary homogeneous reactions, the control parameters are supposed to assume real values, in the following ranges :

$$(3) \quad \xi \gtrless 0 \;\; , \;\; \beta \geqq 0 \;\; , \;\; \lambda \geqq 0 \;\; , \;\; n \gtrless 0$$

Physical interest is only in solutions y of Eq.(1) for which :

$$(4) \quad w(y) > 0$$

Therefore, $y < 1/\xi$ has to be stipulated in case of $\xi > 0$. In consequence of Eqs.(3), (4),

$$(5) \quad \Delta y \leqq 0 \quad \text{in D} \;\; , \;\; y = 0 \quad \text{on } \partial D \; .$$

Due to the maximum principle $\lfloor 7 \rfloor$, a unique maximum y_m of y exists in D, with $y \geqq 0$ in D + ∂D .

Therefore, in order to characterize the internal state of the system, y_m may be adopted as state variable instead of the temperature distribution y. The concept of y_m as a state variable is of particular advantage if the location of y_m within D is fixed due to the existence of a center of symmetry for Eq.(1).

Formally, y_m may be conceived as an independent parameter whereby the solutions of Eq.(1) are sought in parametric representation :

$$(6) \quad (y(x;y_m) \; ; \; \lambda(y_m))$$

Thus, the Frank-Kamenetzki parameter λ is found to be of special importance, as $\lambda(y_m)$ may be used to construct the state surface in a state space spanned by y_m, λ, ξ, β, n . The existence of a stability boundary on that state surface and, in particular, the existence and the type of the singularities in the projection of the stability boundary into the subspace of the independent control parameters λ, ξ, β, n turns out to be governed by the "convexity" of the nonlinear function $w(y)$ (cf. Eq.(2)). An investigation of the stability boundary and its projection may be carried out with the help of lower bounds [8], [9]. The following section presents a discussion of some of the results obtained with this method.

3   Examples

3.1   2-dimensional State Space

If ξ, β, n are kept at a constant value, then a 2-dimensional state space (λ , y_m) results. The state surface is in the form of a curve $\lambda(y_m)$, for which an application of the continuity method and of Green's theorem furnishes the following properties : a stable solution of Eq.(1) is necessarily associated with a point on an increasing section of $\lambda(y_m)$, a critical solution of Eq.(1) is necessarily associated with a stationary point of $\lambda(y_m)$. An example is provided by Fig.1 .

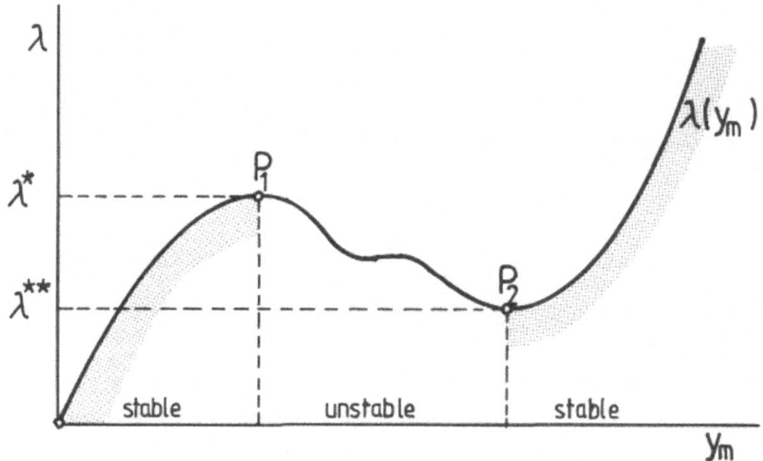

Fig.1   State surface $\lambda(y_m)$ for constant ξ, β, n

The stability boundary is given by the points P_1 , P_2 which upon projection

into the subspace of control parameters (which in this case is just the λ-axis)
produces a visible contour which consists of the two points λ^*, λ^{**} only. P_1,
P_2 designate branching points in the set of all solutions of Eq.(1). Due to y_m
designating the temperature maximum, P_1 (or P_2) serve as jump-off points for
a jump-like temperature increase (or decrease) which may be interpreted as describing an ignition (or extinction) event [6] .

3.2 3-dimensional State Space

If ξ, n are kept at a constant value, a 3-dimensional state space results, which contains a proper 2-dimensional state surface $\lambda(y_m;\beta)$. The stability boundary is found to define a curve C on that state surface, with C
derived pointwise as in 3.1 for stepped constant values of β . Fig.2 displays
a result, with C separating stable solutions of Eq.(1) from unstable ones,
which are found to lie in the shaded region.

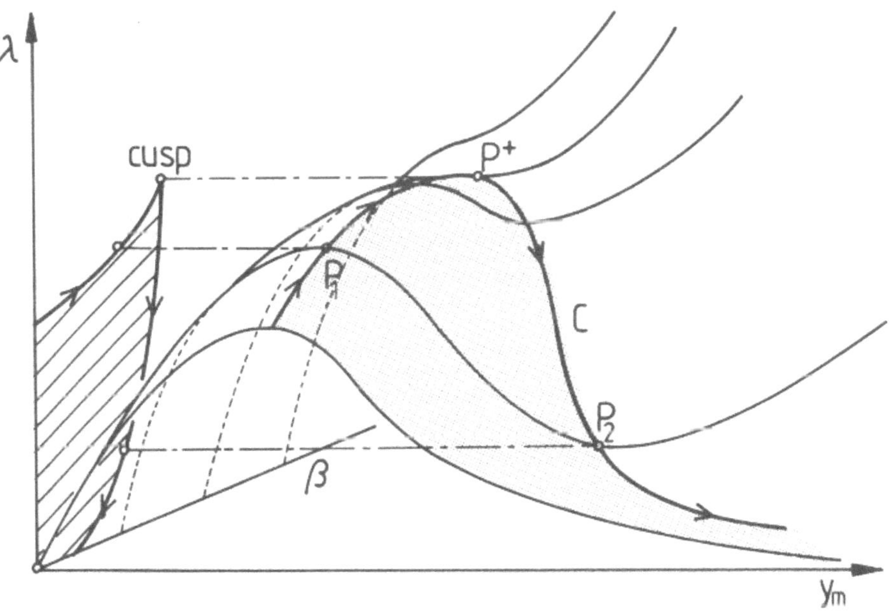

Fig.2   State Surface $\lambda(y_m;\beta)$ for constant ξ, n , with the Region of
Unstable States Marked by Shading

Curve C comprises ignition and extinction points (as exemplified by P_1, P_2) such that a branch of ignition points is separated from a branch of extinction points by a point P^+ of C which is the analogue of a critical point in the theory of phase transitions : certain quasi-stationary processes exist which are given by curves on the state surface such that these curves connect an ignition point P_1 with an extinction point P_2 without passing through the shaded region of unstable states, i.e., without implicating a jump-like passage. For a discussion of phase transitions in equilibrium systems, cf.[1],[1o].

For P^+, a value $\beta = \beta^+$ exists such that the curve $\lambda(y_m;\beta^+)$ possesses a stationary point of inflection . Projection into the subspace of control parameters (λ, β) is along the y_m-axis. The visible contour (i.e., the projection of the curve C) therefore possesses a cusp at the projection image of P^+ because for P^+ the projection is in an asymptotic direction.

Ignition and extinction points represent singularities of lowest order of Eq.(1), whereas a transition point P^+ is of next higher order, which is implied by the cusp in the projection image of C associated with P^+. In the physical interpretation, the transition point P^+ designates the appearance of ignition/extinction phenomena in the following sense : if β approaches β^+ from above, the "convexity" of $w(y)$ is increased with the resulting "birth" of a pair of ignition/extinction branches of C at P^+ .

3.3   4-dimensional State Space

The discussion of the examples given in 3.1 and 3.2 leads to the supposition, that the singularities of Eq.(1) may be classified according to their hierarchical order. This corresponds to the concept of hierarchically ordered phase transitions in equilibrium systems [1], [1o] . In order to provide an example of a singularity of Eq.(1) of higher order, the parameter n is kept constant and the state (hyper)-surface $\lambda(y_m; \xi,\beta)$ is investigated in the resulting 4-dimensional state space (λ, y_m, ξ ,β) . On this state surface, lines of critical points exist which under projection lead to cusp ridges in the subspace of control parameters (λ ,ξ ,β) . A pair of cusp ridges may then again be generated via a "birth event", the birth place designating a singularity P^{++} of Eq.(1) which is of higher order than the singularity associated with P^+ in Fig.2 . Because of its appearance, point P^{++} is referred to as a "swallow-tail" singularity.

Fig.3 gives an impression of such a swallow tail in the projection image of the stability boundary, with n = 2.o and values of β ranging from o.36 to o.21   . Actually, P^{++} in Fig.3 represents a replica in terms of lower bounds of a swallow-tail singularity of Eq.(1), where the bounds turn out to be independent of the particular geometry of D . In order to provide a numerical comparison for a specific geometry, the exact solution for the swallow-tail for an infinite slab (of thickness 1) has been inserted into Fig.3 (marked $\lambda_{exact\ slab}$). The solid lines (highlighted by shading) of $\lambda_{exact\ slab}$ designate a pair of cusp ridges which by themselves are generated in a cusp at the location of the exact swallow-tail singularity. The broken lines which end in the cusp P^{++} provide lower boundsof the cusp ridges for $\lambda_{exact\ slab}$.

According to the concepts discussed in [1o] for infinite equilibrium systems, field-type and density-type variables can be thermodynamically distinguished, with field-type variables (e.g., pressure, temperature, chemical potential) in the role of control parameters. For the example of a van der Waals gas, the investigation of equilibrium phase transitions leads to the uncovering of a swallow-tail singularity in the projection of the state surface into the subspace of control parameters (μ, p, T) (cf.[11], p.15o) which again underscores the analogy between equilibrium and dissipative phase transitions.

For a specific β-value (cf. β=o.21 in Fig.4), four different physical situations result which are given in terms of four values of the parameter ξ. For $\xi = \xi_1$, a pure ignition phenomenon results (ignition point P_1). For $\xi = \xi_2$, a region of stable solutions is found which is embedded in a region of unstable ones. Here, P_1 marks an ignition point, but P_2 and P_3 turn out to be extinction points. For $\xi = \xi_3$ and $\xi = \xi_4$, a region of unstable solutions is found to be embedded in a region of stable ones. Here, P_1 marks an ignition point, P_2 an extinction point, with P_3 for $\xi = \xi_3$ designating an additional ignition point. Thus, a process by which β approaches β^{++} from above leads to the "birth" of a pair of embedded regions of states which differ as to their stability from the states of their surroundings.

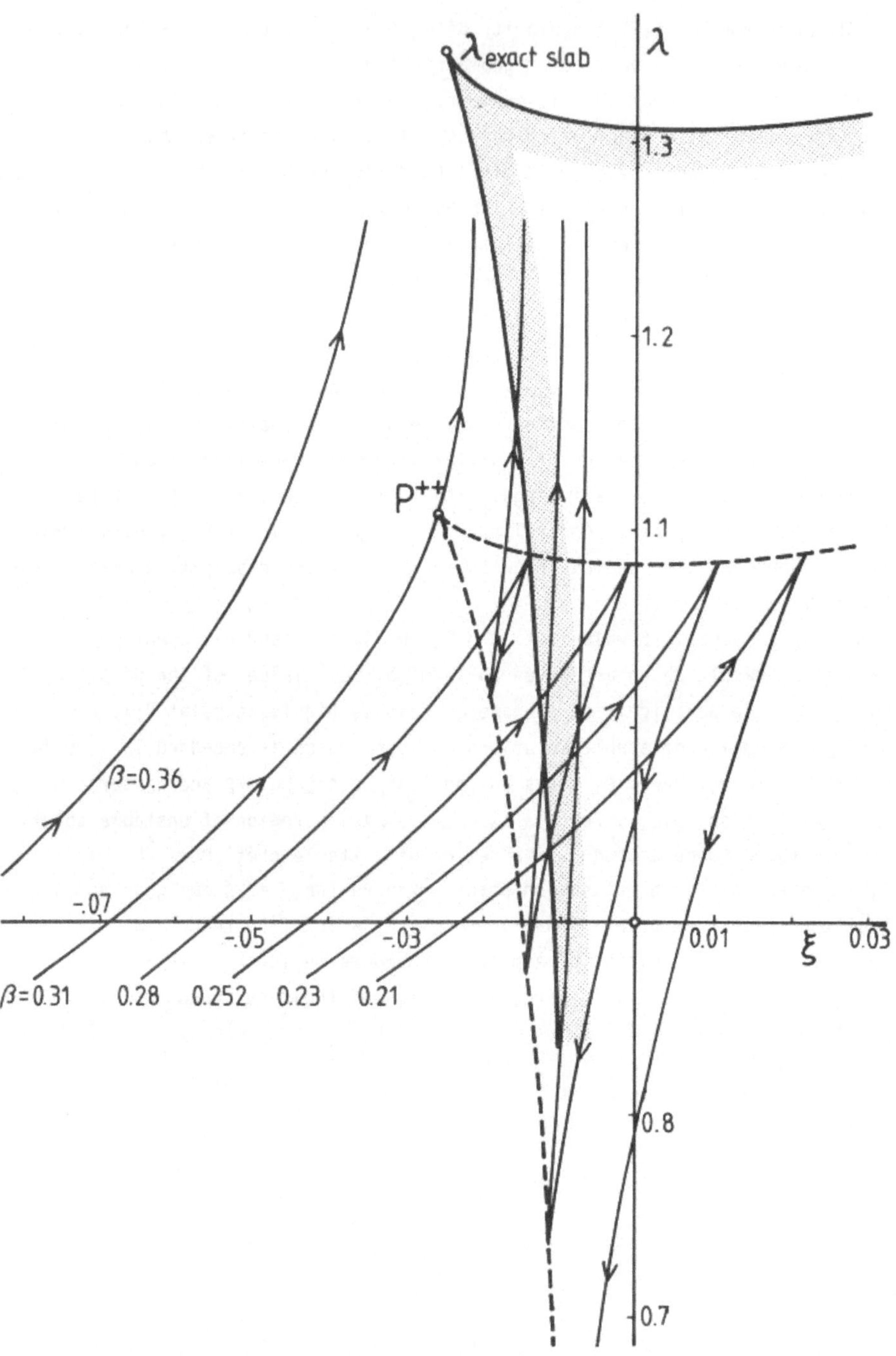

Fig.3   Swallow-Tail Singularity for n=2

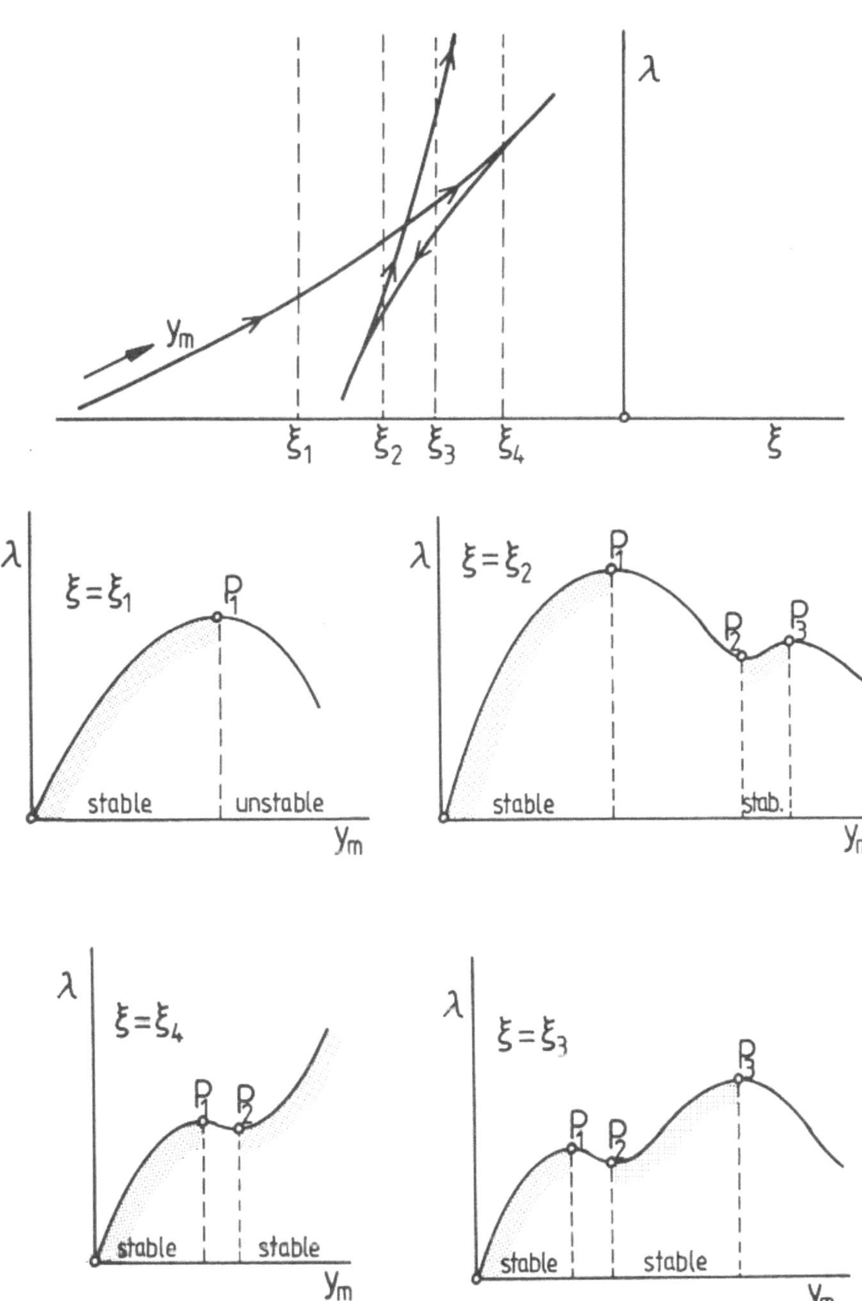

Fig.4   Physical Situations for a Swallow Tail for $\beta=0.21$ and $n=2.0$

4   <u>References</u>

[1]   Landau,L.D.; Lifshitz,E.M.: Statistical Physics. Oxford: Pergamon 198o

[2]   Glansdorff,P.; Prigogine,I.: Thermodynamic Theory of Structure, Stability and Fluctuations. London: Wiley-Interscience 1971

[3]   Rowlinson,J.S.: Liquids and Liquid Mixtures. London: Butterworth 1971

[4]   Kordylewski,W.; Krajewski,Z.: Thermal Ignition of Self-Heating Slab. Comb. Flame $\underline{41}$ (1981) 113-122

[5]   Aris,R.: The Mathematical Theory of Diffusion and Reaction in Permeable Catalysts. Oxford: Clarendon Press 1975

[6]   Gray,P.; Lee,P.R.: Thermal Explosion Theory. Oxydation and Combustion Revs.II, Amsterdam: Elsevier 1967

[7]   Protter,M.H.; Weinberger,H.F.: Maximum Principles in Differential Equations. Englewood Cliffs: Prentice-Hall 1967

[8]   Meinköhn,D.: Disappearance of Criticality in Nonuniform Systems I. SIAM J.Appl.Math. $\underline{48}$ (1988) 536-548

[9]   Meinköhn,D.: Disappearance of Criticality in Nonuniform Systems II: A General Investigation by Upper and Lower Bounds. SIAM J.Appl.Math. $\underline{48}$ (1988) 792-8o7

[1o]   Griffith,R.B., Wheeler,J.C.: Critical Points in Multicomponent Systems. Phys.Rev.$\underline{A2}$ (197o) 1o47-1o64

[11]   Callen,H.B.: Thermodynamics. New York: Wiley 197o

Dirk Meinköhn

Fakultät für Physik

Universität Bielefeld

Postfach 864o

D-48oo <u>Bielefeld</u>

Federal Republic of Germany

ON THE APPROXIMATION OF CONTINUUM STOCHASTIC SYSTEMS BY A DISCRETE STOCHASTIC SYSTEM: A PROBLEM WITH MOVING BOUNDARY

R. Monaco - Dipartimento di Matematica - Politecnico di Torino

1. Introduction

Mathematical systems in stochastic continuum mechanics can be modelled by partial differential equations with stochastic parameters to be joined, in the mathematical formulation of problems, to random initial and/or boundary conditions. An analysis of the stochastic modelling in continuum physics is proposed in paper [6] as well as in the general bibliography quoted in such a paper.

The so-called "stochastic interpolation method" provides a mathematical method, proposed in [3] and developed through papers [4,6], to deal with random initial-boundary value problems. Such a method, preliminary described for some simple cases in Ch.VI of ref.[2], essentially consists in discretizing the space variable into a suitable number of nodal points and in interpolating the solution by a suitable space interpolation (such as the Lagrange type interpolation) in a fashion that the solution itself is equal to the values of the interpolation in the nodal points. This procedure provides, as we shall see in sec.3, the transformation of the original initial boundary value problem into an initial value problem defining the time evolution of the dependent variable in the nodal points in a fashion that the boundary conditions are naturally cast into the interpolation itself, replacing the evolution equation by the values of the boundary conditions into the evolution equations corresponding to the nodal points on the boundaries.

Solving the initial value problem, described by random ordinary differential equations, and applying the space interpolation provides the solution of the original problem so that the relevant statistical measures of the solution itself can be recovered.

One of the crucial points is, as we shall see, determining the gap between the solution of the original problem in continuum mechanics and the one obtained by the discretized system. This correspond to evaluate the distance between a system with an infinite number of degrees of freedom and the corresponding (discretized by the stochastic interpolation method) system with a finite number of degrees of freedom.

This paper developes the mathematical analysis of the problem for the study of moving boundary problems described by parabolic type equations. The paper is in four sections. The second section describes the mathematical model and problem. The third one developes the mathematical method and an application. A discussion follows in the last section to this paper.

2. The mathematical model

Consider a nonlinear model referred to the parabolic heat equation in a one-dimensional random two-phase medium with phase transition at the boundary $s=s(t)\varepsilon]0,1[$. Formally the mathematical model can be written as

$$x\varepsilon[0,s[: \quad u_1=u_1(w,t,x) \quad , \quad \partial u_1/\partial t=\partial/\partial x \{h[u_1,r_1(w,x)]\partial u_1/\partial x\}$$

$$x=s(t): \quad u_1(w,t,s(t))=u_2(w,t,s(t))=u(w,t,s(t)) ,$$
(2.1)
$$L(u,r_s(w))ds/dt=h_2(u_2,r_2(w))\partial u_2/\partial x\big|_{x=s} - h_1(u_1,r_1(w))\partial u_1/\partial x\big|_{x=s}$$

$$x\varepsilon]s,1]: \quad u_2=u_2(w,t,x) \quad , \quad \partial u_2/\partial t=\partial/\partial x \{h_2[u_2,r_2(w,x)]\partial u_2/\partial x\}$$

with initial and boundary conditions, respectively

$$(2.2) \qquad s_0=s(t=0) \quad , \quad u_{10}=u_1(x;t=0) \quad , \quad u_{20}=u_2(x;t=0)$$

$$(2.3) \qquad u_{b0}=u_1(w,t;x=0) \quad , \quad u_{b1}=u_2(w,t;x=1)$$

satisfying the compatibility conditions

$$(2.4) \quad u_{10}(s)=u_{20}(s) \quad , \quad u_{b0}(t=0)=u_{10}(x=0) \quad , \quad u_{b1}(t=0)=u_{20}(x=1)$$

In Eqs.(2.1) the coefficients h are the thermal conductivities of the two phases, whereas L is the latent heat of evaporation.

The random processes r_1 ,r_2 ,r_s ,u_{b0} ,u_{b1} , functions of time and/or space, are defined in a complete probability space $(\Omega,\mathcal{F},\wp)$, where Ω is the space of the elementary events w, $\mathcal{F}$ is a σ-algebra and $\wp$ the induced probability measure. In addition the random processes are assumed to be defined by processes with continuous trajectories (or sample continuous) for every $w\varepsilon\Omega$ in the terms defined in [9], Appendix A.

The randomness characterizing problem (2.1-2.4) can be generated by various reasons, all essentially discussed in ref.[3], so that the general discussion is not repeated here.

The mathematical analysis of the stochastic initial boundary value problem needs to provide the relevant statistical measures of the solutions

$$(2.5) \qquad u_1 = u_1(w,t,x) \quad , \quad u_2 = u_2(w,t,x) \quad , \quad s = s(w,t)$$

of problem (2.1-2.4).

In particular, once the processes (2.5) have been computed, one can evaluate the moments

$$(2.6) \qquad E[u^q](t,x) = \int_\Omega u^q(w,t,x) P(w) dw$$

$$E[s^q](t) = \int_\Omega s^q(w,t) P(w) dw$$

or the second order statistics, i.e. suitable correlation functions such as the auto-correlations

$$(2.7a) \qquad R_u^x(t_1,t_2;x) = \int_\Omega u(w,t_1;x) u(w,t_2;x) P(w) dw$$

$$(2.7b) \qquad R_u^t(x_1,x_2;t) = \int_\Omega u(w,x_1;t) u(w,x_2;t) P(w) dw$$

$$(2.8) \qquad R_s(t_1,t_2) = \int_\Omega s(w,t_1) s(w,t_2) P(w) dw$$

or the auto-covariances

$$(2.9a) \qquad C_u^x(t_1,t_2;x) = \int_\Omega [u(w,t_1;x) - E[u](t_1;x)][u(w,t_2;x) - E[u](t_2;x)] \cdot P(w) dw$$

$$(2.9b) \qquad C_u^t(x_1,x_2;t) = \int_\Omega [u(w,x_1;t) - E[u](x_1;t)][u(w,x_2;t) - E[u](x_2;t)] \cdot P(w) dw$$

$$(2.10) \qquad C_s(t_1,t_2) = \int_\Omega [s(w,t_1) - E[s](t_1)][s(w,t_2) - E[s](t_2)] P(w) dw$$

and so on.

3. Analysis and applications

The mathematical problem described in sec.2 can be dealt with, as already mentioned, by a suitable development of the stochastic interpolation method. Such an application is essentially in three steps:

1) **Rescaling the first and the last equation in (2.1) each with respect to a suitable space interval so that the new space variable spans from zero to one.**
2) **Applying to the solution u(w,t,x) the space interpolation mentioned in the introduction in order to derive an evolution equation, in terms of random ordinary differential equations, for the variable u in each nodal point.**
3) **Solving the system of differential equations and applying again the space interpolation in order to recover u(w,t,x).**

Each step will now be explained in details.

Step 1

The following change of variable is now applied for s strictly larger than zero and less than one

$$(3.1) \qquad x\varepsilon[0,s] : \quad y=x/s \qquad ; \qquad x\varepsilon[s,1] : \quad z=(x-s)/(1-s)$$

which is such that $y\varepsilon[0,1]$, $z\varepsilon[0,1]$ and

$$(3.2) \qquad x\varepsilon[0,s]: \partial/\partial x=(1/s)\partial/\partial y \quad , \quad x\varepsilon[s,1]: \partial/\partial x=[1/(1-s)]\partial/\partial z$$

After this change of variable Eq.(2.1) can be re-written in the new space variables

$$\partial u_1/\partial t = (1/s^2)\, \partial/\partial y \,\{h_1[u_1,r_1(w,y)]\partial u_1/\partial y\}$$

$$(3.3) \qquad ds/dt=\{h_2(u_2,r_2(w))\partial u_2/\partial z\big|_{z=s} -h_1(u_1,r_1(w))\partial u_1/\partial y\big|_{y=s}\}\cdot$$

$$\cdot[s\cdot L(u,r_s(w))]^{-1}$$

$$\partial u_2/\partial t=[1/(1-s)^2]\, \partial/\partial z \,\{h_2[u_2,r_2(w,z)]\partial u_2/\partial z\}$$

to be joined to the initial conditions rescaled in the variables y and z and to the boundary conditions at y=0 and z=1.

Step 2

Let now divide the intervals [0,1] into (N-1) intervals by N equally spaced nodal points using a Lagrange-type interpolation polynomials as in ref.[3] for both variables u_1 and u_2

$$u_1(w,t,y) = \sum_{i=1}^{N} p_i(y)u_{1i}(w,t) \quad , \quad u_2(w,t,z) = \sum_{i=1}^{N} p_i(z)u_{2i}(w,t)$$

where $u_{1i}(w,t)=u_1(w,t;y=y_i)$ and $u_{2i}(w,t)=u_2(w,t;z=z_i)$.

Then, following ref.[3], the space derivatives in each nodal point are given by

$$(3.4) \quad \partial u_{1i}/\partial y = \sum_{j=1}^{N} a_{ij} u_{1j}(w,t) \quad , \quad \partial u_{2i}/\partial z = \sum_{j=1}^{N} a_{ij} u_{2j}(w,t)$$

$$\partial^2 u_{1i}/\partial y^2 = \sum_{j=1}^{N} b_{ij} u_{1j}(w,t) \quad , \quad \partial^2 u_{2i}/\partial z^2 = \sum_{j=1}^{N} b_{ij} u_{2j}(w,t)$$

where the constant coefficients a_{ij} and b_{ij} depend upon the number N and can be computed in the same fashion as indicated in ref.[3]. In addition one imposes the boundary conditions by the positions

$$(3.5) \quad u_{11} = u_{bo}(w,t) \quad , \quad u_{2N} = u_{b1}(w,t) \quad , \quad u_{1N} = u_{21}$$

Setting (3.4) and (3.5) into Eqs.(3.3) one obtains the following set of 2N+1 ordinary differential equations

$$(3.6a) \quad u_{11}(w,t) = u_{bo}(w,t)$$

$$(3.6b) \quad du_{1m}/dt = (1/s^2)\{h_1^*(u_{1m},r_1(w,y_m)) \sum_{k=1}^{N} a_{mk} u_{1k} +$$

$$+ h_1(u_{1m},r_1(w,y_m)) \sum_{k=1}^{N} b_{mk} u_{1k}\} \quad , \quad m=2,\ldots,N$$

$$(3.6c) \quad u_{21}(w,t) = u_{1N}(w,t)$$

$$(3.6d) \quad du_{2j}/dt = [1/(1-s)^2]\{h_2^*(u_{2j},r_2(w,z_j)) \sum_{k=1}^{N} a_{jk} u_{2k} +$$

$$+ h_2(u_{2j},r_2(w,z_j)) \sum_{k=1}^{N} b_{jk} u_{2k}\} \quad , \quad j=2,\ldots,N-1$$

$$(3.6e) \quad u_{2N}(w,t) = u_{b1}(w,t)$$

$$(3.6f) \quad ds/dt = \{[1/(1-s)]h_2(u_{21},r_2(w,z_1)) \sum_{k=1}^{N} a_{1k} u_{2k} -$$

$$- (1/s)h_1(u_{1N},r_1(w,y_N)) \sum_{k=1}^{N} a_{Nk} u_{1k}\}/L(u_{1N},r_s(w))$$

where
$$h_1^* = dh_1/dy \quad , \quad h_2^* = dh_2/dz \quad .$$

Step 3

The third step essentially consists in solving the differential
equations (3.6) and in recovering the statistical measures
indicated in sec.2.

Eq.(3.6) can be regarded as a system of random differential
equations characterized by sample continuous parameters, or, in
other words, by processes characterized by continuous
trajectories. In the following discussion we will call with the
vector $\underline{v}(w,t) = \{u_{11},\ldots,u_{1N},u_{21},\ldots,u_{2N},s\}(w,t)$ the solutions
of the differential system (3.6) and with the vector $\underline{a}(w,t)$ the
set of all the stochastic coefficients which appear in Eqs.(3.6)
(namely r_1, r_2, r_3).

If the parameters $\underline{a}(w,t)$ are defined on the complete
probability space $(\Omega,\mathfrak{F},p)$ with $(w,t)\varepsilon\,\Omega\times T$, then $\underline{a}(w,t)$ is
called sample continuous if its sections $\underline{a}_w(t)$ are continuous
functions of t on $T \subseteq \mathbb{R}_+$ for all $w\varepsilon\Omega$. Consequently the sample
treatment [8] can be applied to Eqs.(3.6) in order to obtain
sample solutions of Eqs.(3.6), namely solutions such that:
i) Almost all trajectories of the processes $\underline{v}(w,t)$ are defined
over some interval $[0,T_o] \subset [0,T]$, where T_o is independent from
the trajectories and the components of $\underline{v}$ are such that Eqs.(3.6)
are satisfied;
ii) At each fixed $t\varepsilon[0,T_o]$ the processes $\underline{v}(w,t)$ are Borel
functions of the random parameters characterizing the system.

Considering that the right-hand-side of Eqs.(3.6) are
bounded Lipschitz functions of their arguments for all $w\varepsilon\,\Omega$ one
can find the solutions which have been mentioned above at least
for some bounded time interval $[0,T_o]$. An example of this kind
of estimates can be found in ref.[7]. Recovering the statistical
measures is then a matter of technical calculations.

In order to show how the method practically works consider
now a system with deterministic initial and boundary conditions
and with stochastic coefficients modelled as follows

$$s_o = 0.5 \quad , \quad u_{10} = x \quad , \quad u_{20} = x$$

$$u_{bo}(t) = 0 \quad , \quad u_{b1}(t) = 1$$

The thermal conductivities, following [3], are chosen such
that

$$h_{1,2} = [1 + r_{1,2}(w,x)u_{1,2}]$$

where the stochastic processes are of separable type

$$r_{1,2}(w,x) = \alpha_{1,2}(w)\sin(2\pi x)$$

with $\alpha_{1,2}(w)$ constant random parameters equally distributed in the intervals $\alpha_1 = [-.5, .5]$ and $\alpha_2 = [-.05, .05]$. In addition, for sake of simplicity, we take L as a constant.

Some results are shown in Figs.1-2 for the motion of the boundary between the two phases: in the first we plot some sample trajectories of the process s(w,t) versus time; in the second we show the behaviour of the variance of s(w,t) versus time.

4. Discussion

The mathematical method developed in the preceding sections can be regarded as a suitable development to moving boundary problems described by partial differential equations of the so-called "stochastic interpolation method" proposed in [3] for random initial-boundary value problems with randomness characterized by separable, Karhunen-Loéve type, processes.

In addition this paper shows how the mathematical model can involve more general stochastic parameters, i.e. random processes with continuous trajectories.

The method is certainly efficient and easily computable. The crucial point still remains the estimate of the distance

$$d = ||\underline{v}^t(w,t,x) - \underline{v}(w,t,x)||$$

between the "true" solution $\underline{v}^t$ of the original initial-boundary value problem and the one $\underline{v}$ obtained by the stochastic interpolation method. In particular the following problems have to be faced:
i) Estimate of the distance between the true space derivatives and the ones obtained by the interpolation (3.4).
ii) Propagation of the error mentioned in (i) through the integration of the ordinary differential equation (3.6).
iii) Error implied by the space interpolation of the solution $\underline{v}$.

The analysis outlined in (i-iii) has to be inserted into the general framework of stochastic interpolation theory [5] and of the treatment of stochastic differential equations [1,2,9];

the method developed in sec.3 is such that the analysis (i-iii) does not differ consistently from the one performed in [4].

Therefore such an analysis is not repeated in this paper; one can simply recall that the solution needs to be sought for in a function space consistent with the stochastic interpolation such as the space

$$\mathbb{C} = \{C_b^o[0,T] \cap C_b^{N+1}[0,1]; L_\infty(\Omega, \mathfrak{J}, p)\}$$

of the functions bounded and continuous with respect to time and (N+1)-differentiable with respect to space and essentially bounded for every $w \varepsilon \Omega$. $\mathbb{C}$ is endowed with the norm

$$||\underline{v}|| = \sup_{(t,x)} ||\underline{v}||_\infty$$

where $\underline{v} = \{v_j\}$, $\underline{v} \varepsilon \mathbb{C}$ and

$$||\underline{v}||_\infty = \max_j \; \operatorname*{ess\,sup}_{w \varepsilon \Omega, \; \forall \gamma \leq N+1} |\partial^\gamma v_j / \partial x^\gamma|$$

Then the estimates indicated in (i) and (iii) can be recovered from bounds on the solutions in $\mathbb{C}$ and the ones mentioned in (ii) by a suitable application of Gronwall's lemma and stability properties. Practical calculations of this kind are the ones developed in paper [4] for a linear problem and in [7] in a nonlinear case.

Acknowledgements: This work has been partially supported by the Ministery for Education and by the National Council for the Research, under grant of P.S. AMTI.

References
[1] Arnold L.: Stochastic Differential Equations, Theory and Application. New York: Wiley 1974.

[2] Bellomo N.; Riganti R.: Nonlinear Stochastic Systems in Physics and Mechanics. Singapore: World Scientific 1987.

[3] Bellomo N; de Socio L.; Monaco R.: On the random heat equation: solutions by the stochastic adaptative method. Comput. Math. with Appl. (in print 1988).

[4] Bellomo N.; Flandoli F.: Stochastic partial differential equations in continuum physics: on the foundations of the stochastic interpolation method for Ito's type equations. Comput. Math. in Simul. (special issue on Stochastic Sistems in print 1989).

[5] Bharucha-Reid T.; Sambandha M.: Random Polynomials. New York: Academic Press 1986.

[6] Bonzani I.; Monaco R.; Zavattaro M.G.: A stochastic model in continuum mechanics: time evolution of the probability density in the random initial boundary value problem. Mathl. Comput. Modelling. 10 (1988) 207-216.

[7] Lachowicz M.; Monaco R.: Existence and quantitative analysis of the solutions to the initial value problem for the discrete Boltzmann equation in all space. SIAM J. of Appl. Math. (in print 1989).

[8] Ruymgoort P.; Soong T.T.: A sample treatment of Langevin-type stochastic differential equations. J. Math. Anal. Appl. 34 (1971) 325-338.

[9] Soong T.T.: Random Differential Equations in Science and Engineering. New York: Academic Press 1973.

Roberto Monaco
Dipartimento di Matematica
Politecnico di Torino
C.so Duca degli Abruzzi 24
10129 Torino - Italy

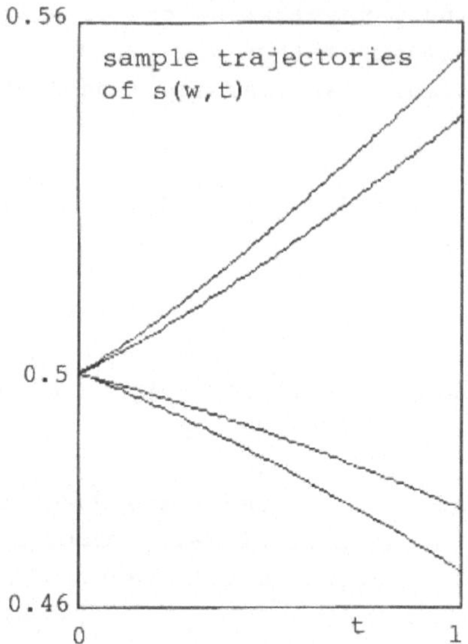

Fig.1 - Random behaviour of the moving boundary

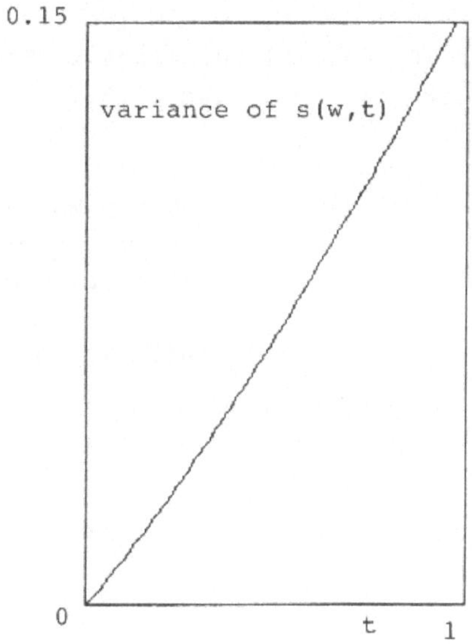

Fig.2 - Random behaviour of the moving boundary

NUMERICAL COMPUTATION OF RAREFIED GAS FLOWS

Aldo Frezzotti

Dipartimento di Matematica del Politecnico di Milano

Piazza Leonardo da Vinci 32 - 20133 - Milano -Italy

Summary: The paper describes a numerical method to solve the Boltzmann Equation in the case of a monatomic gas.The technique is based on a semi-regular method that combines finite differences with the Monte Carlo evaluation of the collision integral.The application of the method to a steady two dimensional supersonic flow is also described.

1.   Introduction

The mathematical difficulties encountered in solving the Boltzmann Equation[1,2] have made the adoption of numerical methods necessary for flows of practical interest.

Since the Direct Simulation Monte Carlo (DSMC) method by Bird[3] has been published in a more or less well established form,the numerical techniques to solve the Boltzmann Equation have been mainly based on statistical methods.

The behavior of the real gas is simulated following the motion of a large number of particles moving through a grid and colliding according to prescribed statistical rules. Various modifications of Bird's DSMC have been proposed in the years.The excellent paper by Nambu[4] is recommended for a comprehensive analysis of the various particle methods and their relation to the Boltzmann Equation.

A direct approach to the solution of the Boltzmann Equation is however possible as described in Refs.(5-7).This class of numerical methods is based on the discretization of the Boltzmann Equation by a proper representation of the distribution function.A finite difference scheme replaces the streaming term,while a Monte Carlo quadrature technique[8] is adopted to evaluate the fivefold collision integral.The details will be described in the next section.The attractive features of this method are its "built-in" consistency with the Boltzmann Equation,its simple structure and the rather natural way in which the algorithm can be organized for vector and/or parallel processing[9].

2.   Description of the numerical method

Let us consider the Boltzmann Equation for a monatomic gas of rigid spheres of diameter d :

$$\frac{DF(\underline{\xi})}{Dt} = \frac{d^2}{2}\int [F(\underline{\xi}^*)F(\underline{\xi}_1^*)-F(\underline{\xi}_1)F(\underline{\xi})]\,|\underline{g}\cdot\underline{k}|\,\sin(\vartheta)\,d\underline{\xi}_1\,d\vartheta\,d\varphi$$

$$(1)$$

In Eq. (1) $F(\underline{\xi})$ is the distribution function of the molecular velocities $\underline{\xi}$, g represents the relative velocity $\underline{\xi}_1-\underline{\xi}$, while $\hat{\underline{k}}\equiv$ ($\sin\theta\,\cos\varphi,\sin\theta\,\sin\varphi,\cos\theta$) is the unit vector that gives the direction of the relative position of the colliding molecules at the time of impact.

Given the distribution function at the time t=0, one can calculate the time evolution of F from Eq. (1). If boundaries are present, their interaction with the gas has to be prescribed giving the probability that a molecule hitting the boundary with velocity $\underline{\xi}_{in}$ is reemitted with velocity $\underline{\xi}_{out}$.

As anticipated in the previous section, the first step toward the construction of the algorithm is the choice of a proper representation of the distribution functions in the velocity space. The simplest choice is to divide a finite domain of the velocity space into a number of cells assuming that the distribution function is constant within each cell. The domain has to be fixed in order to contain most of the particles at any stage of the calculations. Should the distribution function be regular enough in the velocity space higher order approximations could be used to reduce the number of nodes. However, the calculation of the basis functions should not be to expensive in terms of computing time, because thousands of distribution function evaluations are performed during the program execution.

The equation (1) is then integrated over the cell j of the velocity space :

$$\frac{d}{dt}\,N_j = \int_{\nu_j} d\underline{\xi}\;Q(\underline{\xi})$$

$$(2)$$

where :

$$N_j = \int_{v_j} F(\underline{\xi}) \, d\underline{\xi} = \Delta v_j \, F_j \qquad (3)$$

In Eqs.(2,3) N_j represents the number of particles in the j-th cell of the velocity space, while F_j is the mean value of $F(\underline{\xi})$ in the cell j.

The system of differential equations (2) replaces the original system of two Boltzmann equations. An eight fold integral is to be calculated to evaluate the rate of change of the number of particles in a given cell. Since a regular quadrature formula would be too time consuming, a Monte Carlo technique is adopted[8]. In the simplest form of the Monte Carlo integration the right hand side of Eq. (3) is evaluated as:

$$\frac{d}{dt} N_j \cong$$

$$\frac{C_j}{N_{Coll}} \sum_{n=1}^{N_{Coll}} [F^*(n)F_1^*(n) - F(n)F_1(n)] \, |g(n) \cdot \hat{k}(n)| \sin \theta(n) \qquad (4)$$

$$C_j = \pi^2 \Delta v_j V \, d^2$$

In the above equations Δv_j is the volume of the j-th cell, V is the volume of the whole velocity space, N_{Coll} is the number of evaluations of the integrand. A sequence $k(n)$ of unit vectors uniformly distributed on the unit sphere is generated. The values $F^*(n), F_1^*(n), F(n), F_1(n), g(n)$ are defined as :

$$F(n) = F(\underline{\xi}(n))$$
$$F_1(n) = F(\underline{\xi}_1(n))$$
$$F^*(n) = F(\underline{\xi}^*(n))$$
$$F_1^*(n) = F(\underline{\xi}_1^*(n))$$
$$g(n) = \underline{\xi}_1(n) - \underline{\xi}(n)$$

The velocities $\underline{\xi}(n)$ are uniformly distributed in the cell v_j while the velocities $\underline{\xi}_1(n)$ are generated within the whole velocity space $\mathscr{V}$. The value of the expression given in Eq.(4)

is also used to estimate the collision integral in the cells containing $\underline{\xi}_1,\underline{\xi}^*$ and $\underline{\xi}^*$ to increase the number of tests and to reduce the errors on the moments.As is well known the Monte Carlo evaluation of integrals converges to the exact value as $N_{coll}^{-1/2}$,hence the use of variance reduction techniques is strongly recommended[8].A variance reduction method based on the Gaussian generation of some velocity components has been used by Tcheremissine to calculate both the gain and the loss term in the collision integral.The Gaussian was tailored on the average behavior of the distribution function in a number of contiguous cells in the physical space.A few numerical tests by the author indicate that in practice a uniform generator gives often better results because the gain term has not the same behavior of the loss term.A better procedure is probably to calculate the right hand side of Eq. (2) as :

$$\frac{d}{dt}\,N_j \;=\; \int d\underline{\xi} \int d\underline{\xi}_1 \; [\chi_j(\underline{\xi}^*) - \chi_j(\underline{\xi})]F(\underline{\xi}_1)F(\underline{\xi})|\underline{g}\cdot\underline{k}|\sin(\vartheta)$$

$$(5)$$

being $\chi_j(\underline{\xi})$ a function that is equal to one if the argument belongs to the cell j and zero otherwise.In evaluating the integral at the right hand side of Eq.(5),the velocities $\underline{\xi}$ and $\underline{\xi}_1$ should be drawn from F itself.
A drawback of the technique is that owing to the discretization in the velocity space,mass momentum and energy are not exactly conserved.The numerical error is usually small,but tends to accumulate during the the time evolution of the solution. In the case of a simple gas,Aristov and Tcheremissine [6] overcame the difficulty correcting the distribution function in the following way :

$$\tilde{F}^{(n)}(\underline{\xi}) \;=\; F^{(n)}(\underline{\xi}) \; [1+ A +\underline{B}\cdot\underline{\xi} + C\underline{\xi}^2 \,] \tag{6}$$

In Eq.(6) $F^{(n)}(\underline{\xi})$ is the distribution function computed at the n-th time level.A corrected distribution function $\tilde{F}^{(n)}(\underline{\xi})$ is computed multiplying $F^{(n)}(\underline{\xi})$ by the polynomial

$P(\underline{\xi})=1+A+\underline{B}\cdot\underline{\xi}+C\underline{\xi}^2$. The constants $A,\underline{B}$ and C are determined from the conditions :

$$\int \psi(\underline{\xi})\ \tilde{F}^{(n)}(\underline{\xi})\ d\underline{\xi} = \int \psi(\underline{\xi})\ F^{(n-1)}(\underline{\xi})\ d\underline{\xi}\quad ,\psi(\underline{\xi})=1\ ,\ \underline{\xi}\ ,\ \underline{\xi}^2 \tag{7}$$

This method was found to produce satisfactory results,but it is not easily extended to a multi-component gas,since in this case one does not have enough conservation equations to determine the correction coefficients.A possible way out of this further difficulty has been indicated by the author in Ref.(9) to improve the methods proposed by Mausbach and Beylich[10],and by Raines[11].

3.   Applications to 2-D Steady Flows

The numerical method described above was applied to study the steady two-dimensional flow of a monatomic gas past a flat plate at zero angle of attack.Attention was focused on the effects of the gas-surface interaction model on the flow field.

It is assumed that the description of the flow field is obtained from the solution of the steady two-dimensional Boltzmann Equation :

$$\xi_x \frac{\partial F}{\partial x} + \xi_y \frac{\partial F}{\partial y} = \left(\frac{\partial F}{\partial t} \right)_{coll.} \tag{8}$$

The collision term has been calculated either in the form given in Eq.(1) or following the BGK model[12] :

$$\left(\frac{\partial F}{\partial t} \right)_{coll} = \nu(\rho,T)\ \left[\frac{\rho}{(2\pi RT)^{3/2}} \exp \left[-\frac{(\underline{\xi}-v)^2}{2RT} \right] - F \right] \tag{9}$$

Eq. (8) holds in a cartesian frame of reference whose z axis is parallel to the plate which occupies the strip { $(x,y,z)\ \in\ R_3$: y=0,0 < x < L },being L the length of the plate.It is assumed that the distribution function of the gas flowing undisturbed far from the plate is a Maxwellian with number density n_∞, temperature T_∞, and velocity U_∞ parallel to the plate.

The boundary condition for $F(x,y,\underline{\xi})$ on the plate surface is given in the form:

$$(\underline{\xi} \cdot \hat{n})\, F(\underline{\xi}) = \int_{\underline{\xi}' \cdot \hat{n} < 0} R(\underline{\xi},\underline{\xi}')F(\underline{\xi}')|\underline{\xi}'\cdot\hat{n}|\; d\underline{\xi}' \quad , \quad \underline{\xi} \cdot \hat{n} > 0 \tag{10}$$

In Eq.(10) the scattering kernel $R(\xi,\xi')$ gives the probability that a particle hitting the surface element (whose outward pointing normal is $\hat{n}$) with velocity $\underline{\xi}'$ is reemitted with velocity $\underline{\xi}$. Since Maxwell considered the difficult problem of gas-surface interaction in 1879, the most popular choice of the scattering kernel R has been :

$$R(\underline{\xi},\underline{\xi}') = (1-\alpha)\delta(\underline{\xi}'-\underline{\xi}+2\hat{n}[\underline{\xi}\cdot\hat{n}]) + \alpha\, F_w(\underline{\xi})|\underline{\xi}\cdot\hat{n}| \tag{11.1}$$

$$F_w(\xi) = \frac{n_w}{(\,2\pi RT_w\,)^{3/2}}\exp\left(-\frac{\xi^2}{2RT_w}\right) \tag{11.2}$$

In Eqs.(11) n_w , and T_w are the number density and the temperature of the Maxwellian distribution function F_w that is supposed to describe the particles which accommodated to the wall conditions. The wall temperature T_w is usually given, while n_w is to be determined from the mass balance at the surface. A more general model is the one proposed by Cercignani and Lampis [19]:

$$R(\xi,\xi'|\alpha_n,\alpha_t) = \frac{[\alpha_n\alpha_t(2-\alpha_t)]^{-1}}{2\pi(RT_w)^2}\,\xi_n\,\exp\left\{-\frac{\underline{\xi}^2 + (1-\alpha_n)\underline{\xi}'^2}{2RT_w\alpha_n} - \right.$$

$$\left. -\frac{1}{\alpha_t(2-\alpha_t)}\frac{[\xi-(1-\alpha_n)\xi']^2}{2RT_w}\right\} \times I_o\left(\frac{\sqrt{1-\alpha_n}}{\alpha_n RT_w}\xi_n\xi_n'\right) \tag{12}$$

with $\alpha_n \in [0,1]$ and $\alpha_t \in [0,2]$. The function $I_0(x)$ is defined as :

$$I_0(x) = \frac{1}{2\pi} \int_0^{2\pi} e^{x\cos\phi} d\phi \tag{13}$$

It is easily shown that the coefficients α_n and α_t are the accommodation coefficients of the normal kinetic energy $\xi_n^2/2$ and of the tangential momentum ξ_t respectively.

The scattering kernel (12) reduces to Maxwell model with $\alpha=1$ if $\alpha_n=1$ and $\alpha_t=1$. Specular reflection is obtained setting $\alpha_n=\alpha_t=0$, while the choice $\alpha_n \to 0$ and $\alpha_t \to 2$ gives $R(\underline{\xi},\underline{\xi}')=\delta(\underline{\xi}+\underline{\xi}')$.

Eq.(8) has been solved numerically by an iteration procedure to study the sensitivity of the flow field to the gas-surface interaction model. A rectangular region of the physical space enclosing the plate has been divided into a number of cells. The Boltzmann Equation has been replaced by the finite difference expression :

$$\xi_x \Delta y(F_{ij}^{(n+1)} - F_{i-1,j}^{(n+1)}) + \xi_y \Delta x(F_{ij}^{(n+1)} - F_{i,j-1}^{(n+1)}) = \Delta x \Delta y(G_{ij}^{(n)} - L_{ij}^{(n)} F_{ij}^{(n+1)})$$
$$\xi_x, \xi_y > 0 \tag{14}$$

Eq.(13) is clearly based on a first order upwind discretization of the streaming term. The expression is given only positive values of the velocity components, but it is readily written for the general case.

The index n denotes the iteration number, while i and j are spatial index. It is worth noticing that the loss term is treated implicitly to prevent the distribution function to become negative during the calculations. The number of iterations needed to reach the solutions strongly depends on the Knudsen number .

The results of a few test calculation are shown in Figs.(1,2).

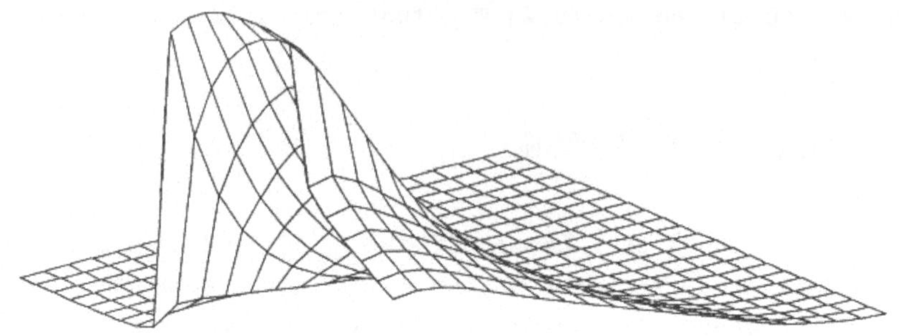

Fig.1 - Qualitative Temperature Field around the plate.Conditions are: Mach Number 4, $T_w/\ T_\infty = 3$,Knudsen Number 1/10, $\alpha_n = 1, \alpha_t = 0.75$.

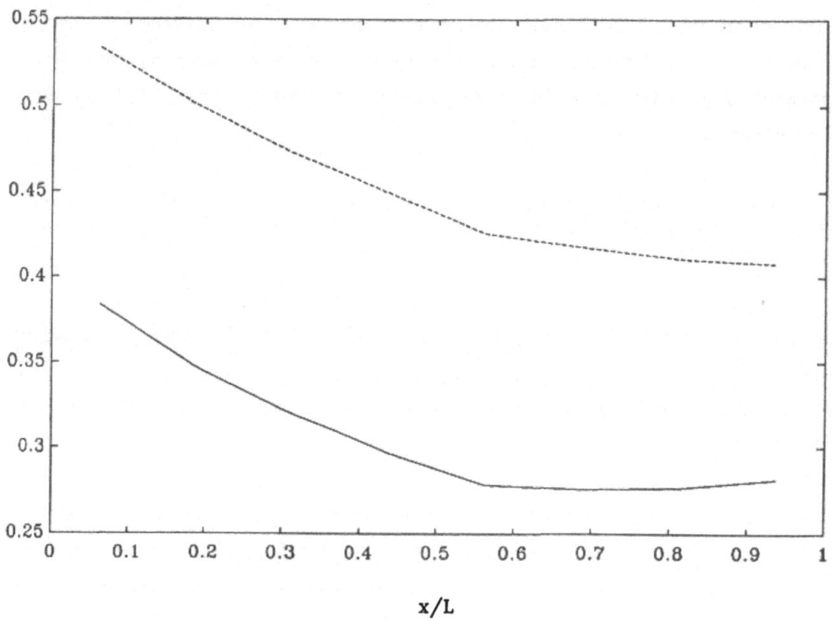

x/L

Fig. 2 -Slip Velocity along the plate .Conditions are : Mach Number 4,Knudsen Number 0.5,Tw/ T_∞=1.Solid line α_t=1 , α_n=1,dashed line α_t=0.75,α_n=1

References

[1] Chapman S.,Cowling T.G.,*The Mathematical Theory of Non-Uniform Gases*,Cambridge University Press,Cambridge,1960.

[2] Cercignani C.,*The Boltzmann Equation and its Applications* , Springer Verlag,N.Y.,1988.

[3] Bird G. A.,*Molecular Gas Dynamics*,Oxford University Press,1976.

[4] Nambu K.,*Theoretical Basis of the Direct Simulation Monte Carlo Method*,Proc. of the 15-th International Symposium on Rarefied Gas Dynamics,edited by V. Boffi and C. Cercignani,1986,pp.369-383.

[5] Nordsieck A.,Hicks B.,*Monte Carlo Evaluation of the Boltzmann Collision Integral*,Rarefied Gas Dynamics,edited by C.L. Brundin,1967,pp.695-710.

[6] Aristov V. V.,Tcheremissine F.G.,*The conservative splitting method for solving the Boltzmann Equation*, USSR Compt. Math. and Math. Phys.,Vol. 20,p.208, 1980.

[7] Tcheremissine F.G.,*Numerical Methods for the Direct Solution of the Kinetic Boltzmann Equation*, U.S.S.R. Comput.Math. Math.Phys., Vol. 25,pp. 156-166,1985.

[8] Kalos M. H.,Whitlock P. A.,*Monte Carlo Methods*,John Wiley & Sons,New York,1986.

[9] A. Frezzotti,R. Pavani,*Numerical study of the homogeneous relaxation in binary mixture of rigid sphere*,presented at III Italian Symposium on Computational Mechanics. Palermo June 7-10,1988.

[10] Mausbach P. Beylich A. E.,*Numerical Solution of the Boltzmann Equation for one-dimensional Problems in Binary Mixtures*,Proc. of the 13-th International Symposium on Rarefied Gas Dynamics edited by O.M. Belotserkovskii et Al. Vol 1 ,1985,p. 285.

[11] Raines A. A.,*Numerical Solution of the Boltzmann Kinetic Equation for the Binary Gas Mixture*, Proc. of the 13-th International Symposium on Rarefied Gas Dynamics edited by O.M. Belotserkovskii et Al. Vol 2 ,1985,pp. 1285-1293.

[12] Bhatnagar P.L.,Gross E. P.,Krook M.,Phys. Rev. 94,511 (1954).

[13] Cercignani, C.,Lampis, M.,*Kinetic Models for Gas-Surface Interactions*, Journal of Stat. Phys., Vol. 1,1971,pp. 101-114.

ON THE NON-LINEAR STABILITY OF PARALLEL SHEAR FLOWS

G.MULONE

1. Introduction

The stability of parallel shear flows for its application in many physical situations (engineering, astrophysics, meteorology, and geophysics) is very important and has been studied by many authors (see, for example [4,8,10] and the references therein).

In some recent papers on the thermal convection of an infinite horizontal layer of an incompressible fluid [12,14,15], we studied the nonlinear stability of the rotating Bénard problem and the magnetic Bénard problem with the Lyapunov direct method and gave a guideline for the choice of a Lyapunov function in order to obtain results as close as possible to the linear results and to the experiments, [12,15]. Applying the aforesaid method, here we study, the non-linear stability of parallel shear flows (plane Couette and plane Poiseuille flows) of an incompressible isothermal fluid with stress-free boundary conditions. These boundary conditions are most appropriate in astrophysics and meteorology. According to the lines given in [12,15] , choosing properly a Lyapunov function, we show that the plane Couette and plane Poiseuille flows are conditionally asymptotically non-linearly stable for *all* Reynolds numbers. Obviously, this result can be expected in this case because of the stabilizing effect of viscosity and the absence of rigid boundaries.

The plan of the paper is as follows: in sect. 2 the basic equations are introduced and some of the classical stability results for the rigid case are summarized. In sect. 3 the linear stability of the plane Couette and plane Poiseuille flows with stress-free boundary conditions is studied through the Lyapunov direct method. In sect. 4 the (conditional) non-linear stability is studied and the behaviour of the initial perturbation is analyzed.

2. <u>Basic equations and stability results in the rigid case</u>

Let $d > 0$, $\Omega = \mathbb{R}^2 \times (-d,d)$ and $Oxyz$ be a cartesian frame of reference with unit vectors i, j, k respectively. The stationary Navier-Stokes system

$$
(2.1) \qquad \begin{cases} U \cdot \nabla U = -\nabla p_1 + \nu \Delta U \\[2mm] \nabla \cdot U = 0 \end{cases} \qquad \text{in } \Omega
$$

with boundary conditions

$$(2.2) \qquad U(x,y,-d) = -Vi, \qquad U(x,y,d) = Vi$$

admits the laminar solutions (parallel shear flows)

$$(2.3) \qquad U = \left\{ \frac{k}{2\nu}(d^2 - z^2) + \frac{Vz}{d} \right\} i, \qquad p_1 = -kx + p_0,$$

where U and p_1 are the velocity and pressure fields, ν is the kinematic viscosity and V, k, p_0 are real numbers.

If $k = 0$ and $V \neq 0$ then we have the

$$(2.4) \qquad \underline{\text{plane Couette flow}} \quad U = \frac{Vz}{d} i, \quad p_1 = p_0,$$

if $k \neq 0$ and $V = 0$ then we have the

$$(2.5) \qquad \underline{\text{plane Poiseuille flow}} \quad U = \frac{k}{2\nu}(d^2 - z^2)i, \quad p_1 = -kx + p_0.$$

Introducing

$$(2.6) \qquad v_0 = \max_{\overline{\Omega}} |U|,$$

and the Reynolds number

211

$$(2.7) \qquad\qquad R= \frac{v_0 d}{\nu} \ ,$$

we obtain the non-dimensional equations for a disturbance   of the basic motion $m_o = (U = U(z)i, \ p_1)$:

$$(2.8) \qquad \begin{cases} u_t + U(z)u_x + wU'(z)i + u \cdot \nabla u = -\nabla p + \frac{1}{R}\Delta u \\[2mm] \nabla \cdot u = 0 \end{cases} \qquad \text{in } I = \mathbb{R}^2 \times (-1,1) \times (0,\infty)$$

with initial condition

$$(2.9) \qquad\qquad u(x,0) = u_o(x) \ , \qquad x \in \mathbb{R}^2 \times (-1,1)$$

and boundary conditions

$$(2.10) \qquad\qquad u(x,y,-1,t) = u(x,y,1,t) = 0, \qquad t \geq 0.$$

In (2.8)-(2.10) $x = (x,y,z)$, $u_x = \frac{\partial u}{\partial x}$, $u_t = \frac{\partial u}{\partial t}$, $u_o(x)$ is an assigned regular field with $\nabla \cdot u_o(x) = 0$ and $U(z) = z$ for <u>Couette flow</u>, $U(z) = 1 - z^2$ for <u>Poiseuille flow</u>.

The linear stability of the basic motion $m_o = (U, p_1)$ has been studied by many authors (see for example the references in [4,10]) with the usual normal - mode analysis solving the corresponding Orr-Sommerfeld equation with numerical and asymptotic expansion methods (see [4,ch.4]).The main results are the following:

a) Couette flow is linear stable for all Reynolds numbers ,[16],

b) Poiseuille flow is linear unstable for any Reynolds number greater than $R_c = 5772$, [13].

The non-linear stability of m_o has been studied by [2,9]. It has been shown that m_o is non-linearly (energy) stable if

$$(2.11) \qquad\qquad R < R_E \ ,$$

with $R_E = 20.7$ for plane Couette flow,

$\qquad R_E = 49.6$ for plane Poiseuille flow.

We also note that a conditional non-linear stability result has been obtained by [16], in the case of plane Couette flow.For other non-linear results and experimental results, see [4, ch.7, § 54].

3. <u>The stress-free boundary case: linear stability</u>

Here we assume that the bounding planes are free, i.e. no tangential stresses act on them. In this case , as it is easy to see, the perturbation (u,p) satisfies the IBVP (2.8),(2.9) with boundary conditions

$$(3.1) \quad w(x,t)=0, \quad u_z(x,t)=v_z(x,t)=0, \qquad \text{on } z= \pm 1, \ t \geq 0 \ .$$

We also assume that the perturbation fields are periodic functions in x and y of periods $\dfrac{2\pi}{a_x}$, $\dfrac{2\pi}{a_y}$, and we require the conditions

$$(3.2) \qquad \int_{\Omega_1} u d\Omega_1 \ = \ \int_{\Omega_1} v d\Omega_1 \ = 0$$

which are necessary for uniqueness, where Ω_1 is the periodicity cell . We observe that, by virtue of the solenoidality condition $(2.8)_2$, the linearized problem associated to (2.8), (2.9),(3.1) may be solved in terms of the variables w and $\xi = k \cdot \nabla \times u$. (see [3 ,ch.II,pag.24]). We call the variables w and ξ the <u>essential variables</u> <u>of the linear problem</u>.

In order to study the stability of m_0 by the Lyapunov direct method, we use the idea given in [12,15]. The Lyapunov function V is the sum of two terms $V_0(t)$ and $V_1(t)$. V_0 is a Lyapunov function for the linear stability problem and depends on the essential variables, while V_1 must dominate the non-linear terms. We need the evolution equations of w and ξ which are:

$$(3.3) \qquad \xi_t + U(z)\xi_x - U'(z)w_y + k \cdot \nabla \times (u \cdot \nabla u) = \frac{1}{R} \Delta \xi \ ,$$

$$(3.4) \qquad \Delta w_t + U(z)\Delta w_x - U''(z)w_x - k \cdot \nabla \times (\nabla \times (u \cdot \nabla u)) = \frac{1}{R} \Delta \Delta w \ .$$

In this section we solve the linear stability problem. For

this, we consider the Lyapunov function

(3.5) $$V_o(t) = \frac{1}{2} [\ \|\xi\|^2 + \beta \ \|\Delta w\|^2 \ ] \ ,$$

where β is a positive constant that shall be chosen later. First, we observe that, because of the boundary conditions (3.1) it follows that

(3.6) $$\xi_z = 0 \ , \ \Delta w = 0 \qquad \text{on } z = \pm 1, \quad t \geq 0 \ .$$

Now we write the evolution equation of $V_o(t)$. First we linearize the equations (3.3) and (3.4), i.e. we neglect the terms $k \cdot \nabla \times (u \cdot \nabla u)$ and $k \cdot \nabla \times (\nabla \times (u \cdot \nabla u))$, then we multiply (3.3) by ξ, (3.4) by Δw and integrate over Ω_1. Because of the boundary conditions, we have

(3.7) $$\dot{V}_o = I_o - D_o$$

where

(3.8) $$I_o = - \int_{\Omega_1} U'(z) w_y \xi d\Omega_1 \quad ,$$

(3.9) $$D_o = \frac{1}{R} [\ \|\nabla \xi\|^2 + \beta \ \|\nabla \Delta w\|^2 \ ].$$

Now we use the following relations:

(3.10) $$\|\xi\| \leq \frac{2}{\pi} \|\nabla \xi\| \qquad \text{(Wirtinger inequality, see [7])},$$

(3.11) $$\|w\| \leq \frac{2}{\pi} \|\nabla w\| \ , \ \|\Delta w\| \leq \frac{2}{\pi} \|\nabla \Delta w\| \qquad (\text{ Poincaré inequality)},$$

(3.12) $$ab \leq \frac{a^2}{2\varepsilon} + \frac{\varepsilon}{2} b^2 \quad , \ a,b \in \mathbb{R} \ , \ \varepsilon > 0 \ ,$$

the identity

(3.13) $$\int_{\Omega_1} w \Delta w d\Omega_1 = - \ \|\nabla w\|^2 \quad ,$$

and the Schwarz inequality to obtain the following estimate:

(3.14) $$I_o \leq m_1 \ \|w_y\| \ \|\xi\| \leq \frac{8}{\pi^3} m_1 \|\nabla \Delta w\| \ \|\nabla \xi\| \leq$$

$$\leq \frac{32}{\varepsilon \pi^6} m_1^2 \|\nabla \Delta w\|^2 + \frac{\varepsilon}{2} \|\nabla \xi\|^2,$$

with $\varepsilon > 0$ and $m_1 = \max_{[-1,1]} |U'(z)|$. From (3.8) and (3.14), we obtain:

$$(3.15) \qquad \dot{V}_o \leq [\ \frac{32}{\varepsilon \pi^6}\ m_1^2\ -\ \frac{\beta}{R}\ ]\|\nabla\Delta w\|^2 + (\frac{\varepsilon}{2} - \frac{1}{R})\|\nabla\zeta\|^2 \quad .$$

Choosing

$$(3.16) \qquad\qquad \varepsilon = \frac{1}{R} \quad , \qquad \beta = \left(\ \frac{8Rm_1}{\pi^3}\ \right)^2 \quad ,$$

we have

$$(3.17) \qquad \dot{V}_o \leq -\frac{1}{2R}\ [\ \|\nabla\zeta\|^2 + \beta\ \|\nabla\Delta w\|^2\ ] \leq -\frac{\pi^2}{4R}\ V_o \quad .$$

Integrating this last inequality, it follows

$$(3.18) \qquad\qquad V_o(t) \leq V_o(0)\ \exp\ (-\frac{\pi^2}{4R}\ t) \quad .$$

So we have proved the

Theorem 3.1.   The basic motion (plane Couette or plane Poiseuille flow) is linear stable for all Reynolds numbers.

4. Non-linear stability

Here we study the non-linear (conditional) stability of m_o and analyze the behaviour of the intial perturbation. First of all we observe that if we apply the classic energy method then it is easy to see that the critical Reynolds number R_E is finite.

Instead of this classical method, according to the lines given in sec. § 3, we now define the Lyapunov function

$$(4.1) \qquad\qquad V(t) = V_o(t) + b\ V_1(t) \quad ,$$

where $V_o(t)$ is given by (3.5),

$$(4.2) \qquad\qquad V_1(t) = \frac{1}{2}\ [\ \|\nabla u\|^2 + \|\nabla(\nabla\times u)\|^2\ ] \quad ,$$

and b is a positive constant.

In order to study the evolution equation of V(t), we multiply $(2.10)_1$ by $-\Delta u$ and integrate over Ω_1. Then we take the curl of $(2.10)_1$, multiply the equation so deduced by $-\Delta(\nabla\times u)$, integrate over Ω_1 and add. We obtain the evolution equation

(4.3) $\qquad \dot{V}(t) = I_o - D_o + N_o + bI_1 - bD_1 + bN_1 + bB_1$

with I_o and D_o given by (3.8) and (3.9),

(4.4) $\qquad N_o = \int_{\Omega_1} [-k\cdot\nabla\times(u\cdot\nabla u)\xi + \beta\nabla\times\nabla\times(u\cdot\nabla u)\Delta w \,]d\Omega_1$,

(4.5) $\qquad I_1 = - \int_{\Omega_1} \{ U'(z) [u_z\cdot u_x + \nabla u\cdot\nabla w + w_{xyz}(w_y - v_z) + v_{zz}w_{xy}$

$\qquad\qquad + u_{zz}w_{xx} + w_{xxz}(w_x - u_z) - \xi_x\xi_z - w_{zzz}u_z - v_{zz}\xi_z - u_x\cdot\nabla(\Delta w)$

$\qquad\qquad + \Delta\xi v_x - \Delta(\nabla\times u\cdot j)w_z + (w_z j - w_y k)\cdot\Delta(\nabla\times u)] + U''(z) [u_z w$

$\qquad\qquad + \Delta w w_x + w j\cdot\Delta(\nabla\times u)] \} \, d\Omega_1$,

(4.6) $\qquad\qquad D_1 = \frac{1}{R} [\; \|\Delta u\|^2 + \|\Delta(\nabla\times u)\|^2 \;]$,

(4.7) $\qquad N_1 = \int_{\Omega_1} \{ u\cdot\nabla u\cdot\Delta u + [\nabla\times u \cdot\nabla u - u\cdot\nabla(\nabla\times u)]\cdot\Delta(\nabla\times u) \} \, d\Omega_1$,

finally B_1 is a boundary integral which vanishes for the boundary conditions and the periodicity.

By virtue of the boundary conditions and the periodicity, it is easy to see that $B_1 = 0$.

From (3.17) and (4.3) we have

(4.8) $\qquad\qquad \dot{V}(t) \leq - D_o' + N_o + bI_1 - bD_1 + bN_1$,

where

(4.9) $\qquad\qquad D'_o = \frac{1}{2} D_o$.

By using the Schwarz inequality, (3.11) - (3.13), the inequality

(4.10) $\qquad \|w_{x_i x_j}\| \leq \|\Delta w\| \qquad (i,j = 1,2,3, \; x_i, \; x_j$ stand for x,y,z)

and the definitions (3.9),(4.6), we obtain the following estimate of the term I_1:

$$I_1 \leq \frac{8}{\pi} A_o \|\nabla\Delta w\| \; \|\Delta u\| + 2B_o \|\nabla\xi\| \; \|\Delta u\| + \frac{4}{\pi^2} \|\nabla\Delta w\| \; \|\Delta(\nabla\times u)\|$$

$$\leq \frac{2}{\pi\varepsilon} (2A_o^2 + \frac{C_o^2}{\pi}) \|\nabla\Delta w\|^2 + \varepsilon(\frac{4}{\pi} + 1) \|\Delta u\|^2 + \frac{B_o^2}{\varepsilon} \|\nabla\xi\|^2 + \frac{2\varepsilon}{\pi^2} \|\Delta(\nabla\times u)\|^2$$

where $A_o = m_1 (\frac{3}{\pi^2} + 2) + \frac{m_2}{\pi} (\frac{1}{\pi} + 1)$, $B_o = m_1 (\frac{2}{\pi^2} + \frac{2}{\pi} + 1) + \frac{m_2}{\pi}$,

$$C_o = 3m_1 + m_2 \quad , \text{ with } \quad m_1 = \max_{-1 \leq z \leq 1} |U'(z)| \quad , \quad m_2 = \max_{-1 \leq z \leq 1} |U''(z)| \quad .$$

Choosing

$$(4.11) \quad \varepsilon = \frac{1}{2R\left(\frac{4}{\pi} + 1 + \frac{2}{\pi^2}\right)} \quad , \quad b = \frac{1}{8R^2\left(\frac{4}{\pi} + 1 + \frac{2}{\pi^2}\right)\left[\frac{2}{\pi\beta}\left(2A_o^2 + \frac{1}{\pi}C_o^2 + B_o^2\right)\right]}$$

we have

$$(4.12) \qquad\qquad bI_1 \leq D_2$$

where

$$(4.13) \qquad\qquad D_2 = \frac{1}{2}\left[D_o' + D_1\right] \quad .$$

In a similar way it is possible to estimate the non-linear terms:

$$(4.14) \quad N_o + bN_1 \leq \left(\frac{2}{b}\right)^{1/2} CR\left[5 + \frac{4\left(1 + 2\sqrt{\beta}\right)}{\sqrt{b}}\right] D_2 V^{1/2} \quad ,$$

where C is a positive computable constant in the following inequalities

$$(4.15) \qquad \sup_{\Omega_1} |u| \leq C\|\Delta u\| \quad , \quad \sup_{\Omega_1} |\nabla \times u| \leq C\|\Delta(\nabla \times u)\| \quad ,$$

and β is given by $(3.17)_2$. (For the proof of (4.16) see [6], Appendix and [5], lemmas A.2 - A.3; a value of C is given in [6], A.15).If now we put

$$(4.16) \qquad\qquad A = \left(\frac{2}{b}\right)^{1/2} CR\left[5 + \frac{4\left(1 + 2\sqrt{\beta}\right)}{\sqrt{b}}\right] \quad ,$$

then we have

$$(4.17) \qquad\qquad \dot{V} \leq -D_2\left(A - V^{1/2}\right) \quad .$$

This last inequality implies the following non-linear stability theorem.

THEOREM 4.1. *Let* $V(0) < A^{-2}$, *with A given by (4.16), then*

$$(4.18) \qquad\qquad V(t) \leq V(0) \exp\left\{-\frac{\pi^2}{8R}\left[1 - AV(0)^{1/2}t\right]\right\} \quad .$$

Proof. The condition $V(0) < A^{-2}$ and the inequality (4.17) assure that $\dot{V}(0) < 0$. Therefore, from (3.11),(3.12) and (4.17), by a

217

recursive argument, we have:

$$\dot{V} \leq - D_2 \left[A - V(0)^{1/2} \right] \leq - \frac{\pi^2}{8R} V[1-AV(0)^{1/2}] \quad .$$

The last inequality implies (4.18).

The following lemma holds:

LEMMA 4.1. *If* u *is a regular field, periodic in x and y of periods* $\frac{2\pi}{a_x}$, $\frac{2\pi}{a_y}$ *respectively, with* $\nabla \cdot u = 0$ *and such that each of its component* ψ *statisfies one of the following boundary conditions* $\psi = 0$ *or* $\psi_z = 0$ *on* $z = \pm 1$, *then the following identity*

(4.19) $$\|\nabla(\nabla \times u)\|^2 = \|\nabla\nabla u\|^2$$

holds.

The proof easily follows from the solenoidality and the boundary conditions.

THEOREM 4.2. *In the hypothesis of theorem 4.1 we have the following punctual (exponential) decay:*

$$\sup_{\mathbb{R}^2 \times [-1,1]} |u(x,t)| \quad 0 \text{ as } t \quad \infty .$$

The proof is consequence of Lemma 4.1, of the well known imbedding theorem, [1]:

$$\sup_{\Omega_1} |u(x,t)| \leq c \, \|u\|_{W_2^2(\Omega_1)} ,$$

of the inequalities $\pi^2 \|u\| \leq \pi \|\nabla u\| \leq \|\Delta u\| \leq \|\nabla\nabla u\|$, and of (4.19).

References

[1] Adams, R.: *Soboler spaces*, Academic Press: New York,1976.

[2] Busse, F.H., *Z.Angew.Math.Phys.*, <u>20</u>,(1969) 1-14 .

[3] Chandrasekhar, S.:*Hydrodynamic and hydromagnetic stability*, Oxford: Clarendon Press, 1961.

[4] Drazin, P,G.;Reid, W.H.: *Hydrodynamic stability*, Cambridge

Univerity Press (1981).

[5] Galdi, G.P., *Arch. Rational Mech. Anal.* <u>87</u>,(1985) 167-186 .

[6] Galdi, G.P.; Straughan B., *Proc.Soc. Lond. A* <u>402</u> (1985), 257-283.

[7] Hardy,G.H.; Littlewod, J.E.; Polya, G.: *Inequalities*, Cambridge University Press,1959.

[8] Joseph, D.D.: *Stability of fluid motions* (2 vols.) Springer Tracts in Natural Phylosophy, vols. 27 and 28. Berlin: Springer - Verlag, 1976.

[9] Joseph,D.D.; Carmi, S.,*Quart.Appl.Math.*,<u>26</u> ,(1969) 575-599.

[10] Lin, C.C.: *The theory of hydrodynamic stability,* Cambridge University Press, 1955.

[11] Mulone, G.: On the nonlinear stability of the magnetic Bénard problem with rotation,(*to appear*).

[12] Mulone, G.; Rionero, S.:On the non-linear stability of the rotating Bénard problem via the Lyapunov direct method *J. Mat.Anal App.*,(*to appear*).

[13] Orszag, S.A., *J. Fluid Mech.* <u>50</u>, (1971) 689-703 .

[14] Rionero, S.: On the choice of the Lyapunov function in the fluid motion stability, *lecture given at the meeting* "Energy stability and convection",Capri (1986).

[15] Rionero, S.; Mulone, G.:A nonlinear stability analysis of the magnetic Bénard problem via the Lyapunov direct method, *Arch.Rational Mech.Anal*,(*to appear*).

[16] Romanov, V.A., *Functional Anal. App.* <u>7</u>,(1973) 137-146 .

Dipartimento di Matematica, viale A.Doria, 6

95125 CATANIA - ITALY

A rigorous Onsager-Machlup formulation of nonequilibrium thermodynamics

Paolo Dai Pra*
Mathematics Department
Rutgers University
New Brunswick, NJ 08903
USA

Michele Pavon
Dipartimento di Elettronica e Informatica
Universita' di Padova
and
LADSEB-CNR
35100 Padova
ITALY

Abstract : let $p(x,t)$ be the probability density of a general diffusion process $\{x(t); 0 \le t \le T\}$. A variational path-integral representation for the square root of p, and of $(p/\bar{p})$, where $\bar{p}$ is the density of the invariant measure, is derived. This is accomplished by showing that these functions are optimal performances of stochastic controls problems, where the controlled equation evolves backward in time. This provides a rigorous counterpart of the formal Onsager-Machlup formulation of nonequilibrium thermodynamics.

* The research of this author was conducted at LADSEB-CNR, Padua, ITALY, with support provided by a CNR postgraduate fellowship.

220

1. Introduction.

In a very influential paper [1], Onsager and Machlup proposed a most-appealing approach to nonequilibrium thermodynamics. Their work, and later generalizations [2]-[4], was aimed at deriving a functional integral representation for the transition density of a diffusion process, which models the macroscopic dynamics of a system. A simple maximization would then yield "the most probable path" of the process. This approach is considered particularly attractive from both the theoretical and computational viewpoint by physicists. Consider a diffusion process with drift b and constant diffusion coefficent a. The Onsager-Machlup theory asserts that the measure induced by the process $\{x(t) ; 0 \leq t \leq T\}$ on path space admits a density, with respect to a (fictitious) uniform measure, which is proportional to the following exponential

$$\exp \left\{ -(1/2) \int_0^T L(x(t),\dot{x}(t),t)dt \right\}$$

where the Lagrangian is

$$(1.1) \qquad L(x,\dot{x},t) = \frac{1}{2a}\| \dot{x} - b(x,t) \|^2 + \frac{1}{2} \nabla \cdot b(x,t).$$

As the paths of a diffusion process are nowhere differentiable with probability one, evaluating this Lagrangian on them, as required in the Onsager-Machlup theory, has to be regarded as purely formal.The Lagrangian, however, appears in a rigorous asymptotic formula for the probability that the paths of the diffusion lie in a small tube about a given path *of class* C^2 , i.e. in the mathematical theory of the most probable path [5],[6] . The problem of describing the dynamics through a path-integral representation of the density of the diffusion was given a rigorous formulation in [7], where -logp(x,t) was shown to be the value function of a stochastic control problem. The same interpretation was obtained in [8] for $\log(p/\bar{p})(x,t)$, where $\bar{p}$ is the density of the invariant measure. These functions may be seen as local versions of the *entropy* and of the *Helmholtz free energy* . In [9] similar results are obtained for $p^{1/2}$ and $(p/\bar{p})^{1/2}$. A peculiar feature of these control problems is that the controlled equation evolves backward in time. In this paper we summarize some of the main results of [9], discussing in more detail the case of general diffusions. The key tools are stochastic control theory [10] and Nelson's kinematics of stochastic processes [11]. In the first variational principle, the Lagrangian is actually the Onsager - Machlup function (1.1), the nonexisting derivative being replaced by a control field. The second representation concernes the way the equilibrium measure is approached by p(x,t)dx. In both results the optimal drift has an appealing physical meaning within Nelson's kinematics. Moreover, both results lend themselves naturally to computational methods.   The paper is outlined as follows. In Section 2 we collect some basic results and concepts from Nelson's kinematics of stochastic processes. Section 3 contains our main results: we obtain a variational path-integral representation for $p^{1/2}$ and $(p/\bar{p})^{1/2}$. For simplicity, we consider first the case of nondegenerate diffusion processes with constant diffusion coefficent.The variational principle for $p^{1/2}$ is extended to general diffusions in Section 4.

2. Preliminaries.

Consider a probability space $(\Omega, F, \mathbf{P})$ and let $\{F_t ; 0 \le t \le T]$ be an increasing filtration of sub-σ-algebras of F. Let $x(t)$ be an n-dimensional, mean-square continuous, $\{F_t\}$-adapted stochastic process defined in the interval $[0,T]$.   According to Nelson [11], the process $x(t)$ is said to be *mean-forward differentiable* with respect to the filtration (F_t) if the limit

$$D_+x(t) := \lim_{h \to 0^+} E\left\{\frac{x(t+h)-x(t)}{h} \ / \ F_t\right\} \qquad t \in (0,T)$$

exists and forms a continuous curve in $L_n^2(\Omega, F, \mathbf{P})$. In this case, the process

$$x(t) - x(0) - \int_0^t D_+x(s)ds$$

is an (F_t)-martingale [11]. Similarly, if $x(t)$ is adapted to a decreasing family (G_t) of sub-σ-algebras of F, we say that $x(t)$ is *mean-backward differentiable* with respect to (G_t) if

$$D_-x(t) := \lim_{h \to 0^+} E\left\{\frac{x(t)-x(t-h)}{h} \ / \ G_t\right\} \qquad t \in (0,T)$$

exists and forms a continuous curve in $L_n^2(\Omega, F, P)$. Then

$$x(t) - x(T) + \int_t^T D_-x(s)ds$$

is a backward (G_t)-martingale. Suppose now that $x(t)$ is the strong solution of the stochastic differential equation:

(2.1) $$dx(t) = b_+(x(t),t)dt + \sigma \, dw_+(t), \quad x(0) = x_o,$$

where w_+ is a standard Wiener process and x_o is independent of w_+. We assume that the drift b^+ is a continuously differentiable function. By the uniform ellipticity of the generator, the probability density of $x(t)$ is a smooth and everywhere positive solution of the Fokker-Planck equation

(2.2)
$$\frac{\partial p}{\partial t} + \nabla \cdot (b^+ p) - \frac{1}{2} a \, \Delta p = 0, \quad p(x,0) = p_o.$$

where p_o is the density of x_o and $a = \sigma^2$. Nelson considered first the possibility of a backward representation of $x(t)$ .More precisely, this means to find a function $b_-(x,t)$ and $aG_t = \sigma\{x(s): s \geq t\}$-adapted Brownian motion w_- such that $x(t)$ is a solution of the following stochastic differential equation (going backward in time):

(2.3)
$$dx(t) = b_-(x(t),t)dt + \sigma \, dw_-(t).$$

Nelson proved that, if such a representation holds, then necessarily:

$$b_-(x,t) = b_+(x,t) - a\nabla \log p(x,t), \qquad dw_- = dw_+ + \sigma \, \nabla \log p(x(t),t).$$

Carlen [12], and Haussmann and Pardoux [13] recently showed that the process $x(t)$, given as the solution of (2.1), is a weak solution of the S.D.E. (2.3), i.e. for t fixed and f bounded and smooth, the process

$$f(x(t)) - f(x(s)) + \int_s^t L^- f(x(\tau))d\tau \qquad s \leq t$$

is a backward G_t-martingale, where $L^- = b_- \cdot \nabla - \frac{1}{2}\sigma^2\Delta$ .So L^- can be viewed as the *backward generator* of the diffusion process $x(t)$, just like $L^+ = b_+ \cdot \nabla + \frac{1}{2}\sigma^2 \Delta$ is the *forward generator* of $x(t)$ (the result actually holds for general, possibly degenerate diffusions, under weak assumptions). Let $f(x,t)$ be of class $C^{2,1}$. Ito's rule corresponding to (2.1) and (2.3) yields

$$D_+ f(x(t),t) = \frac{\partial f}{\partial t}(x(t),t) + L^+ f(x(t),t),$$
$$D_- f(x(t),t) = \frac{\partial f}{\partial t}(x(t),t) + L^- f(x(t),t).$$

Of course, D_+ and D_- may be identified with the differential operators $\frac{\partial}{\partial t} + L^+$ and $\frac{\partial}{\partial t} + L^-$.

Moreover we have the backward Fokker-Planck

(2.4)
$$\frac{\partial p}{\partial t} + \nabla \cdot (b_- p) + \frac{1}{2} a \, \Delta p = 0.$$

Combining (2.2) and (2.4) we get

(2.5)
$$D^* p = \frac{\partial p}{\partial t} + \nabla \cdot (vp) = 0,$$

where D is given by

$$Df(x(t),t) = [(D_+ + D_-)/2]f(x,t) := \frac{\partial f}{\partial t}(x(t),t) + v \cdot \nabla f(x(t),t).$$

Here $v := (b_+ + b_-)/2$ is the *current velocity* [11]. The partial differential operator D has the form of a substantial derivative, i.e. a derivative along stream lines. (2.5) should be compared to the *continuity equation* of fluid dynamics . As in [7], we define the *local entropy* $S(x,t)$ by

(2.6) $$S(x,t) = -k \log p(x,t),$$

where k is Boltzmann's constant. When x models the macroscopic dynamics of a system coupled to a heat bath, the macroscopic entropy $S(t)$ satisfies Boltzmann's relation

$$S(t) = \int S(x,t)p(x,t)dx = E\{S(x(t),t)\}.$$

Next we assume that x possesses an invariant probability measure with density $\bar{p}$, which corresponds to thermal equilibrium. We define the *local internal energy* by

$$U(x) = - kT \log \bar{p}(x),$$

where T is the absolute temperature of the fluid, and the *local (Helmholtz) free energy* by

$$\psi(x,t) = U(x) - TS(x,t) = kT \log(p/\bar{p})(x,t).$$

When x models the macro-evolution of a thermodynamical system, the internal energy is

$$U(t) = = \int U(x)p(x,t)dx = E\{U(x(t))\},$$

and the Helmholtz free energy is

$$\psi(t) = U(t) - TS(t) = kT \int \log(p/\bar{p})(x,t)\, p(x,t)dx.$$

The latter, neglecting the constant kT, may be recognized as the *Kullback-Leibler pseudo-distance* between the measures pdx and $\bar{p}$ dx [14]. Both $S(x,t)$ and $\psi(x,t)$ may be shown to be value function of stochastic control problems in [7] and [8], respectively. In both cases the (backward) dynamics on the interval [0,t] is

$$dx(s) = u\, ds + \sigma\, dw_- \, , x(t) = x,$$

and an optimal control is the backward drift of $\{x(t)\}$, namely $u^* = b_-$. It is a classical result of thermodynamics [15] that for systems with constant volume under isothermal conditions the function $\psi(t)$ is nonincreasing, and it becomes constant only in equilibrium. The Hamilton-Jacobi-Bellman (H.J.B.) equation

(2.7) $$D_-\psi = -(\sigma^2/2kT) \| \nabla\psi \|^2$$

associated to the variational principle leads to a much stronger local result ,which extended a result proven in [7] for the Ornstein-Uhlenbeck process in phase space. Indeed, it follows that $\psi(x(t),t)$ is a backward submartingale with respect to the natural filtration $G_t = \sigma\{x(s): s \geq t\}$. This gives a local specification of thermodynamical stability, namely

$$D_- U(x(t)) \leq T\, D_- S(x(t),t) , \qquad \text{with probability one.}$$

3.Stochastic variational principles.

In this section we derive variational principles alternative to those mentioned in the previous section, by applying a different increasing concave function to $p(x,t)$ and $p(x,t)/\bar{p}(x)$, i.e. the square root instead of the logarithm.The following theorem describes the way equilibrium is approached.

Theorem 1:*The ratio* $(p(x,t)/\bar{p}(x))^{1/2}$ *is the value function of the following stochastic control problem*

$$(3.1) \qquad (p(x,t)/\bar{p}(x))^{1/2} = \max_{u \in \mathcal{U}} E_{x,t} \left\{ (p_0(x(0))/\bar{p}(x(0)))^{1/2} \exp\left[-\int_0^t \frac{1}{2a} \| \bar{b}_- - u \|^2 ds \right] \right\},$$

$$dx = u\, dt + \sigma\, dw_- \,,$$

where the maximum is attained at $u^* = (b_- + \bar{b}_-)/2$.

Proof: From (2.7), observing that $(p/\bar{p})^{1/2} = \exp(\psi/2kT)$, we have

$$D_-\left((p/\bar{p})^{1/2} \right) = - \frac{3a}{2(2kT)^2} \| \nabla\psi \|^2 (p/\bar{p})^{1/2} \qquad \Rightarrow$$

$$\frac{\partial (p/\bar{p})^{1/2}}{\partial t} + \left(b_- + \frac{a}{2kT}\nabla\psi \right) \cdot \nabla (p/\bar{p})^{1/2} - \frac{1}{2} a\, \Delta(p/\bar{p})^{1/2} = - \frac{a}{2(2kT)^2} \| \nabla\psi \|^2 (p/\bar{p})^{1/2}$$

Observing that $\nabla(p/\bar{p})^{1/2} = \frac{1}{2kT} \nabla\psi\, (p/\bar{p})^{1/2}$, this equation can be rewritten in dynamic-programming form:

$$(3.2) \qquad \frac{\partial (p/\bar{p})^{1/2}}{\partial t} - \frac{1}{2} a\, \Delta(p/\bar{p})^{1/2} = \max_u \left\{ - u \cdot \nabla (p/\bar{p})^{1/2} - \frac{1}{2a} \| b_- - u \|^2 (p/\bar{p})^{1/2} \right\}$$

where the max is attained at $u^* = \bar{b}_- - \frac{a}{2kT} \nabla\psi$. Note that (3.2) has the form of the dynamic-programming equation of a control problem with exponential cost . A verification theorem for (3.2) can be obtained just like in the classical case ([16]). //

The representation (3.1) provides information on the rate at which the measure $p(x,t)dx$ tends to the invariant measure. Numerical methods designed for stochastic control problems may then be employed to compute this rate. A similar variational principle may be established for $p^{1/2}$.

Theorem 2: *The square root of the density satisfies the following variational principle*

$$(3.3) \qquad p(x,t)^{1/2} = \max_{u \in \mathcal{U}} E_{x,t} \left\{ p_0(x(0))^{1/2} \exp\left[- \int_0^t \frac{1}{2a} \| b_+ - u \|^2 + \frac{1}{2} \nabla \cdot b_+ \, ds \right] \right\}$$

$$dx = u\, dt + \sigma\, dw^-$$

and the maximum is achieved at $u^* = v$.

Proof: It readily follows from (2.2) that S satisfies the H.J.B. equation

$$D_- S = \frac{a}{2} \| \nabla S \|^2 + \nabla \cdot b_+$$

Observing that $S = - 2 \log p^{1/2}$, we get

$$\frac{\partial p^{1/2}}{\partial t} + v \cdot \nabla p^{1/2} - \frac{1}{2} a \, \Delta p^{1/2} = - \left[\frac{1}{2a} \left\| \frac{a \, \nabla S}{2} \right\|^2 + \frac{1}{2} \nabla \cdot b_+ \right] p^{1/2}$$

and also

$$\frac{\partial p^{1/2}}{\partial t} - \frac{1}{2} a \, \Delta p^{1/2} = \max_{u} \left\{ - u \cdot \nabla p^{1/2} - \left[\frac{1}{2a} \| b_+ - u \|^2 + \frac{1}{2} \nabla \cdot b_+ \right] p^{1/2} \right\}.$$

The maximum is attained at $u^* = b_- - \frac{1}{2} a \, \nabla S$. We conclude as in Theorem 2. $//$

As for equation (3.1), it seems possible to use (3.3) to solve numerically (2.2).

Remark : The Lagrangian of the control problem (3.3)

$$L(x,u) = - \frac{1}{2a} \| b_+ - u \|^2 - \frac{1}{2} \nabla \cdot b_+$$

can be seen as an Onsager-Machlup functional acting on the space of diffusions having the form

$$dx = u \, dt + \sigma \, dw^-$$

rather than on the space of differentiable paths. Hence (3.3) is a rigorous representation of the density obtained through an Onsager-Machlup functional. Indeed, L(x,u) has the same form of the Lagrangian of the deterministic variational problem giving the Most Probable Path . An alternative representation for the density involving Wiener measure was obtained in [17].

Example : let w(t) be a Wiener process starting at t=0 with a distribution $p_o(x)$. In this case $b_+ = 0$, a = 1, and the invariant measure is Lebesgue measure (which implies $\bar{b}_- = 0$). Hence, Theorems 1 and 2 give the same representation for the density of this process:

$$\sqrt{p(x,t)} = \max_{u \in \mathcal{U}} E_{x,t} \left\{ \sqrt{p_o(x(0))} \exp \left[- \int_0^t \frac{1}{2} \| u \|^2 ds \right] \right\}$$

$$dx = u \, dt + \sigma \, dw_-.$$

4. General nondegenerate diffusions.

In this section we extend Theorem 2 to the case of general nondegenerate diffusion. We refer to [18] for a thorough discussion of diffusion processes on a manifold. Consider the n-dimensional diffusion process defined by the following stochastic differential equation:

$$dx(t) = b^+(x(t),t)\, dt + \sigma(x(t),t)\, dw(t), \quad x(0) = x_0 .$$

Here $\sigma_{ij}(x,t)$ is a $C^{2,1}$-function for any (i,j). We assume that $a(x,t) = \sigma(x,t)^T\sigma(x,t)$ is positive definite for all x and t. Given a positive definite matrix $\alpha \in \mathbf{R}^{n\times n}$, we use the following notation

$$\| v \|_\alpha^2 = \sum_{i\,j}\alpha_{ij}v_i v_j , \qquad v \in \mathbf{R}^n .$$

The backward drift of the process x(t) assumes the more general form

$$b_- = b_+ - p^{-1}\sum_j \frac{\partial a_j p}{\partial x_j} ,$$

where a_j is the jth column of a. Now recall that $a^{-1}(x,t)$ may be used as a time-varying metric tensor to define a Riemannian geometry on $\mathbf{R}^n$. Let

$$|\sigma(x,t)| := \det(a(x,t))^{1/2}, \quad \mathfrak{b}_i := b_i - \tfrac{1}{2}|\sigma|\sum_j \frac{\partial(a_{ij}/|\sigma|)}{\partial x_j} , \quad \nabla_t.f := |\sigma|\sum_i \frac{\partial(f/|\sigma|)}{\partial x_i}$$

where f is a differentiable function. The latter expression is the divergence with respect to the metric induced by $a^{-1}(x,t)$, whereas $\mathfrak{b}$ is the natural drift of x(t) associated with the Riemannian structure. In fact, if Δ denotes the Laplace-Beltrami operator on the manifold M, the generator of x(t) can be written as

$$\mathfrak{b}\cdot\nabla_t + \tfrac{1}{2}\,\Delta$$

Furthermore, we have the measure on $\mathbf{R}$ $d\xi = |\sigma|^{-1}dx$. Thus $q(x,t) = |\sigma(x)|p(x,t)$ is the density with respect to the measure $d\xi$. It soves the Fokker-Planck equation:

$$(4.1)\qquad \frac{\partial q}{\partial t} + \nabla_t \cdot (\mathfrak{b}\, q) - \tfrac{1}{2}\,\Delta q = 0, \quad q(x,0) = q_0(x) := |\sigma(x)|p_0(x).$$

Let u(x,t) be a smooth field, and let $x^u(t)$ be the solution of the backward s.d.e.

$$dx^u(t) = u(x^u(t),t)dt + \sigma(x^u(t))dw_-(t).$$

Then the backward generator of $x^u(t)$ can be written in the form

$$\mathfrak{u} \cdot \nabla_t - \frac{1}{2}\Delta, \qquad \text{where} \quad \mathfrak{u}_i := u_i + \frac{1}{2}|\sigma| \sum_j \frac{\partial(a_{ij}/|\sigma|)}{\partial x_j} \ .$$

Theorem 3: $q(x,t)$ admits the following representation:

$$(4.2) \qquad q^{1/2}(x,t) = \max_{u \in \mathfrak{U}} E\left\{ q_0^{1/2}(x^u(0))\exp\left[-\frac{1}{2}\int_0^t \left(\|b - \mathfrak{u}\|_a^2 1 + \nabla_t \cdot b \right) ds \right] \right\}$$

where $x^u(s)$ solves the backward equation:

$$dx^u(s) = u(x^u(s),s)ds + \sigma(x^u(s))dw_-(s)$$

$$x^u(t) = x$$

The optimal control is given by $\mathfrak{u}^*_i = b_i - \frac{1}{2}p^{-1}\sum_j \frac{\partial a_{ij}p}{\partial x_j}$, that is the *current velocity* of

$x(t)$. In other words $q(x,t)$ solves the continuity equation

$$\frac{\partial q}{\partial t} + \nabla_t \cdot (\mathfrak{u}^* q) = 0$$

The proof of Theorem 3 may be found in [19]. It rests on a long manipulation of (4.1) in order to derive the appropriate Hamilton-Jacobi-Bellman equation for $q^{1/2}$ in analogy to the proof of Theorem 2. The Lagrangian (4.2) has essentially the same form of the Onsager-Machlup function for general diffusions [2],[6], with two main differences. First, the Lagrangian is evaluated on diffusion processes given as solutions of backward stochastic differential equations, rather than on differentiable paths. The derivative of the path is replaced by the mean backward derivative (backward drift) of the diffusion process. Secondly, (4.2) does not contain a scalar curvature term.

REFERENCES

[1] L.Onsager and S.Machlup, Fluctuations and Irreversible Processes, *Physical Review* 91, 1953, 1505-1515.

[2] R.Graham, Path integral formulation of general diffusion processes, *Zeitschrift fur Physik B* 26, 1977, 281-290.

[3] K.Yasue, A simple derivation of the Onsager-Machlup formula for one-dimensional nonlinear diffusion processes, *J. Math. Phys.* 19, 1978, 1671-1673.

[4]   K.Yasue, The role of the Onsager-Machlup Lagrangian in the theory of stationary diffusion processes, *J. Math. Phys.* 20, 1979, 1861-1864.

[5]   D.Dürr and A.Bach, The Onsager-Machlup Functional as Lagrangian for the most probable path of a diffusion process, *Comm. Math. Physic* 60, 1978, 153-170.

[6] Y.Takahashi ans S.Watanabe, The probability functionals (Onsager-Machlup functions) of diffusion processes, Springer-Verlag, *Lecture Notes in Mathematics* 851, 1980, 433-463.

[7] M.Pavon, Stochastic Control and Nonequilibrium Thermodynamical Systems, *Applied Mathematics and Optimization* (to appear).

[8] F.Guerra and M.Pavon, Stochastic Variational Principles for Dissipative Processes, *Proceeding 1987 MTNS Conference* , Phoenix, Arizona (to appear).

[9] P.Dai Pra and M.Pavon, Variational path-integral representations for the density of a diffusion process, *Stochastics* (in press).

[10] W.H.Fleming and R.W.Rishel, *Deterministic and Stochastic Optimal Control* , Springer-Verlag, 1975.

[11] E.Nelson, *Dynamical Theories of Brownian Motion* , Princeton University Press, 1967.

[12] E.Carlen, Conservative diffusions, *Comm.Math.Phys.* 94 ,1984, 293-315.

[13] U.G.Haussmann and E.Pardoux, Time reversal of diffusions, *The Annals of Probability* 14, 1986, 1188-1205.

[14] S.Kullback, *Information Theory and Statistics* , Dover, New York, 1968.

[15] H.B.Callen, *Thermodynamics* , Wiley, New York, 1960.

[16] P.R.Kumar and J.H.van Schuppen, On the optimal control of stochastic systems with an exponential-of-integral performance index, *J. Math. Anal. Appl.* 80, 1981, 312-332.

[17] V.E.Benes and L.A.Shepp, *Theor. Prob. Appl.* 13, 1968, 475.

[18] N.Ikeda and S.Watanabe, *Stochastic Differential Equations and Diffusion Processes* , North-Holland, Amsterdam, 1981.

[19] P.Dai Pra and M.Pavon, A new approach to the Feynman-Kac formula and related results, preprint June 1988, submitted for publication.

Paolo Dai Pra
Mathematics Department
Rutgers University
New Brunswick, NJ 08903
USA

Michele Pavon
Dipartimento di Elettronica e Informatica
Universita' di Padova
and
LADSEB-CNR
35100 Padova
ITALY

CONSTITUTIVE EQUATIONS FOR MASONRY-LIKE MATERIALS

Anna Maria Gennai*, Cristina Padovani*

Introduction

It is well-known that the constitutive equation of elastic materials not supporting tension has a unique solution ([2],[5]). For isotropic materials it is easy to calculate the solution for the constitutive equation, as the material's elastic constants and the deformation tensor vary. On the contrary,the study of anisotropic materials is more complex. The aim of this paper is to propose a method for calculating the solution for transversely isotropic materials ,which can be used in the case of plane strain state.

Notation

Let $\mathcal{V}$ the three-dimensional inner product space.

Let $\{e_1,e_2,e_3\}$ be an orthonormal basis of $\mathcal{V}$,fixed once and for all;if $v \in \mathcal{V}$,the quantities $v_i = v \cdot e_i$, $i=1,2,3$ are the components of v with respect to the basis $\{e_1,e_2,e_3\}$. Let us write $v \leqslant 0$ $(v \geqslant 0)$ provided $v_i \leqslant 0$, $i=1,2,3$ $(v_i \geqslant 0$, $i=1,2,3)$. We denote by Lin the space of all linear transformations on $\mathcal{V}$.

Moreover let us denote by Sym the subspace of Lin consisting of all symmetric tensors;by Sym^+ and and Sym^- the subsets of Sym constituted,respectively,by positive semidefinite and negative semidefinite tensors.

$A,B \in$ Sym are said to be coaxial if there is an orthonormal basis $\{u_1,u_2,u_3\}$ of $\mathcal{V}$ such that

$$A=\sum_{i=1}^{3} a_i u \otimes u \quad \text{and} \quad B=\sum_{i=1}^{3} b_i u_i \otimes u_i \ .$$

where $\otimes$ indicates the tensor product. It is known ([8])

that A and B are coaxial if and only if they commute, e.g. AB=BA. As usual we consider Lin with the inner product defined by $A \cdot B = tr(A^t B)$, $A, B \in$ Lin, and we say that A and B are orthogonal if $A \cdot B = 0$. In what follows, if $A \in$ Sym we shall write $A \geqslant 0$, (resp. $A \leqslant 0$) provided $A \in Sym^+$ (resp. $A \in Sym^-$). Let us denote by $\mathscr{C}^*$ the subset of the set of fourth order tensors, consisting of symmetric and positive definite elements. It is clear that the elements of $\mathscr{C}^*$, restricted to Sym, are invertible.

1. The formulation of the problem

We shall consider the linearized strain tensor E and we shall suppose E to be the sum of an elastic part E^e and an inelastic part E^a. We shall also assume that the Cauchy-stress tensor T is linearly dependent on the elastic deformation E^e by means of the elasticity tensor $\mathbb{C}$ which will be always supposed to be symmetric and positive definite. The inelastic part of the deformation is further characterized by the relations $E^a \geqslant 0$, $E^a \cdot T = 0$; finally, the hypothesis of no reaction to tractions is expressed by the condition $T \leqslant 0$.

Given $E \in$ Sym and $\mathbb{C} \in \mathscr{C}^*$, we then intend to determine explicitly two tensors, $T, E^a \in$ Sym, satisfying the following relations:

$$(1.1) \quad \begin{cases} T = \mathbb{C} \, [E - E^a] \\ T \cdot E^a = 0 \\ \\ E^a \geqslant 0 \\ T \leqslant 0 \end{cases}$$

The existence of the solution of the problem (1.1) is

confirmed by a classical theorem of convex analysis ([3]) and the solution is unique because $\mathbb{C}$ is positive definite. The aim of the following paragraphs is to give a method to construct the solution of the problem (1.1) for isotropic materials and in one case of transversely isotropic material which is compatible with a plane strain state.

2. The isotropic case

This section deals with the solution of the problem (1.1) in the case in which T is isotropically dependent on the deformation's elastic part E^e . (1.1)$_1$ thus becomes

$$(2.1) \qquad T = 2\mu E^e + \lambda\, tr(E^e)I \qquad ,$$

where λ and μ are the Lame' moduli of the material satisfying ([8])

$$(2.2) \quad \mu > 0 \ , \ 2\mu + 3\lambda > 0,$$

since $\mathbb{C} \in \mathcal{C}^*$. First we shall prove that, given $E \in$ Sym and the numbers λ and μ verifying the relations (2.2), to find T and E^a satisfying (1.1) is equivalent to resolving a linear complementarity problem, whose solution's existence and unicity are assured by a known optimization theorem. Successively we shall determine the explicit espressions of the inelastic deformation E^a as E, λ , μ vary. Let us begin by proving a preliminary result whose proof is omitted.

Proposition 2.1

Let $A \in$ Sym$^+$ and $B \in$ Sym$^-$ be such that $A \bullet B = 0$.

Then $AB = BA = 0$ (A and B are coaxial).

From this proposition and from (1.1)$_2$,(1.1)$_3$ and (1.1)$_4$ it follows that T and E^a are coaxial. On the other hand, in virtue of the material's isotropy,we have $0=TE^a-E^aT=\mu(EE^a-E^aE)$, $TE-ET=-\mu(E^aE-EE^a)=0$, and therefore T,E,E^a are coaxial.

Let us now consider the orthonormal basis of $\mathcal{V}$, $\{u_1,u_2,u_3\}$ consisting of eigenvectors of E,so that $E=\sum_{i=1}^{3}e_i\,u_i\otimes u_i$; it is also possible to set $T=\sum_{i=1}^{3}t_i\,u_i\otimes u_i$, $E^a=\sum_{i=1}^{3}a_i\,u_i\otimes u_i$, $E^e=\sum_{i=1}^{3}(e_i-a_i)\,u_i\otimes u_i$, where $t_i,a_i,i=1,2,3$ are quantities to be determined. The conditions $E^a\geqslant 0$, $T\leqslant 0$ and $T\cdot E^a=0$ are equivalent to have, respectively $a_i\geqslant 0$, $t_i\leqslant 0$ $i=1,2,3$ and $\sum_{i=1}^{3}a_it_i=0$. At this point,considering the vectors $\mathbf{t}=(t_1,t_2,t_3)$ $\mathbf{a}=(a_1,a_2,a_3)$, $\mathbf{e}=(e_1,e_2,e_3)$ constituted by the principal stresses,inelastic deformations and total deformations respectively,from (2.1) we obtain $\mathbf{t}=D(\mathbf{e}-\mathbf{a})$, where the matrix D having components $d_{ij}=\lambda$ if $i\neq j$, $d_{ii}=2\mu+\lambda$ is positive definite by (2.2). Therefore we can reformulate the problem (1.1) as follows :

Given the principal deformations $\mathbf{e}=(e_1,e_2,e_3)$ and the matrix D, to find two vectors $\mathbf{t}$ and $\mathbf{a}$ so that

$$(2.3)\quad\begin{cases}\mathbf{e}=D^{-1}\,\mathbf{t}+\mathbf{a}\\[4pt]\mathbf{t}\leqslant 0\\[4pt]\mathbf{a}\geqslant 0\\[4pt]\mathbf{t}\cdot\mathbf{a}=0\end{cases}$$

holds. This last is a linear complementarity problem which admits only one solution because the D matrix is positive definite ([4]). Since $(2.3)_4$ holds if and only if $a_i\,t_i=0$, $i=1,2,3$, to resolve the problem (2.3) is equivalent to find a solution (a_1,a_2,a_3) of the system

$$a_i[2\mu(e_i-a_i)+\lambda(e_1+e_2+e_3-a_1-a_2-a_3)]=0 \ , \ i=1,2,3$$

which satisfies the conditions $a_i \geqslant 0$ and

$$2 \mu (e_1 - a_1) + \lambda (e_1 + e_2 + e_3 - a_1 - a_2 - a_3) \leqslant 0 \ , \ i=1,2,3.$$

The solution of the problem (2.3) is given in the Table 2.1 (where $\alpha = \lambda/\mu$), which represents a subdivision of $\mathcal{V}$ in eight regions; once one has determined the region to which $\mathbf{e}$ belongs, one can read the corresponding principal values of the inelastic deformation $\mathbf{a}$.

3. A transversely-isotropic case

Let us examine a transversely-isotropic material having elastic constants ν_1 , E_1 and ν_2 , E_2 associated respectively with the behaviour in the x-z plane and with the behaviour in the direction y normal to this plane ([9]). Let us assume that the constants ν_1 and ν_2 are positive and they satisfy the following inequalities

$$(3.1) \quad \nu_2^2 < 1/n \ , \ 1 - \nu_1 - 2n\nu_2^2 > 0 \ ,$$

where $n = E_1/E_2$. Moreover let us consider the plane deformation E with $e_{ij} = 0$ if $i=2$ or $j=2$. The constitutive law is expressed by the relations ([9]) :

$$e_{11} = t_{11}/E_1 - \nu_2 t_{33}/E_2 - \nu_1 t_{22}/E_1 + a_{11} \quad ,$$

$$e_{22} = -\nu_1 t_{11}/E_1 - \nu_2 t_{33}/E_2 + t_{22}/E_1 + a_{22} = 0 \quad ,$$

$$e_{33} = -\nu_2 t_{11}/E_2 + t_{33}/E_2 - \nu_2 t_{22}/E_2 + a_{33} \quad ,$$

$$(3.2)$$

$$2e_{12} = 2(1+\nu_1)t_{12}/E_1 + 2a_{12} = 0 \quad ,$$

$$2e_{13} = t_{13}/G_2 + 2a_{13} \quad ,$$

$$2e_{23} = t_{23}/G_2 + 2a_{23} = 0 \ ,$$

where $t_{ij}, a_{ij}, i,j=1,2,3$ are the components of tensors T and E^a respectively, and where we have set $G_2 = E_2/[2(1-\nu_2)]$. It is easy to prove that

$$(3.3) \quad t_{12} = t_{23} = a_{12} = a_{23} = a_{22} = 0$$

hold. In fact from the proposition 2.1 it follows that $T\widetilde{E}=0$ and therefore

$$(3.4) \qquad t_{12}a_{12}+t_{23}a_{23}+t_{22}a_{22}=0 \quad .$$

This relation,$(1.1)_3$ and $(1.1)_4$ imply

$$(3.5) \qquad t_{12}a_{12}+t_{23}a_{23} \geqslant 0 \quad .$$

The hypothesis follows from relations $(3.2)_2,(3.2)_4,$ $(3.2)_6$ (3.4) and (3.5),taking into account the positiveness of the quantities $2(1+\nu_1)/E_1$ and G_2. In virtue of $(3.2)_1$, $(3.2)_2$, $(3.2)_3$ and $(3.2)_5$ from (3.3) the relation $t = C(e-x)$, follows,where we have set $e=(e_{11},e_{33},2e_{13})$, $x=(a_{11},a_{33},2a_{13})$, $t=(t_{11},t_{33},t_{13})$ and where the matrix C having components $c_{13}=c_{23}=0$ and

$$c_{11}=nE_2(1-n\nu_2^2)/[(1+\nu_1)(1-\nu_1-2n\nu_2^2)] \quad ,$$

$$c_{22}=E_2(1-\nu_1)/(1-\nu_1-2n\nu_2^2) \qquad ,$$

$$c_{33}=E_2/[2(1-\nu_2)] \qquad ,$$

$$c_{12}=n\nu_2 E_2/(1-\nu_1-2n\nu_2^2) \qquad ,$$

is positive definite because of (3.1). Moreover we have from $(3.2)_2$, $t_{22}=\nu_1 t_{11}+n\nu_2 t_{33}$. From this last relation and from the hypothesis of the positiveness of ν_1 and ν_2, it follows that when t_{11} and t_{33} are both non positive, t_{22} is also non positive. Solving the problem (1.1) is equivalent to determining a vector x which satisfies

$$[C(e-x)]_1 \leqslant 0$$

$$[C(e-x)]_2 \leqslant 0$$

$$(3.6) \qquad [C(e-x)]_1[C(e-x)]_2-[C(e-x)]_3^2 \geqslant 0 \quad .$$

$$x_1 \geqslant 0$$

$$x_2 \geqslant 0$$

$$x_1 x_2-x_3^2/4 \geqslant 0$$

$$C(e-x)\cdot x = 0$$

The problem (3.6) has a unique solution because it is equivalent to the problem (1.1). Let us consider the functional $\Omega(x)=(x\cdot Cx)/2 - x\cdot Ce$ defined on the domain $\mathcal{D}=\{\ x\in V\ :\ x_1\geqslant 0\ ,\ x_2\geqslant 0\ ,\ x_1x_2-x_3^2/4\geqslant 0\ \}$. Because the domain is closed and convex and the quadratic form

$$x \longrightarrow x\cdot Cx$$

is symmetric ,coercive and continuous,the functional Ω has a unique minimum $x\in\mathcal{D}$. In virtue of a John' s theorem ([6]),there are $\theta\in\mathbb{R}$ and $1\in\mathbb{R}^3$ such that

$$(3.7)\quad\begin{cases} \theta\ \mathrm{grad}\ \Omega(x)-1\mathrm{gradh}(x)=0 \\[4pt] 1\cdot h(x)=0 \\[4pt] 1\geqslant 0 \\[10pt] \theta\geqslant 0 \\[4pt] (\theta,1)\neq 0 \\[4pt] h(x)\geqslant 0 \end{cases}$$

where $h(x)=(x_1,x_2,x_1x_2-x_3^2/4))$. The following theorem relates the solution of the problem (3.6) to the solution of the system of inequalities (3.7).

<u>Theorem 3.1</u>

i) let x be the solution of (3.6), with $x\neq 0$, thus there are $\theta\in\mathbb{R}$ and $1\in\mathbb{R}^3$ such that $(x,\theta,1)$ is a solution of (3.7).

ii) Vice versa if $(x,\theta,1)$ with $x\neq 0$ is a solution of (3.7),then x is the solution of (3.6).

We now propose to calculate explicitly all solutions of (3.6) as $e,\gamma_1,E_1,\gamma_2,E_2$ vary. We can observe that (3.6) admits the zero solution if and only if the inequalities $(3.6)_1$, $(3.6)_2$, $(3.6)_3$) are verified with $x=0$. We can then suppose that x=0 does not verify these inequalities. By means of the theorem 3.1 the solution of (3.6) can be obtained as the solution of the problem (3.7). If we suppose $l_3=0$, we obtain the first, the second and the third solution of the table 3.1. Let us now suppose $l_3> 0$. Easy calculations show that the supposition $(l_1, l_2) \neq 0$ leads to values of x already determined in the case $l_3=0$. Then by supposing $l_1 = l_2 = 0$ we obtain the solution five and six of the table 3.1, where we have set $C = c_{11}e_{11}+c_{12}e_{33}$, $B = c_{12}e_{11}+c_{22}e_{33}$,

$$K= -\frac{C}{\sqrt{c_{11}c_{22}}} + \sqrt{\frac{C^2}{c_{11}c_{22}} + \frac{16c_{33}^2 e_{13}^2 c_{11}}{\sqrt{c_{11}c_{22}}(2c_{33}+c_{12}+ \sqrt{c_{11}c_{22}})^2}} \, .$$

and $A=c_{11}c_{22}-c_{12}^2$. The multiplier α_3 is a root of the polynomial $P(y)=a_4y^4+a_3y^3+a_2y^2+a_1y+a_0$ with coefficients

$a_4=BC-4c_{33}^2 e_{13}^2$,

$a_3=4BCc_{33}+e_{33}AB+e_{11}AC+16c_{12}c_{33}^2 e_{13}^2$,

$a_2=4BCc_{33}^2 +e_{11}e_{33}A^2+4c_{33}e_{33}AB+4c_{33}e_{11}AC+$

$+8Ac_{33}^2 e_{13}^2 -16c_{12}^2 c_{33}^2 e_{13}^2$,

$a_1=4c_{33}(e_{11}e_{33}A^2+c_{33}e_{33}AB+c_{33}e_{11}AC-4Ac_{12}c_{33}e_{13}^2)$,

$a_0=4c_{33}^2 A^2(e_{11}e_{33}-e_{13}^2)$.

The existence and unicity of the solution of (3.6) and the theorem 3.1 guarantee that if the data belong to the region R_6, complementary of $R = \bigcup_{i=1}^{5} R_i$, where

$R_1=\{e: e_{11} \geqslant 0 , e_{33} \geqslant 0 , e_{11}e_{33}-e_{13}^2 \geqslant 0 \}$

$R_2=\{e : e_{33} \leqslant 0 , e_{11}+c_{12}e_{33}/c_{11} \geqslant 0 , e_{13}=0 \}$

$R_3=\{e : e_{11} \leqslant 0 , e_{33}+c_{12}e_{11}/c_{22} \geqslant 0 , e_{13}=0 \}$

$R_4=\{e: C \leqslant 0 , B \leqslant 0 , CB-4c_{33}^2 e_{13}^2 \geqslant 0 \}$

$R_5=\{e : c_{11}B+\sqrt{c_{11}c_{22}} C=c_{22}C+\sqrt{c_{11}c_{22}} B=0 \}$

there is a positive root α_3 of the polynomial $P(y)$ so that

237

the sixth (x_1,x_2,x_3) given in the table 3.1 is the solution
of (3.7).

Table 2.1

$2(1+\alpha)e_1+\alpha e_3 \geq 0$	$a_1=e_1+\alpha e_3/[2(1+\alpha)]$	$(2+\alpha)e_1+\alpha e_2+\alpha e_3 \leq 0$	$a_1=0$
$2(1+\alpha)e_2+\alpha e_3 \geq 0$	$a_2=e_2+\alpha e_3[2(1+\alpha)]$	$\alpha e_1+(2+\alpha)e_2+\alpha e_3 \leq 0$	$a_2=0$
$e_3 \leq 0$	$a_3=0$	$\alpha e_1+\alpha e_2+(2+\alpha)e_3 \leq 0$	$a_3=0$
$\alpha e_2+2(1+\alpha)e_1 \leq 0$	$a_1=0$	$e_1 \geq 0$	$a_1=e_1$
$2(1+\alpha)e_2+\alpha e_1 \leq 0$	$a_2=0$	$e_2 \geq 0$	$a_2=e_2$
$(2+\alpha)e_3+\alpha(e_1+e_2) \geq 0$	$a_3=e_3+\alpha(e_1+e_2)/(2+\alpha)$	$e_3 \geq 0$	$a_3=e_3$
$(2+\alpha)e_1+\alpha(e_2+e_3) \geq 0$	$a_1=e_1+\alpha(e_2+e_3)/(2+\alpha)$	$e_1 \leq 0$	$a_1=0$
$\alpha e_3+2(1+\alpha)e_2 \leq 0$	$a_2=0$	$2(1+\alpha)e_2+\alpha e_1 \geq 0$	$a_2=e_2+\alpha e_1/[2(1+\alpha)]$
$2(1+\alpha)e_3+\alpha e_2 \leq 0$	$a_3=0$	$2(1+\alpha)e_3+\alpha e_1 \geq 0$	$a_3=e_3+\alpha e_1/[2(1+\alpha)]$
$\alpha e_1+2(1+\alpha)e_3 \leq 0$	$a_1=0$	$2(1+\alpha)e_1+\alpha e_2 \geq 0$	$a_1=e_1+\alpha e_2/[2(1+\alpha)]$
$2(1+\alpha)e_1+\alpha e_3 \leq 0$	$a_2=e_2+\alpha(e_1+e_3)/(2+\alpha)$	$e_2 \leq 0$	$a_2=0$
$(2+\alpha)e_2+\alpha(e_1+e_3) \geq 0$	$a_3=0$	$2(1+\alpha)e_3+\alpha e_2 \geq 0$	$a_3=e_3+\alpha e_2/[2(1+\alpha)]$

Table 3.1

$e \in R_1$	$x_1=e_{11}$ $x_2=e_{33}$ $x_3=2e_{13}$	$e \in R_4$	$x_1=0$ $x_2=0$ $x_3=0$
$e \in R_2$	$x_1=e_{11}+\dfrac{c_{12}e_{33}}{c_{11}}$ $x_2=0$ $x_3=0$	$e \in R_5$	$x_1=\dfrac{K\sqrt{c_{11}c_{22}}}{2c_{11}}+\dfrac{C}{r_{11}}$ $x_2=\dfrac{K}{2}$ $x_3=\dfrac{4c_{33}e_{13}}{2c_{33}+c_{12}+\sqrt{c_{11}c_{22}}}$
$e \in R_3$	$x_1=0$ $x_2=e_{33}+\dfrac{c_{12}e_{11}}{c_{22}}$ $x_3=0$	$e \in R_6$	$x_1=\dfrac{e_{11}A+\alpha_3B}{A-\alpha_3^2+2c_{12}\alpha_3}$ $x_2=\dfrac{e_{33}A+\alpha_3C}{A-\alpha_3^2+2c_{12}\alpha_3}$ $x_3=\dfrac{4c_{33}e_{13}}{2c_{33}+\alpha_3}$

References

[1] Abruzzese, D.; Grimaldi, A.; Sacco, E. , Indagine numerica su alcuni problemi di materiali non resistenti a trazione , Rapporto n.14 Dipartimento di Ingegneria Civile Edile II Universita' di Roma,Scienza e Tecnica delle Costruzioni , 1987.

[2] Anzellotti, G.: "A Class of non Coercive Functionals and Masonry-like Materials" , Ann. Inst. H.Poincare', $\underline{2}$ (1985)

[3] Cea, J.: Lectures on Optimization.Theory and Algoritms Tata Institute of Fundamental Research , Bombay , 1978 ;

[4] Cottle,R.W.; Giannessi, F.; Lions, J.L.: Variational Inequalities and Complementarity Theory and Applications , Wiley ,1980 ;

[5] Del Piero, G.: Private Communication , 1987 ;

[6] Giannessi, F.: Metodi matematici della programmazione. Metodi lineari e non lineari , Quaderni dell' U.M.I. n. 23 , 1982 ;

[7] Giaquinta, M.; Giusti, G.: "Researches on the Equilibrium of Masonry Structures" , Arch. Rat. Mech. Analysis , $\underline{88}$ (1985)

[8] Gurtin, M.E.: An Introduction to Continuum Mechanics , Academic Press,1981

[9] Zienkiewicz, O.C.: The Finite Element Method , Mc Graw-Hill , 1977 .

* Istituto CNUCE (CNR) , Via Santa Maria,36

56100 Pisa ITALY

AMICS:A MULTIFUNCTIONAL ASSISTENT FOR STATE ACCOUNTING QUERIES

G.A.M. Perrotta, C. Squitieri
Italsiel S.p.A. - PAC1 Branch

1. Introduction

Using new processing systems often leads to problems concerning user´s capability to understand and easely use the new system he has to work with. Most of the times training courses are insufficient to override such problems, for they should enable people to communicate with computers by using a new restricted language so different from the natural speaking one. This leads to pay more attention on ´how´ to query the system rather than ´what´ to ask it.

For such reasons many A.I. researching areas have been studying human ways of communication, trying to automize such processes. As a consequence, new ´friendly´ systems have been created to facilitate data extraction and processing.

Following these new attempts, AMICS is a sw/hw system explicitly created to consult budget information stored on mainframe. It is actually used by the Italian Treasury Department and has been developed by Italsiel and Tecsiel s.p.a, both companies belonging to the IRI Finsiel Group.

AMICS main facilities are:

* decision support tool

* easy use and approach

* fast and efficient processing

* communication with mainframe host
  containing relational databases

Such objectives have been realized implementing a sort of ´electronic desk´ working on an IBM PC/AT connected to a mainframe host. The system is able to retrieve and postelaborate information using a) two interconnected data-retrieval systems that permit query specification by menus or natural language (Italian) and b) a postelaboration system allowing specific data displays, graphics, statistics and formatted printouts.

Natural language querying is usefull especially for users whose knowledge about the State budget structure is not too technical ,but who require fast answers to specific requests; on the other hand, menu-driven interaction is more adeguate for people who must analyse different data types contemporary (in these cases natural language queries would be too slow to formulate).
Anyway, the only knowledge requested is budget information, while no technical skill is needed.

2. The working domain

Amics databases contain month and year historical accounting data of the Italian State budget refered to the last five years. Common users are people working for the National political economy or experts in economics who are interested in previsional models.
The State budget includes information such as:

* State expenditures for institutions

* money destination

* State institutional objectives

    etc...

It is possible to analyse the National income trend and expenditure cycle by requesting accounting data related to the single budget items or intermingled viewpoints. In fact, budget data is information organized in relational databases where logical schema of budget items and viewpoints are represented by an hierarchical structure of codes.

3. AMICS architecture

AMICS architecture is based on four principal functions:

* user´s interface

* system action planning

* data manipulation

* output management

A supervisor system (Driver) activates each function
by request and without any predetermined order. In fact,
Driver provides transition from one state to another on
the bases of user´s choices, functional constraints and
component return codes.
    Data manipulation and output management include five
modules: interface processing, table browser, graphics
manager, statistics package and printing manager. These
components use one single memory environment (´Mailbox´)
to unify parameter using.

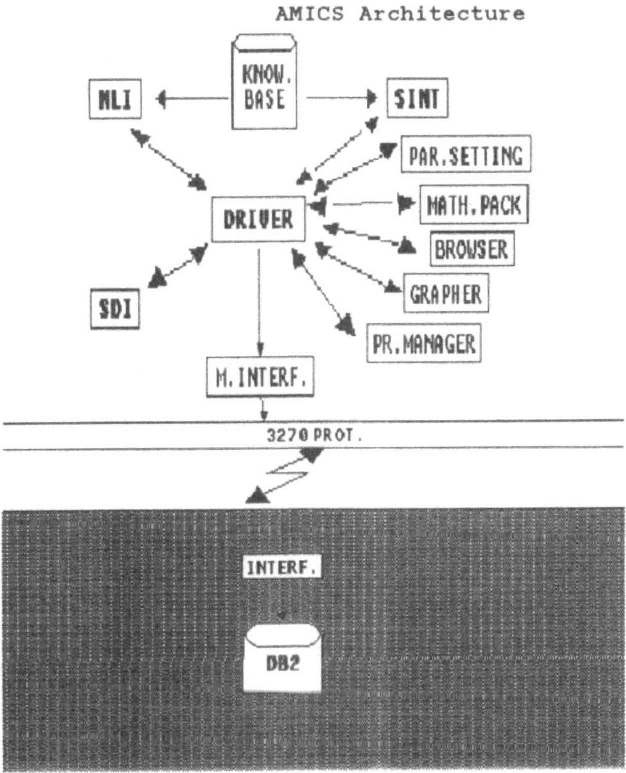

AMICS Architecture

    User´s interfacing   is realized by   other two
components: NLI (Natural Language Interface) and SDI

(System Driven Interface), while action planning is performed by SINT, an SQL command builder.

NLI performs a ´primary´ grammatical and semantic analysis for simple and frequent requests. It includes also a ´nonstandard´ treatment for sentence error recovery, for it is a selfdiagnostic expert system and a therapeutic planner.

SDI component is a menu-driven system that uses windowing techniques and its use requires familiarity with the budget data structure. As in NLI, the user must choose to work in the expendure domain or in the revenue one. A series of interconnected menus guide him to formulate his request, specifying data organization conditions (e.g. appropriation of Treasury Department) and temporal/numerical ones on accounting data (e.g. which year, month or numerical limit the data are refered to).

As soon as meaningful information is gathered, a window appears on the screen to verify at each step the actual query formulation. The outline of a specific request could look like the following:

```
SHOW            appropriation
RELATIVE TO     Dept = 4
OF              year = 1985
WITH            payments > 1000 billions
```

The information specified can be divided into four types: the first one contains information about accounting data, the second what the accounting data must refer to, while the third and the fourth respectively temporal conditions and numerical constraints.

Once the request is made, its correctness is checked and is translated into SQLIF, a formal language processed by the SINT component.

In fact both NLI and SDI cooperate with SINT, an SQL command planner supporting Universal relational system. User´s queries arrive to Sint in SQLIF and are then translated into sequences of SQL ´SELECT´ commands. These are sent to the mainframe host, where they are executed, and the retrieved data is sent back to the user´s working station, ready for statistical manipulation, output management, graphic and table generation.

On the whole, as natural language interface represents one of AMICS main feature, the system modules involved in such implementation (NLI, SINT) are considered the most important.

4. Natural Language Interface: NLI

NLI subsytem accepts user´s queries in natural languages (Italian), analyses then for morphological/ semantic correctness and transforms them into SQLIF commands sent to the SINT component.

In fact, first of all NLI analyses the morfological structure of the request: each word is searched in a vocabolary containing the roots of Italian words necessary to formulize budget queries. A vocabolary entry contains the following information:

* the root of the word it refers to

* its ending rules (noun) and/or
  archetypical pattern (verb)

* category (the semantic role of the
  word in the domain)

optionally:

* semantic value (proper noun)

* ´abstract´ attribute (see further details)

* type of the connective or comparative
  operator

* etc...

Once words have been recognized, NLI creates one or more and/or semantic graphes, using a set of context-free grammar rules. Rules contain information about constraints and the functional role of the semantic category in the sentence; they are of the type:

$$< NT_k > \quad -----> \quad \{< NT_j >\} \; T_k \; \{< NT_m >\}$$

The T (terminal symbols) are the mentioned semantic categories and the NT (non terminal symbols) are structured groups of semantic categories representing a specific semantic role in the sentence.

Each rule has an associated function used for SQL translation and it is evaluated with a bottom-up strategy on the semantic tree.

After all possible semantic trees are built, they are presented to the user as paraphrases of the original request, with the help of a natural language generator. When the correct interpretation is choosen, the corresponding SQLIF translation is sent back to the SINT module which is immediately activated.
NLI module can detect different kinds of query errors (so called ´non-standard´ situations): meanless words, structural or functional errors, anaphoras and ellipses. In such cases the expert system, which is self-diagnostic, is able to provide different responses according to the particular situation it has to handle. In fact it can send diagnostic messages ,or interact with the user for further details ,or automatically solve the odd situation if it exactly knows what caused it.

Natural Language Processing

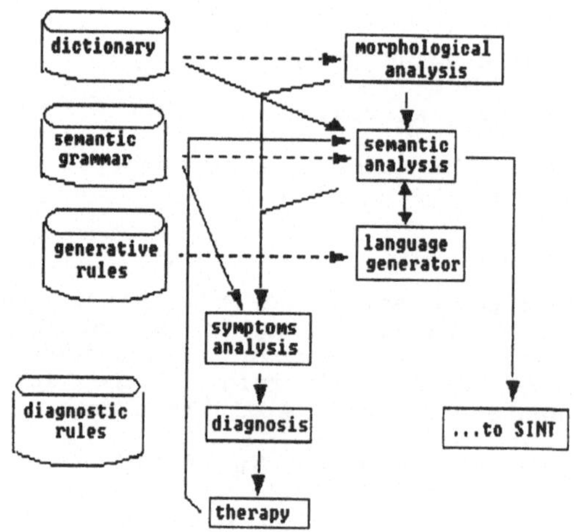

Non-standard treatment is assigned to the TNS module (´Trattamento non standard´), which is activated by primary analysis as soon as symptoms are known, such as morfological ones (e. g unknown word) or interpretation ones (e.g. unpossible function attachment).When semantic errors are revealed, the sentence is instead segmented and primary analysis is applied to each segment for sympton research.

TNS treats word ambiguity creating different copies of the input sentence refered to all possible combinations. For each copy the TNS does the following steps:

      a) sympton detection to find out which
         rule has failed
      b) diagnostics
      c) recovery ('therapy')

As TNS is a self-diagnostic expert system, it is composed of an inferential engine, a knowledge base system (production rules) and a blackboard memory. In fact, the inferential engine uses a depth first strategy on the knowledge base, containing rules able to generate diagnostic trees, and uses also a blackboard memory where the current state is always memorized. Each rule is related to a particular therapy module associated to the diagnostic tree (at present the TNS draws only one diagnostic tree for each sentence, even if there could be more than one odd situation in a sentence). Terminal nodes of a tree are possible symptoms, while non-terminal nodes are error situations able to generate subtrees.

When a diagnostic tree is generated, recoverable situations are distinguished from unrecoverable ones. The former are submitted to primary analysis, which makes all possible interpretations and sends SINT the one choosen by the user. The latter instead cause the display of diagnostic messages, and explanations can be request by on-line help.

5. Intelligent database interface: SINT

SINT component receives the analysed user's request transformed into SQLIF language and translates it into SQL 'SELECT' commands, corresponding to DBMS actions. In this way the user's abstract attributes are related to their corresponding database real attributes directly or indirectly, according to the amount of information in the SQLIF request. That is why the user is not obliged to know what the data base relations are, but only how logical notions are interweaved with each other. Missing information and soundness of the generated SELECTs are also controlled by SINT.

For example, if we consider the two following examples:

1) **"What are the initial fund appropriations
   of the Treasury department?"**
   ---> DAMMI (stanziamento iniziale di competenza)
        (MINISTERO = tesoro)

2) **"What are the payments of the
   departments?"**
   ---> DAMMI (pagato) (MINISTERO = tutti)

The system retrieves directly the attribute ´payment´ if it exists in the db (example 1.), otherwise it is able to calculate it (example 2.) on the ground of the virtual relations the attribute has with the real ones.

Another important feature is SINT´s capability of adeguating the extensional meaning of the words to the user´s particular depth knowledge. For this reason AMICS consists of two knowledge bases, relative to the user´s different specification level: synthetic and analytic. Such level is choosen interactively at the beginning of a working session, so that SINT can relate the user´s attributes with the real ones depending on his specific point of view.

SINT knowledge base system consists of a set of frame-like structures and production rules for the SQL command planning activity. In fact, rules are used in order to:

    a) manage frame-like structures able to
       organize information about abstract
       attributes

    b) manage ´from´ fields and ´join´ con-
       ditions of SELECT commands

    c) introduce attributes in the SELECT
       commands for more readable output

    d) plan sequences of SELECT commands if
       more than one relation is involved

    e) insert default conditions when a
       request is uncomplete.

6. Conclusions.

AMICS has been in use since March 1987 and up to now it has been well accepted by users, for it reduces the cost and difficulty of transferring accounting

information from the data processing offices to the top
manager offices.
    Actually new developments are made to let the system
handle   other   types   of   natural   language   requests
concerning:

* information about semantics and features
  of the domain

* information about the system capabilities
  (metarequest)

* mathematical and statistical processing
  on extracted data

References

[1]   A.Bisseret "Psycology   for   man   computer
         cooperation in knowledge processing", in
         Mason (ed.) "Information processing 83",
         North-Holland, 1983
[2]   E.Rich "Natural language interfaces"
         Computer Sep. 1984
[3]   A.K.Joshi "Varieties of cooperative
         responses in Question-Answer systems" in
         Kiefer "Questions and Answers" Riedel
         Publishing Company 1983
[4]   P.J.Hayes, J.G.Carbonell "Multy-strategy
         parsing and its role in robust man-machine
         communication", CMU May 1981
[5]   D.Maier et al., "Toward a logical data
         independence: a relational query language
         without relations", Proceedings 1982 Int.
         Conf. on VLDB, 1983
[6]   D.Maier, J.D.Ullman, M.J.Vardi, "On the
         foundation of the universal relation
         databases", ACM TODS, 9,2,1984
[7]   M.Jarke, Y.Vassiliou, "A framework for
         choosing a database query language",
         ACM Comp. Surv., 17,3,1985
[8]   A.D´Altri and F.Ricci "L´interpretazione
         d´interrogazioni ambigue su basi di dati
         statistiche", Proceedings of Giornata di
         studi su basi di dati statistici, Roma
         1987

TOWARDS A THEORY OF
NONLINEAR STOCHASTIC SYSTEMS

L. Arnold, Bremen

Summary: We describe certain concepts related to the multiplicative ergodic theorem which enable us to develop a genuine theory of nonlinear stochastic systems, in particular a stochastic bifurcation theory.

1. Introduction

Numerous applications in science and engineering lead to the following situation: Let

$$(1.1) \qquad \dot{x} = f(x,\alpha), \quad x(0) = x_0 \in IR^d,$$

be a nonlinear differential equation whose right-hand side depends on a parameter α. Often this parameter is perturbed by noise. In this case (1.1) turns into a stochastic differential equation

$$(1.2) \qquad \dot{x} = f(x,\alpha+\sigma\xi(t)), \quad x(0) = x_0 \in IR^d,$$

where $\xi(t)$ is some stationary stochastic process (possibly white noise, see section 2), and σ is a strength parameter. Typically, a steady state of (1.1) will survive as a stationary solution $x^0(t)$ of (1.2). The following questions arise naturally:

(i)   Does $x^0(t)$ inherit the stability behaviour from the original steady state? Can noise stabilize an unstable steady state?

(ii)   When is (1.2) chaotic in the sense that (almost) every point of the state space is a saddle point (so-called hyperbolicity)? This means in particular that the system exhibits sensitive dependence on initial conditions. Let us mention that today hyperbolicity is the only mathematical explanation of chaotic behaviour.

(iii)   How are the bifurcation scenarios modified in the presence of noise? In what direction does noise shift the critical parameter value? Is there a stochastic center manifold? Are there stochastic normal forms? What kind of solutions bifurcate if $\sigma \neq 0$?

We will discuss and partly answer these questions in the following sections. Let us emphasize that the basic technique for tackling the above problems in the deterministic case is as follows: One linearizes (1.1) at the steady state x^0,

$$(1.3) \qquad \dot{v} = A(\alpha)v,\ v(0) = v_0 \in {I\!R}^d,\ A(\alpha) = \frac{\partial f}{\partial x}(x^0,\alpha) = \text{Jacobian matrix},$$

and tries to draw conclusions from the linear system (1.3) about the original system (1.1). The key fact is that a matrix has eigenspaces and eigenvalues. The dynamical relevance of this fact for (1.3) is that they provide invariant subspaces and exponential growth rates for its solutions. Those invariant subspaces for (1.3) have their nonlinear versions for (1.1) (e.g. the stable/unstable/center manifolds) which are fundamental objects of nonlinear dynamics.

We claim that it is possible to transfer the above approach to the stochastic case. We try to explain this to the non-specialist and thus put our emphasis on concepts and instructive examples rather than on completely listing all assumptions under which our statements are true.

2. Stochastic flows

Remember that an ordinary differential equation (1.1) generates, by assigning to each initial value x the solution $\varphi(t;x)$ at time $t \in {I\!R}$, a flow of diffeomorphisms (smooth bijections) $(\varphi(t))_{t \in {I\!R}}$ of the state space ${I\!R}^d$, "flow" meaning

$$\varphi(t+s) = \varphi(t) \cdot \varphi(s) \quad \text{for all } t,s \in {I\!R}.$$

Consider the following three classes of stochastic dynamical systems:

White noise case:

$$(2.1a) \qquad dx(t) = f_0(x(t))dt + \sum_{i=1}^{m} f_i(x(t)) \cdot dW_i(t),\ x(0) = x \in {I\!R}^d,\ t \in {I\!R}$$

("$\cdot$" means Stratonovich stochastic integral).

250

<u>Real noise case:</u>

(2.1b) $\dot{x}(t) = f(x(t),\xi(t,\omega))$, $x(0) = x \in IR^d$, $t \in IR$

$(\xi(t,\omega) = $ (ergodic) stationary stochastic process$)$.

<u>Discrete time case   (products of random mappings):</u>

(2.1c) $x(t+1) = f(x(t),\xi(t,\omega))$, $x(0) = x \in IR^d$, $t \in \mathbb{Z}$

$(\xi(t,\omega) = $ (ergodic) stationary sequence$)$.

All three cases are particular examples of the following general scheme:

Given a probability space (Ω,P) and a flow $\vartheta = (\vartheta_t)_{t \in T}$, $T = IR$ or $\mathbb{Z}$, of measure preserving bijections of Ω, i.e. $\vartheta_0 = $ id, $\vartheta_{t+s} = \vartheta_t \cdot \vartheta_s$. Assume for simplicity that ϑ is ergodic. In the cases (2.1b) and (2.1c) just use $\Omega = $ set of sample paths of $\xi(t,\omega)$, $P = $ distribution of $\xi(\cdot,\omega)$ on Ω, $\vartheta_t = $ shift defined by $\vartheta_t\omega(\cdot) = \omega(\cdot + t)$. In the white noise case (2.1a) use $\Omega = \{\omega: IR \to IR^m$ continuous, $\omega(0) = 0\}$, $P = $ Wiener measure, and $\vartheta_t\omega(\cdot) = \omega(\cdot + t) - \omega(t)$.

For more details for the real noise case see Crauel ([11],[12]), for the white noise case see Arnold [1] and Boxler [8], chapter 2.

The basic observation is now as follows:

By assigning to each initial value $x \in IR^d$ the solution $\varphi(t,\omega,x)$ of (2.1) at time $t \in T$, $x \to \varphi(t,\omega,x)$ we define a random mapping $\varphi(t,\omega)$ of IR^d into itself which, under certain conditions, is a random diffeomorphism of IR^d. Moreover, with P–probability 1 (2.1) generates a <u>cocycle</u> (or <u>flow</u>) of diffeomorphisms of IR^d over ϑ, i.e. we have the cocycle property

$$\varphi(t+s,\omega) = \varphi(t,\vartheta_s\omega) \cdot \varphi(s,\omega) \quad \text{for all } s,t \in T .$$

On $\Omega \times IR^d$, $(\omega,x) \to (\vartheta_t\omega,\varphi(t,\omega,x)) = \Theta_t(\omega,x)$ defines a <u>(skew–product)flow.</u>

Given a probability space (Ω,P) and a flow $\vartheta = (\vartheta_t)_{t\in T}$, $T = \mathbb{R}$ or $\mathbb{Z}$, of measure preserving bijections of Ω, i.e. $\vartheta_0 = \mathrm{id}$, $\vartheta_{t+s} = \vartheta_t \cdot \vartheta_s$. Assume for simplicity that ϑ is ergodic. In the cases (2.1b) and (2.1c) just use $\Omega = $ set of sample paths of $\xi(t,\omega)$, $P = $ distribution of $\xi(\cdot,\omega)$ on Ω, $\vartheta_t = $ shift defined by $\vartheta_t\omega(\cdot) = \omega(\cdot + t)$. In the white noise case (2.1a) use $\Omega = \{\omega: \mathbb{R}\to\mathbb{R}^m \text{ continuous}, \omega(0) = 0\}$, $P = $ Wiener measure, and $\vartheta_t\omega(\cdot) = \omega(\cdot + t) - \omega(t)$.

<u>Example: Ornstein-Uhlenbeck flow</u>

Let $A \in \mathbb{R}^{d\times d}$, $B \in \mathbb{R}^{d\times m}$ and

$$(2.2) \qquad dx(t) = Ax(t)dt + BdW(t), \quad x_0 = x \in \mathbb{R}^d .$$

The cocycle generated by (2.2) is given by the affine mappings

$$x \to \varphi(t,\omega,x) = e^{tA}x + g(t,\omega), \quad g(t,\omega) = \int_0^t e^{(t-s)A}BdW(s).$$

If $\mathrm{Re}\,\lambda_i(A) \neq 0$ for all i, there exists a unique invariant μ on $\Omega \times \mathbb{R}^d$. It takes the form

$$\mu(d\omega,dx) = \delta_{x_0(\omega)}(dx)\, P(d\omega) ,$$

where

$$x_0(\omega) = (x_0^-(\omega),x_0^+(\omega)) = \left(\int_{-\infty}^0 e^{-sA^-}(BdW)^-, -\int_0^{-\infty} e^{-sA^+}(BdW)^+ \right).$$

Here we have assumed w.l.o.g. that $A = \mathrm{diag}\,(A^-,A^+)$, where A^- and $-A^+$ are stable, and we write $BdW = ((BdW)^-,(BdW)^+)$. The process $\varphi(t,\omega,x_0(\omega))$ is stationary on $\mathbb{R}$, but only if x_0^+ is not present the distribution ρ of x_0 solves the Fokker–Planck equation for (2.2) on $\mathbb{R}^+$.

More examples of stochastic flows, also on manifolds, can be found in Arnold [1].

3. The multiplicative ergodic theorem

Given the situation of section 2: $\varphi(t,\omega)$ is a cocycle over ϑ, and μ an invariant probability.

We linearize (differentiate) the diffeomorphism $\varphi(t,\omega)$ at $x \in IR^d$:

$$T\varphi(t,\omega,x): T_x IR^d \rightarrow T_{\varphi(t,\omega,x)} IR^d .$$

The identification $T\,IR^d \cong IR^d \times IR^d$ makes $T\varphi(t,\omega,x)$ a cocycle of linear isomorphisms of IR^d over the base flow Θ_t on $\Omega \times IR^d$, i.e.

$$T\varphi(t+s,\omega,x) = T\varphi(t,\Theta_s(\omega,x)) \cdot T\varphi(s,\omega,x) \quad \text{for all } t,s \in IR, \ x \in IR^d .$$

Of course, the linearized flow $v(t) = T\,\varphi(t,\omega,x)$ is generated by the linearized version of (2.1), for example in case (2.1b) by

$$\dot{v}(t) \quad = A(\xi(t,\omega), \varphi(t,\omega,x))v(t), \ v(0) = v \in IR^d ,$$

$$A(\xi,x) = \frac{\partial f}{\partial x} (\xi,x) .$$

The <u>Lyapunov exponent</u> of the linear cocycle $T\varphi(t,\omega,x)$ at $x \in IR^d$ under $\omega \in \Omega$ in the tangent direction $v \in IR^d$, $v \neq o$, is defined by

$$\lambda(\omega,x,v) = \lim_{t \to \infty} \frac{1}{t} \log \|T\varphi(t,\omega,x)\| .$$

<u>Theorem 3.1</u>   (<u>Multiplicative ergodic theorem of Oseledec</u> [14]). Given a cocycle φ over ϑ and an ergodic invariant probability μ . Assume

$$\log^+ \sup_{0 \leq t \leq 1} \|T\varphi(t,\omega,x)^{\pm 1}\| \ \in \ L^1(\Omega \times IR^d,\mu) .$$

Then there is a Θ_t–invariant subset $\Gamma \subset \Omega \times IR^d$ with $\mu(\Gamma) = 1$, r real numbers

$\lambda_1 > ... > \lambda_r$ $(1 \leq r \leq d)$ and naturals $d_i, \overset{r}{\underset{1}{\Sigma}} d_i = d$, such that for all $(\omega,x) \in \Gamma$:

(i)   *There is a measurable invariant splitting*

$$IR^d = E_1(\omega,x) \oplus \ldots \oplus E_r(\omega,x), \ \dim E_i(\omega,x) = d_i ,$$

$$E_i(\Theta_t(\omega,x)) \ = \ T\varphi(t,\omega,x) \, E_i(\omega,x) .$$

(ii)   *The Lyapunov exponent exists for each* $v \neq 0,$ *and*

$$\lambda(\omega,x,v) = \lim_{t\to\pm\infty} \frac{1}{t} \, \log\|T\varphi(t,\omega,x)v\| \ = \ \lambda_i \ \text{ iff } \ v \in E_i(\omega,x) .$$

<u>Remark.</u>   The Lyapunov exponents λ_i are independent of (ω,x) and thus global constants (real parts of "eigenvalues") of the cocycle under μ. In contrast, the $E_i(\omega,x)$, the stochastic analogues of the eigenspaces, depend in general on (ω,x). The "message" of Theorem 3.1 is that the situation for a cocycle is as nice as for a deterministic steady state.

4. Invariant manifolds, stochastic stability, hyperbolicity

We would like to have nonlinear analogues $M_i(\omega,x)$ of the invariant linear subspaces $E_i(\omega,x)$. More generally define for $\varnothing \neq \Lambda \subset \{\lambda_1,\ldots,\lambda_r\}$

$$E_\Lambda(\omega,x) = \bigoplus_{\lambda_i \in \Lambda} E_i(\omega,x), \ \dim E_\Lambda(\omega,x) = \sum_{\lambda_i \in \Lambda} d_i = d_\Lambda .$$

If Λ consists of all $\lambda_i < 0$ ($>0, =0$) we obtain the stable (unstable, center) subspace $E_s(\omega,x)$ ($E_u(\omega,x)$, $E_c(\omega,x)$).

This is still work in progress. Here is a general statement.

<u>Theorem 4.1</u> <u>(invariant manifolds)</u>. Assume the situation of Theorem 3.1. Then (under further conditions) for each $(\omega,x) \in \Gamma$ *there is a smooth submanifold* $M_\Lambda(\omega,x)$ *of* IR^d *tangent to* $E_\Lambda(\omega,x)$ *at* x *which is invariant w.r.t.* φ:

$$\varphi(t,\omega)M_\Lambda(\omega,x) = M_\Lambda(\Theta_t(\omega,x)) \ .$$

For more details see the forthcoming Ph.D.thesis of Dahlke [13]. The stable (unstable, center) manifolds $M_s(\omega,x)$ $(M_u(\omega,x), M_c(\omega,x))$ are particular cases. A rather complete center manifold theory has been developed by Boxler [8] (see also Arnold and Boxler ([2],[3])).

Theorem 4.2 (*stochastic stability*). *Assume the situation of Theorem 3.1. Suppose* $\lambda_1 < 0$. *Then the cocycle* φ *is* *stable* *in the following sense: For each* $(\omega,x) \in \Gamma$ *the stable submanifold* $M_s(\omega,x)$ *is an open neighborhood of* x *such that for all* $y \in M_s(\omega,x)$

$$\|\varphi(t,\omega,x) - \varphi(t,\omega,y)\| \to 0 \quad (t\to\infty)$$

exponentially fast. If $\lambda_1 > 0$ *then* φ *is unstable.*

See Carverhill [9] and Ruelle [15] for the particular case of a μ coming from the Fokker-Planck equation.

The main task of nonlinear stochastic stability theory is, therefore, to check whether $\lambda_1 < 0$ and estimate the (random!) domain of attraction $M_s(\omega,x)$ of a solution starting at x.

Example: Consider the scalar differential equation

$$(4.1) \qquad \dot{x} = \xi(t,\omega)x + (1 + \xi(t,\omega))x^2 \ ,$$

where the stationary ergodic real noise $\xi(t,\cdot)$ satisfies $-1 < a \le \xi(t,\omega) \le b, \ E\xi(t,\omega) = \lambda < 0$. Since $x = 0$ is a solution of (4.1) the measure

$$\mu(d\omega,dx) = P(d\omega) \otimes \delta_0(dx)$$

is invariant and ergodic. The Lyapunov exponent λ_1 of (4.1) corresponding to this measure is the exponential growth rate of the linearized equation $\dot{v} = \xi(t,\omega)v$, i.e.

$$\lambda_1 = \lim_{t\to\infty} \frac{1}{t} \log|v(t,\omega)| = \lim_{t\to\infty} \frac{1}{t} \int_0^t \xi(s,\omega)ds = E\xi = \lambda < 0.$$

By Theorem 4.2 there is a stable manifold $M_s(\omega)$ of $x = 0$. Since (4.1) can be solved explicitly, $M_s(\omega)$ turns out to be

$$M_s(\omega) = (-\infty, \beta(\omega)) \, , \quad \beta(\omega) = \frac{1}{\int_0^\infty (1+\xi(t,\omega)) \exp\left(\int_0^t \xi(s,\omega)ds\right)dt} \, .$$

Observe that $M_s(\omega)$ depends on the whole future of $\xi(\cdot,\omega)$. The probability

$$p_y = P(y \in M_s(\omega)) = P(y < \beta(\omega))$$

that the solution of (4.1) starting at y is exponentially attracted by $x = 0$ is equal to 1 if $y \le 0$ and can be made arbitrarily close to 1 if $y > 0$ is small enough.

For hyperbolicity all we have to check is whether $\lambda_r < 0 < \lambda_1$. This is "generically true" (see Baxendale [7], Carverhill [10], Arnold and Kliemann [4]).

5. Stochastic bifurcation

It is now quite obvious what the "correct" definition of stochastic bifurcation should be: Assume $\sigma \neq 0$ in (1.2). Then the invariant probability $\mu = \mu(\alpha)$ and the corresponding Lyapunov exponents $\lambda_1(\alpha) > ... > \lambda_r(\alpha)$ depend on α. The simple observation that the $\lambda_i(\alpha)$ are the stochastic analogues of Re $\lambda_i(A)$ opens the door to stochastic bifurcation theory.

We will expect bifurcation of new (stationary and other) solutions if α moves in such a way that $\lambda_1(\alpha)$ (say) changes sign at a critical value α_0. Near bifurcation point it suffices to study the cocycle on the center manifold (which exists, see section 4 and Boxler [8]). There it can be transformed into normal form, see the forthcoming Ph.D. thesis of Xu [16].

This approach to stochastic bifurcation is in sharp contrast to the one studied so far by physicists who have mainly looked for qualitative changes of $\mu(\alpha)$.

<u>**Example   (stochastic pitchfork bifurcation):**</u>

Consider the scalar differential equation

(5.1) $\dot{x} = (\alpha + \xi(t,\omega))x - x^3$

which, without noise, undergoes a pitchfork bifurcation at $\alpha = 0$. Assume $E\xi(t,\omega) = 0$.

Since $x = 0$ solves (5.1) for all $\alpha \in \mathbb{R}$, (5.1) has the invariant probability $\mu(d\omega,dx) = P(d\omega) \otimes \delta_0(dx)$ which does not depend on α. The corresponding Lyapunov exponent is

$$\lambda_1 \;=\; \lim_{t\to\infty} \frac{1}{t} \int_0^t (\alpha + \xi(s,\omega))ds \;=\; \alpha\,,$$

i.e. it changes sign at $\alpha = 0 =$ deterministic bifurcation point. (5.1) can be explicitly solved and yields for $x \neq 0$

$$\varphi(t,\omega,x) \;=\; \pm\, \frac{\exp \displaystyle\int_0^t (\alpha+\xi(s,\omega))ds}{\sqrt{\dfrac{1}{2x^2} + \displaystyle\int_0^t \exp\!\left(2\int_0^s (\alpha+\xi(u,\omega))du\right)ds}}\,.$$

For $\alpha < 0$ $\varphi(t,\omega,x) \to 0$ $(t\to\infty)$ exponentially fast for every x, i.e. the stable manifold of $x = 0$ is $\mathbb{R}^1$. For $\alpha = 0$ a more detailed analysis gives again $\varphi(t,\omega,x) \to 0$. For $\alpha > 0$ $x = 0$ is unstable, and its unstable manifold is

$$M_u^\alpha(\omega) = (-x_0^\alpha(\omega), x_0^\alpha(\omega)),\;\; x_0^\alpha(\omega) \;=\; \frac{1}{\sqrt{2\displaystyle\int_{-\infty}^0 \exp\!\left(2\int_0^s (\alpha+\xi(u,\omega))du\right)ds}}\,.$$

Since $M_u^\alpha(\omega)$ is invariant, i.e. $\varphi(t,\omega)M_u^\alpha(\omega) = M_u^\alpha(\vartheta_t\omega)$, $\pm x_0^\alpha(\omega)$ are invariant, i.e.

$$\varphi(t,\omega,\pm x_0^\alpha(\omega)) \;=\; \pm x_0^\alpha(\vartheta_t\omega)\,.$$

In other words, $\pm x_0^\alpha(\omega)$ are the initial values of two new stationary solutions. Since for

$\alpha \downarrow 0 \quad x_0^\alpha(\omega) \to 0$ with probability 1, those solutions bifurcate from the trivial one at

$\alpha = 0$. We thus have the familiar pitchfork scenario, where now two new stable stationary solutions (= invariant probabilities) bifurcate from the original one.

Stochastic Hopf bifurcation requires the concept of rotation number in higher dimensions (see Arnold and San Martin [5]).

References

[1]   Arnold, L.: Lyapunov exponents of nonlinear stochastic systems. In: Ziegler, F.; Schueller, G.I. (eds.): Nonlinear stochastic dynamic engineering systems. IUTAM Symposium Innsbruck 1987, Springer 1988, 181-201

[2]   Arnold, L.; Boxler, P.: Eigenvalues, bifurcation and center manifolds in the presence of noise. In: Dafermos, C.; Ladas, G. (eds.): Proceedings of EQUADIFF 87, M. Dekker, New York 1988

[3]   Arnold, L.; Boxler P.: Elimination of fast variables in the presence of noise: stochastic center manifold theory. In: Velarde, M. (ed.): Synergetics, Order and Chaos, World Scientific, Singapore 1988

[4]   Arnold, L.; Kliemann, W.: Large deviations of linear stochastic differential equations. In: Engelbert, H.; Schmidt, W. (eds.): Stochastic Differential Systems, Springer Lecture Notes in Control and Information Sciences Vol. 96, 117-151 (1987)

[5]   Arnold, L.; San Martin, L.: A multiplicative ergodic theorem for rotation numbers. Dynamics and Differential Equations 1 (1988)

[6]   Arnold, L.; Wihstutz, V. (eds.): Lyapunov exponents, Springer Lecture Notes in Mathematics Vol. 1186 (1986)

[7]   Baxendale, P.: Lyapunov exponents and relative entropy for a stochastic flow of diffeomorphisms. Preprint, University of Aberdeen (1986)

[8]   Boxler, P.: Stochastische Zentrumsmannigfaltigkeiten. Ph.D. thesis Universität Bremen (1988)

[9]   Carverhill, A.: Flows of stochastic dynamical systems: ergodic theory. Stochastics 14 (1985), 273-317

[10]   Carverhill, A.: Furstenberg's theorem for nonlinear stochastic systems. Prob. Th. Rel. Fields 74 (1987), 529-534

[11] Crauel, H.:   Lyapunov exponents and invariant measures of stochastic systems on manifolds, in [6], 271-291

[12] Crauel, H.:   Random dynamical systems: positivity of Lyapunov exponents and Markov systems. Ph.D. thesis Universität Bremen (1987)

[13] Dahlke, S.:   Invariante Mannifaltigkeiten für stochastische Systeme. Ph.D.thesis Universität Bremen (1989)

[14] Oseledec, V.I.:   A multiplicative ergodic theorem. Lyapunov characteristic numbers for dynamical systems. Trans. Moscow Math. Soc. 19 (1968), 197-231

Ludwig Arnold
Institut für Dynamische Systeme
Universität Bremen
D-2800 Bremen 33
Fed. Rep. of Germany

STOCHASTIC AGGREGATION

Giorgio Picci[†]

Dipartimento di Elettronica e Informatica
Universita' di Padova
Via Gradenigo 6/A, 35131 Padova, Italy

ABSTRACT. This paper surveys recent work by the author on a theory of Aggregation of Dynamical Systems by Stochastic models. Aggregation is a concept relevant to various areas of Engineering, Physics (in particular to Statistical Mechanics) and Economics. The mathematical formulation proposed here owes to diverse areas of Applied Mathematics such as Systems Theory (in particular Stochastic Realization Theory), Topological Dynamics, Ergodic Theory, Hamiltonian Mechanics and general wisdom from the basic paradigms of Statistical Mechanics. The development of the Theory is yet in a very preliminary stage although some very interesting facts have already emerged. Here we discuss especially necessary conditions for aggregability in a general nonlinear context. In the linear case much more specific results are available.

1. Introduction and Basic Definitions

There are two basic ingredients in a Dynamic Aggregation Problem,

— **The microscopic dynamics.** This is to be thought of as the model of a (proverbially "very large") aggregate of interacting systems, possibly describing particles, economic units, subsystems or a large network etc. The aggregate is collectively described by an ordinary differential equation

$$\dot{z}(t) = F(z(t)) \quad z(t) \in M \tag{1.1}$$

evolving on a smooth phase space M. We could take $M = \mathbf{R}^N$ but more generally (1.1) could be a local description of a flow $\Phi(t)$ on a smooth N-dimensional manifold. We shall always assume that the flow is defined for all $t \in \mathbf{R}$.

The variable $z(t)$ is called the **microscopic state** of the system at time t.

— **The observables.** Normally $z(t)$ is not accessible to direct observation and, in any case, it is not the variable of immediate interest for describing the physical phenomenon

---

[†] On leave at the Department of Mathematics, Arizona State University, Tempe, AZ 85287

being modeled. The relevant variables which are immediately meaningful to describe the physics or the economy of the system are some m (with $m < N$ and usually $m << N$) functions of the microscopic state, called the observables,

$$y_k(t) = h_k(z(t)) \quad k = 1 \ldots m \tag{1.2}$$

We shall make the assumption that the dynamic equation (1.1) is "completely coupled" to the system of observables $y_1 \ldots y_m$ in the sense that there is no smooth smaller dimensional microscopic state dynamics which can also be used to represent the observables. To this purpose we introduce the notion of *irreducibility*. Hereafter by a *smooth dynamical system* Σ we shall always mean a triple $(M, \Phi(t), h)$ where $\Phi(t)$ is a flow on a smooth manifold M and h is a smooth function $M \to \mathbf{R}^m$.

DEFINITIONS 1.1

a) *We say that* Σ *"covers"* $\hat{\Sigma}$ *(notation:* $\Sigma > \hat{\Sigma}$*), or* Σ *is* **reducible** *and* $\hat{\Sigma}$ *is a* **reduction** *of* Σ *if there is a smooth submersion* $\pi : M \to \hat{M}$ *intertwining the flows* Φ *and* $\hat{\Phi}$*, i.e.* $\pi\Phi(t) = \hat{\Phi}(t)\pi$ *and making the diagram*

$$
\begin{array}{ccc}
M & \xrightarrow{\ h\ } & \mathbf{R}^m \\
{\scriptstyle \pi}\downarrow & \nearrow & \\
\hat{M} & &
\end{array}
\tag{1.3}
$$

commutative.

b) *The Dynamical system* Σ *is* **irreducible** *if* $\Sigma > \hat{\Sigma}$ *implies* $\hat{\Sigma} > \Sigma$*. In this case it can be shown that the map* π *is actually a* **diffeomorphism***. Systems* $\Sigma, \hat{\Sigma}$ *related by the commutative diagram* (1.3) *where* π *is a diffeomorphism (intertwining the flows* $\Phi, \hat{\Phi}$*) are called* **equivalent***.*

c) *A weakening of irreducibility is obtained by defining* **weak covering** *as in* (a) *the only difference being that* π *is only required to map an* **open dense subset** $\tilde{M} \subset M$ *onto* $\hat{M}$*.* **Weak irreducibility** *is defined accordingly.* **Weakly equivalent** *systems are related by a commutative diagram like* (1.3) *where* π *is a diffeomorphism of an open dense subset of* M *onto an open dense subset of* $\hat{M}$*.* ∎

Irreducibility means in particular that there is no smaller dimensional differential equation

$$\dot{\hat{z}}(t) = \hat{F}(\hat{z}(t)) \ , \ \hat{z}(t) \in \hat{M}$$

$\hat{z}(t) = \pi(z(t))$, describing the observables (1.2) also as,

$$y_k(t) = \hat{h}_k(\hat{z}(t)) \quad k = 1 \ldots m.$$

In other words, the microscopic dynamics (1.1) brought in to describe the dynamics of the observables was taken of minimal complexity. All the degrees of freedom are necessary to determine the temporal evolution of $y_1, \ldots, y_m$.

In the following we shall assume that the system (1.1), (1.2) is irreducible.

We wold now like to **define** the **aggregation problem** for the dynamical system (1.1–1.2) as that of finding a simpler, in fact the simplest possible, dynamical model sufficient to describe the time evolution of the macroscopic variables $y_1, \ldots, y_m$. Ideally, we would like to represent the temporal evolution of the observables $y_1(t), \ldots, y_m(t)$ **exactly** (We will not be interested in approximations in this paper) by an **aggregate model** of much simpler structure; specifically, by a phase space of much smaller dimension than N in (1.1) (1.2).

In view of the irreducibility postulated above, this hope is clearly illusory unless the meaning of the term "aggregate model" is somehow broadened up to allow for something different than deterministic ordinary differential equations.

The problem can be made meaningful by introducing **probabilistic** representations of the observables. In order to do this we shall have to introduce a few more concepts.

DEFINITION 1.2

Assume there is an invariant probability measure μ for the flow $\Phi(t)$ defined by the differential equation (1.1). The quadruple $(M, \Phi(t), h, \mu)$ is called a randomization *of the (deterministic) system (1.1)(1.2). A randomization is* smooth *if μ is absolutely continuous and* singular *if it is singular with respect to Lebesgue measure.*

Randomization is a typical device of Statistical Mechanics. In this context a smooth randomization may correspond to an invariant distribution on the phase space of the underlying microscopic Hamiltonian system of the **canonical** type. A singular randomization would occur instead when the system is endowed with a microcanonical distribution (a measure concentrated on a surface of constant energy). See e.g. [7] p. 110.

We can assume without loss of generality that the support of any smooth invariant measure μ is the whole state manifold M. Indeed $M_0 := \text{supp}(\mu)$ is always an invariant set for the flow generated by (1) (because of continuity of the solutions with respect to initial data) and thus, in case M_0 is a proper subset, the randomization can for all purposes be considered as obtained from the subsystem induced by (1.1)(1.2) on the invariant set M_0.

Note that a smoothly randomized dynamical system as defined above can also be viewed as a "Classical" Dynamical System in the sense of [1] with the adjuction of output variables (observables).

The Observables of a randomized system become a stationary **stochastic process** $y : \mathbf{R} \times M \to \mathbf{R}^m$ defined on the probability space $\{M, \mathcal{B}, \mu\}$ (here $\mathcal{B}$ is the Borel σ-algebra of subsets of M) as

$$y(t, z) := h(\Phi(t)z) \tag{1.4}$$

Note that in this context $\Phi(t)$ plays just the role of the measure preserving transformation in the theory of stationary processes (see e.g. [13] p. 149).

The following definition introduces a very general class of stochastic models.

DEFINITION 1.3

A **Stochastic Dynamical System** *consists of*

- *a probability space* $\{\Omega, \mathcal{A}, \mu\}$
- *A Markov process* $x = \{x(t)\}$ *defined on* $\{\Omega, \mathcal{A}, \mu\}$ *with values in some measurable space* $(X, \mathcal{X})$ *called the* **state space** *of the system*
- *A Borel function* $c : x \to \mathbf{R}^m$.

The $\mathbf{R}^m$*-valued stochastic process y with*

$$y(t) = c(x(t)) \tag{1.5}$$

is called the **output** *process while x is the* **state** *process of the system.*

A stochastic system is stationary *if there is a measure preserving group* $\Phi(t) : \Omega \to \Omega$ *with respect to which* x *(and hence* y*) is a stationary process.*

The system is finite dimensional *if* X *is a finite dimensional manifold with* $\mathcal{X}$ *the relative Borel* σ*-algebra; dim* X *is called the* order *or* dimension *of the system. A finite dimensional system is* smooth *if* x *is a diffusion process (possibly degenerate) ([3]) and* c *is a smooth map.*

The state process of a smooth system with say $X = \mathbf{R}^n$ can be represented by a stochastic differential equation of the form

$$dx(t) = f(x(t))dt + \sigma(x(t))dw(t) \tag{1.6}$$

where $\{w(t)\}$ is a p-dimensional Wiener process (see [2] [5]).

If $\{x(t)\}$ is *Gaussian* and *stationary* the above simplifies to

$$dx(t) = Ax(t)dt + Bdw(t) \tag{1.7}$$

where A and B are $n \times n$ and $n \times p$ real matrices. The representation of smooth Markov processes is well established. In particular one can give constructive procedures which allow to build A, B (or f, σ) and $\{w(t)\}$ from $\{x(t)\}$. We shall not touch on these questions now, just stress the fact that smooth finite dimensional stochastic dynamical systems admit a differential equation representation of the form

$$\begin{aligned} dx(t) &= f(x(t))dt + \sigma(x(t))dw(t) \\ y(t) &= c(x(t)) \, . \end{aligned} \tag{1.8}$$

These are the *dynamical equations* of the system.

A very natural question to ask is which processes can be represented by dynamical equations of the form (1.8), i.c. which processes can be described as the output of a smooth finite dimensional stochastic system. This is essentially the **Stochastic Realization** (also called **Markovian Representation**) **problem** which has been considered in various places in the literature [10] [11] [8] [14] [9] [17].

DEFINITION 1.4

A "strong" Stochastic Realization of a (stationary) random process y defined on $(\Omega, \mathcal{A}, \mu)$ is a (stationary) stochastic Dynamical System defined on the same probability space with output process η almost surely equal to y i.e.

$$y(t) = \eta(t) \quad \text{a.s.} \quad t \in \mathbf{R} \tag{1.9}$$

The attribute "strong" (which will be omitted in the following) refers to an alternative possibility of representing y in a weaker sense than a.s. equality as in (1.9), namely by only requiring equality in **distribution** of η and y. Also in such a **weak sense stochastic realization** the system "realizing" y need not be defined on the same probability space. We will not consider this alternative as our main interest here will be to construct stochastic systems which reproduce y **samplewise**. This is intended to mean that for (almost) all $\omega \in \Omega$ the trajectories $t \to \eta(t, \omega)$ and $t \to y(t, \omega)$ are the same function of time. For smooth, e.g. sample continuous, processes this is clearly guaranteed by condition (1.9). Also we shall only consider the problem in the stationary setting. By **stationarity of a realization** we mean stationarity with respect to the same measure preserving group $\Phi(t)$ w.r. to which y is stationary.

At this point we return to aggregation. The following definition contains the basic idea.

DEFINITION 1.5

The Dynamical system (1.1)(1.2) is **aggregable** *if it admits a smooth randomization whose output process y has a smooth stationary stochastic realization of dimension $n < N$.*

An **aggregation** *of the system (1.1)(1.2) is then a finite dimensional smooth stationary Stochastic Dynamic System defined on $(M, \mathcal{B}, \mu)$, where μ is a smooth $\Phi(t)$-invariant measure. This stochastic system generates ($\mu - $ a.s.) the same output trajectories as the deterministic system.*

A Dynamical System Σ could be aggregable and yet some "subsystem" preserve the same dimension. To exclude this we introduce the stronger concept of *complete aggregability.*

DEFINITION 1.6

A Dynamical System $\sum$ is decomposable if $M = M_1 \times M_2$, $\Phi = \Phi_1 \times \Phi_2$ with M_i smooth manifolds of dimension $N_i > 0$ and $\Phi_i(t) : M_i \to M_i$ smooth flows. Whenever a decomposable system is randomized it is understood that the invariant measure has the form $\mu = \mu_1\mu_2$ with μ_i smooth and Φ_i-invariant, $i = 1, 2$.

$\sum$ is **completely aggregable** *if for any such decomposition there are smooth Markov processes x_i on $(M_i, \mathcal{B}_i, \mu_i)$ of dimensions $n_i < N_i$ and a smooth function c, such that*

$$h(\Phi_1(t)z_1, \Phi_2(t)z_2) = c(x_1(t), x_2(t))$$

for μ-almost all $(z_1, z_2) \in M_1 \times M_2$ and all $t \in \mathbf{R}$. ■

REMARKS

Whenever the flow generated by (1.1) admits a smooth invariant measure μ, there automatically is a smooth N-dimensional stochastic realization of y with state process the process ς defined by

$$\varsigma(t, z) = \Phi(t)z \tag{1.10}$$

Note that ς is a very degenerate (in fact purely deterministic) type of Markov process as its present say $\varsigma(0)$, determines the future and past evolution exactly. **Aggregability** means that this trivial stochastic realization of y is reducible i.e. there is some other smooth Markov process x with values on a smaller dimensional state space X and some smooth function $c : X \to \mathbf{R}^m$ such that

$$h(\varsigma(t)) = c(x(t)) \quad \mu - \text{a.s.} \tag{1.11}$$

for all $t \in \mathbf{R}$.

When will then a deterministically irreducible system become reducible in the stochastic sense defined above? This is the central issue of aggregability that will be taken up in the next section.

2. Necessary Conditions for Aggregability

In this section we shall show that aggregability is possible only if the observables of the (randomized) system behave in a strongly stochastic manner (they should form a *purely non deterministic process*). This condition makes contact with Chaotic Dynamics, in particular with the notion of a *K-system*, see e.g. [1]. It gives a first hint on how aggregation may be possible. Indeed, the reduction of dimension in the stochastic representation could be roughly explained by imagining that the white noise forcing term $\frac{dw}{dt}$ in the stochastic differential equation (1.8) is an equivalent substitute for the nasty "chaotic portion" of the deterministic (unforced) dynamics of the microscopic phase variables $z(t)$. Note that there is no "stochastic dynamic equation" which can describe a white noise process. In a sense (that can be made precise) white noise is a *memoryless* i.e. zero dimensional, signal. The synthesizing power of probabilistic modeling really lies in this capability of describing most complicated temporal behaviours by rather simple (stochastic) models.

In order to characterize aggregability as roughly indicated above one has first to see under what condition a smooth system like (1.1)(1.2) can produce "chaotic" (i.e. p.n.d.) observables. It turns out that, generically at least, this can happen *only if the phase space is infinite dimensional*. So aggregation of *smooth* systems in the "exact" sense of Definition 1.5, could only be possible for microscopic system with infinitely many degrees of freedom. We emphasize the word "smooth" since examples abound of chaotic systems on finite dimensional manifolds (e.g. Anosov flows). It is to be appreciated however that in all these examples, the only functions of the flow which can generate purely non deterministic observables are *finitely valued*, i.e. highly discontinuous. These functions essentially have to describe certain special *finite* partitions of the phase space. The corresponding output processes (if p.n.d.) has then to be *finitely valued*. The goal here is instead to model *continuous* variables and the smoothness of the observation function h cannot be given up.

The necessity of dim $M = \infty$ has to do with general beliefs (e.g. [4]) that as the dimension N of the microscopic state grows (i.e. the system becomes "larger") then (and only then) "statistics" should become applicable to describe the behaviour of the system. Various "Thermodynamic limit Theorems" can be found in the literature dealing with particular instances of this transition, but general arguments about *existence* of a limit behaviour are never given. We will have something more to say about this point at the end of the section.

Everything which follows will take place in a probability space $(M, \mathcal{B}, \mu)$ with μ a smooth measure, equipped with a measure preserving flow $\Phi(t)$. All σ-algebras will be μ-complete. We shall use the notions of **purely deterministic** (p.d.) and **purely non deterministic** (p.n.d.) (also called "regular", [13] p. 178) processes both forward and backward in time. A stochastic realization will be called purely deterministic (or purely non deterministic) if the state process x is p.d. (or p.n.d.).

LEMMA 2.1

A purely deterministic smooth realization of y cannot have smaller dimension than the trivial realization (M, ς, h) defined in (1.10).

Proof:

Let x be a forward [resp. backward] p.d. Markov process and let $\mathcal{X}$ be the σ-algebra induced by $\{x(t); t \in \mathbf{R}\}$. By definition the past [future] σ-algebra generated by x is constant, i.e.

$$\mathcal{X}_t^- = \mathcal{X} \quad [\mathcal{X}_t^+ = \mathcal{X}]$$

for all t. Since the present $\mathcal{X}_t = \sigma\{x(t)\}$ makes $\mathcal{X}_t^-$ and $\mathcal{X}_t^+$ conditionally independent we have $\mathcal{X}_t \supset \mathcal{X}_t^- \cap \mathcal{X}_t^+ = \mathcal{X}_t^+[\mathcal{X}_t^-]$ and hence $\mathcal{X}_t = \mathcal{X}_t^+[\mathcal{X}_t^-]$ for all t.

Let x be also a smooth diffusion with values on some finite dimensional Borel space X. Then the limit as $\Delta \to 0+$ of $1/\Delta$ times $E[x(t + \Delta) - x(t)|\mathcal{X}_t^-] = E[x(t + \Delta) - x(t)|\mathcal{X}_t] = x(t + \Delta) - x(t)$ exists and is some smooth function of $x(t), g(x(t))$. Since the conditioning is trivial we see that $x(t)$ is actually samplewise differentiable (from the right) and the sample paths satisfy $D_+ x(t) = g(x(t))$ (D_+ denotes derivative from the right). Now, as the sample paths are continuous by assumption, the right hand side is continuous in t so the right derivative is an ordinary derivative and x satisfies the ordinary differential equation

$$\dot{x}(t) = g(x(t)) \tag{2.1}$$

If x is backward p.d. just let g be the lim as $\Delta \to 0-$ of $1/\Delta$ $E[x(t + \Delta) - x(t)|\mathcal{X}_t^+]$

and we reach the same conclusion.

Finally suppose that (X, x, c) is a stochastic realization of y. We assume $\dim X < \infty$ (otherwise there is nothing to be proven) and x smooth and p.d. (either forward or backward). By the above, almost all trajectories of y can be described by the dynamical system

$$\dot{x}(t) = g(x(t))$$
$$\eta(t) = c(x(t)) \tag{2.2}$$

where η is the output process of the realization which satisfies $\eta(t, z) = y(t, z)\ \mu - \text{a.s.}$ on M for all $t \in \mathbf{R}$. Since both η and y are sample continuous we have $\eta(\cdot, z) = y(\cdot, z)$ for all $z \in M$ except on some μ-null set N. Clearly then the realization (12) cannot have dimension smaller than N as this would contradict irreducibility.

Hence there can be reduction in dimension (i.e. aggregation) only if the state process is non deterministic. In fact, in view of the result above it is to be expected that for *complete* aggregability to be possible, there should exist a Markovian representation of y without deterministic components at all. So one is really led to look for **purely non deterministic** realizations. In Statistical Mechanics, this requirement can even be justified directly on physical grounds.

Let $\mathcal{Y}$ be the σ-algebra generated by h (i.e. $y(0)$) on $(M, \mathcal{B})$, let $\mathcal{Y}_t := \sigma\{y(t)\} = \sigma\{h(\Phi(t))\}$ and $\mathcal{Y}_t^{\pm}$ the future (past) histories of the output process at time t. If $t = 0$ the subscript is usually dropped. For any stochastic realization with state process x inducing (at time zero) the σ-algebra $\mathcal{X}$, we have $\mathcal{Y} \subset \mathcal{X}$. Hence

$$\mathcal{Y}^- \subset \mathcal{X}^-,\ \mathcal{Y}^+ \subset \mathcal{X}^+ \tag{2.3}$$

and it follows that $\mathcal{Y}^- \vee \mathcal{Y}^+ \subset \mathcal{X}^- \vee \mathcal{X}^+ \subset \mathcal{B}$.

Now, by irreducibility we have $\mathcal{Y}^- \vee \mathcal{Y}^+ = \mathcal{B}$ (this actually follows from injectivity of the map $h_{\mathbf{R}}$, see below) so if y is, say, forward-p.d., we have $\mathcal{Y}^- = \mathcal{B}$ and then x is necessarily also forward-p.d. as $\mathcal{X}^- = \mathcal{B}$ as well. What we have discovered is the following basic fact.

269

THEOREM 2.1

A necessary condition for aggregability of an irreducible system is that the output y be a non deterministic process (i.e that neither y_t^+ nor y_t^- be constant in time).

For complete aggregability as roughly indicated above, we will have to strengthen this condition requiring y to be p.n.d. A key step is then to answer the following

QUESTION: *Consider a smooth randomization of the dynamical system* (1.1)(1.2). *Under what conditions will it produce a p.n.d. output process?*

Note that this condition can be read as a condition of **chaotic behaviour** of the "classical" dynamical system $(M, \Phi(t), \mu)$. In fact, very much in accordance with the spirit of Kolmogorov's definition, y being (say, forward) p.n.d. is equivalent to $(M, \Phi(t), \mu)$ being a K-**system** ([1] p. 32), with contracting family of σ-algebras equal to $\{y_t^-\}$. This terminology tends to induce the impression that the chaotic behaviour of the system (1.1)(1.2) as understood in our present context (i.e. p.n.d. output) depends on structural features of the dynamic group $\Phi(t)$ alone (countable Lebesgue spectrum etc.). In fact, before all chaotic behaviour depends on the **observation map** $h : M \to \mathbf{R}^m$ and on the **observability** properties of the system. That **observability** is the crucial concept to understand questions of chaotic behaviour of dynamical systems, has also been argued by T. Taylor ([15]). For convenience here the standard definition of observability will be weakened a bit.

DEFINITION 2.1

Let $h : M \to \mathbf{R}$ be a Borel measurable function. The System $(M, \Phi(t), h)$ is **observable** *on the time set $T \subset \mathbf{R}$ if the map $h_T : M \to C(T; \mathbf{R}^m)$ where*

$$h_T(z) : t \to h(\Phi(t)z) \; ; \; t \in T \tag{2.4}$$

is injective (at least) on a dense open subset of M.

$[\Sigma = (M, \Phi(t), h)$ is to be called **exactly observable** on the set T if h_T is injective on the whole of M.]

It can be shown that if $(M, \Phi(t), h)$ is observable on the subset T then the σ-algebra generated by the functions $\{h(\Phi(t)); t \in T\}$ is equal to the Borel σ-algebra $\mathcal{B}$ of M modulo μ-null sets (for any μ equivalent to Lebesgue measure).

For irreducible systems observability is interesting only on *proper* subsets of the time axis $\mathbf{R}$. For, we have the following result

LEMMA 2.2

If $(M, \Phi(t), h)$ is irreducible then $h_{\mathbf{R}}$ is injective on an open dense $M' \subset M$. In fact, this condition is equivalent to weak irreducibility (as defined in (c), Def. 1.1).

The proof is based on standard arguments in Nonlinear Systems Theory and is therefore skipped (see [6]).

We shall say that $\sum$ is **finite time-observable** if it is observable on a time set T contained in a bounded interval $[t_0, t_1]$. If it is observable on T but **not finite time observable** and, i) $\sup\{t \in T\} = +\infty$, $\inf\{t \in T\} > -\infty$ or, ii) $\sup\{t \in T\} < +\infty$, $\inf\{t \in T\} = -\infty$, we shall say that $\sum$ is **observable in the infinite future** (case i) or **in the infinite past** (case ii).

The following simple observation ties together observability and randomness. The basic idea of the proof elaborates on Taylor's paper [15].

PROPOSITION 2.2

Let $(M, \Phi(t), h, \mu)$ be a smooth randomized dynamical system with output process $y(t) = h(\Phi(t)\cdot)$, $t \in \mathbf{R}$. Then,

i) *if $(M, \Phi(t), h)$ is observable in the infinite future then,*

$$y_t^+ = \mathcal{B} \quad \forall t \tag{2.5}$$

ii) *if $(M, \Phi(t), h)$ is observable in the infinite past then,*

$$y_t^- = B \quad \forall t \tag{2.6}$$

iii) *if $(M, \Phi(t), h)$ is finite time observable both (2.5) and (2.6) hold and the process y is purely deterministic both in forward and backward sense.*

Proof:

Let $T + s$ be the observation set T translated by s i.e. $T + s = \{t + s; t \in T\}$. By $h_{T+s} = h_T \cdot \Phi(s)$ and bijectivity of $\Phi(s)$ for all real s, observability on the time set T is equivalent to observability on any translated set $T + s$. Now, observability on T implies that the σ-algebra $\sigma\{h_T\} = \sigma\{y(t); t \in T\}$ is equal to the Borel σ-algebra B of M. Then observability of T implies that $\sigma\{h_{T+s}\} = B$ for any finite s. Assume $(M, \Phi(t), h)$ is observable in the infinite future. Then for any t there is an s such that $[t, +\infty] \supseteq T + s$ and therefore

$$y_t^+ \supseteq \sigma\{h_{T+s}\} = B$$

which is (2.5). Dually for observability in the infinite past. Also, it is obvious that finite time observability implies **both** conditions (2.5) and (2.6). ∎

REMARK

In discrete time the implications (i), (ii) do not hold unless $\Phi(t)$ is **invertible**. In this case the evolution of the system is also defined for all $t \in \mathbb{Z}$. For a non invertible map $\Phi \equiv \Phi(1)$, observability in the infinite future is compatible with a purely non deterministic output process (i.e. with chaotic behaviour). The classical "Tent" map $\Phi : [0, 1] \to [0, 1]$, graph (Φ) = segments joining $(0,0)$, $(1/2,1)$, $(1,0)$, and observation function $h(z) = 0$ if $0 \leq z < 1/2$; $h(z) = 1$ if $1/2 \leq z \leq 1$ is observable in the infinite future although

$$y_t^+ \supset y_{t+1}^+ \text{ strictly}$$

and in fact $\cap y_t^+$ is the trivial σ-algebra. Note that h in this example is finitely valued and hence discontinuous. This is essential in order to rule out finite time observability which (as we shall recall below) is a generic property of smooth systems.

PROPOSITION 2.3 (Aeyels, Takens [18], [19])

Finite time observability is a generic property of smooth finite dimensional systems. In fact, for a generic set of smooth systems, h_T can be rendered injective by choosing T as almost any finite set of $2N + 1$ distinct time instants.
(See especially [18] Theorems 1 and 2 for details.)

Although this statement is of a rather weak type (because of genecicity), it shows quite clearly the role played by the dimension of the phase space N. As soon as $N < \infty$, finite time observability is automatic, so that *smooth finite dimensional dynamical systems generically produce purely deterministic output* processes. Therefore, they are not aggregable.

It is believed that a much stronger statement can be made: *No smooth finite dimensional system is aggregable.* We are currently working to prove this conjecture. (A claim of this type is contained in [20], but the proof contains an error). At the stage at which things are now, the discussion can be concluded with the following claim.

THEOREM 2.4

Generically, at least, smooth systems can be aggregable only if they are infinite dimensional (*i.e. $N = \dim M = \infty$*).

Of course whenever talking about necessary conditions for solvability of a certain problem one feels pressed to show that the necessary conditions at hand actually do apply to a nonempty set of situations. For the case in point, indeed, *there is* at least *one* interesting class of aggregation problems which is explicitly solvable.

We shall very sketchily report here on aggregability of *linear Hamiltonian systems*. The material which is referred to is to be found in refs. [12] and [21], a more complete exposition being under preparation.

It is shown in [21] that, by a suitable normalization, all *nonsingular linear Hamiltonian systems* (linearly observed) can be represented as a Dynamical System $(M, \Phi(t), h)$ where M is a real Hilbert space, $\Phi(t)$ is a continuous unitary group on M and h is a linear map $M \to \mathbf{R}^m$, i.e. a collection of m linear functionals $h_k(z) = \langle h_k, z \rangle$, $k = 1, \ldots, m$. The interesting case, according to Theorem 2.4, is when $\dim M = \infty$ (M will be always separable) and $\Phi(t)$ is strongly continuous. In this case randomization involves cylinder measures in Hilbert spaces [22], but, for the case at hand, the invariant measure can be chosen very naturally as the ∞-dimensional Gaussian distribution [22]. Then, it is shown in [21] that the output of a randomized linear Hamiltonian system is just *the most general m-dimensional stationary Gaussian process.*

Irreducibility amounts to asking that M coincides with the $\Phi(t)$-reducing subspace H spanned by the vectors $\{h_1, \ldots, h_m\}$ representing the observables,

$$H := \overline{\text{span}}\{\Phi(t)h_k; k = 1, \ldots, m, t \in \mathbf{R}\}.$$

This is a quite natural condition which can be met just by restricting $\Phi(t)$ to H and taking $M = H$. Note that anything in the phase space M which is orthogonal to H will be invisible to an external observer having access only to the observables $y_k(t, z) = \langle h_k, \Phi(t)z \rangle$.

The delicate question is observability (say, on $T = \mathbf{R}_+$). Note that the orthogonal complement of the nullspace of $h_{\mathbf{R}_+}$ is

$$[H^+]^\perp := [\ker h_{\mathbf{R}_+}]^\perp = \overline{\text{span}}\{\Phi^*(t)h_k; k = 1, \ldots, m, t \geq 0\}$$

and for "most" vectors $h_1, \ldots, h_m$, H^+ will not coincide with H. In fact, the complete story is told by the following theorem.

THEOREM 2.5 [12]

An irreducible linear Hamiltonian system $(H, \Phi(t), h, \mu)$ with μ normalized Gaussian measure, generates a (forward) purely non deterministic output process if and only if,

i) *H is infinite dimensional,*

ii) *The infinitesimal generator of the group $\Phi(t)$ has Lebesgue spectrum (of multiplicity $\leq m$).*

iii) *The spectral density matrix $M = [M_{jk}]$*

$$M_{ij}(\lambda) = \frac{d}{d\lambda}\langle h_j, E(\lambda)h_k \rangle \quad j, k = 1 \ldots m$$

where E is the spectral measure of $\Phi(t)$, is factorizable, i.e. admits matrix spectral factors $W(\lambda)$ satisfying,

$$M(\lambda) = W^*(\lambda)W(\lambda)$$

which are **analytic** *i.e. the columns of W belong to the m-dimensional Hardy space H_m^2 of the half plane.*

The system is **aggregable** *iff $M(\lambda)$ is a rational function of λ. In this case the dimension of any minimal Markovian representation of the observables is $n = \frac{1}{2}$ McMillan degree of $M(\lambda)$. Any rational analytic spectral factor $W(\lambda)$ originates a finite dimensional Markovian representation*

$$dx(t) = Ax(t)dt + Bdw(t)$$

$$y(t) = Cx(t)$$

where the matrices C, A, B can be computed from $W(\lambda) = C(\lambda I - A)^{-1}B$ and $\{w(t)\}$ is a suitable p-dimensional Wiener process.

REMARKS ON "THERMODYNAMIC LIMIT"

Roughly, we have seen that **exact** matching of the trajectories of the observables of the system (1.1)(1.2) may be obtained by a stochastic model of dimension strictly smaller than N only if $N = \infty$ (i.e. M is infinite dimensional).

Suppose now that the infinite dimensional system $\sum$ is approximated in the limit as $N \to \infty$, in some appropriate topology, by a sequence of finite dimensional systems $\{\sum_N\}$ of increasing "complexity".

Assume also that a (**finite dimensional**) **aggregate stochastic model** exists for the infinite dimensional limit $\sum$ of the sequence of microscopic dynamics. This stochastic model will then describe in an "approximate sense" the observables produced by each finite dimensional model $\sum_N$. Various notions of **approximate (stochastic) aggregation** of the observables can then be introduced. With a right choice of topology, the approximate aggregation will become exact in the *thermodynamic limit $N \to \infty$*. We should say that this seems to us a far more clean picture than what is currently found in the literature.

In any case the problem deserves a much longer treatment and we shall not discuss it any further in this paper.

References

[1] ARNOLD, V.I., AVEZ, A. (1986) *Ergodic Problems of Classical Mechanics*, Benjamin.

[2] DOOB, J. L. (1953) *Stochastic Processes*, Wiley.

[3] DYNKIN, E. B. (1965) *Markov Processes*, Vol. 1, Springer Verlag.

[4] FORD, J. (1983) How Random is a Coin Toss? *Physics Today* April 1983, 40–47.

[5] GIKHMAN, I. I., SKOROKHOOD, A. N. (1965) *Introduction to the Theory of Random Processes*, Saunders.

[6] ISIDORI, A. (1985) *Nonlinear Control Systems: An Introduction*, Springer Lect. Notes Control and Inf. Sciences **72**.

[7] KINTCHINE, A. (1949) *Mathematical Foundations of Statistical Mechanics*, Dover.

[8] LINDQUIST, A., PICCI, G. RUCKEBUSCH (1979) On Minimal Splitting Subspaces and Markovian Representation, *Math, Syst. Theory* **12**, 271–279.

[9] LINDQUIST, A., PICCI, G. (1985) Realization Theory of Multivariate Stationary Gaussian Processes, *SIAM J. Control Optimiz.* **23**, 809–857.

[10] PICCI, G. (1976) Stochastic Realization of Gaussian Processes, *Proc. IEEE 65*, 112–122.

[11] PICCI, G. (1977) On the Internal Structure of Finite State Stochastic Processes, *Springer Lect. Notes Econom. Math. Systems* **162**, 288–304.

[12] PICCI, G. (1986) "Application of Stochastic Realization Theory to a Fundamental Problem of Statistical Physics" in *Modelling Identification and Robust Control*, C. J. Byrnes and A. Lindquist eds., North Holland.

[13] ROZANON, Y. A. (1967), *Stationary Random Process*, Holden Days.

[14] RUCKEBUSCH, G. (1976) Representations Markoviennes de Processus Gaussiens Stationnaires, *C. R. Acad. Sci. Paris Series A* **282**, 649–651.

[15] TAYLOR, J.S.T. (1987) On Observations of Chaotic Dynamical Systems and Randomness, preprint.

[16] TAYLOR, J.S.T. (1987) An Example of Global Observability of a Chaotic System, *Proc. Dec. Control Conference*, Los Angeles, CA.

[17] TAYLOR, J.S.T., PAVON, M. (1987) On the Nonlinear Stochastic Realization Problem to appear on *Stochastics*.

[18] AEYELS, D. (1981) Generic Observability of Differentiable Systems, *SIAM Journal Control & Optimiz.* **19**, 595–603.

[19] TAKENS, F. (1981) Detecting Strange Attractors in Turbulence, in: *Dynamical Systems in Turbulence*, Springer Lect. Notes Math #898, 366–387.

[20] PICCI, G. (1988) Stochastic Aggregation, in: *Linear Circuits, Systems and Signal Processing, Theory and Applications*, North Holland, 493–501.

[21] PICCI, G. (1988) Hamiltonian Representation of Stationary Processes, in: *Operator Theory Advances and Applications*, vol. 35 Bir häuser,

[22] KUO, H. (1975) *Gaussian Measures in Banach Spaces*, Springer Lect. Notes Math. #463, Springer Verlag.

INVERSE PROBLEMS IN MEDICINE

A.K. Louis, TU Berlin

The research of the author was supported by the Deutsche Forschungsgemeinschaft under grant Lo 310/2-4.

Summary : In this paper we present some inverse problems in technical medicine. They are stemming from diagnostic methods like imaging. The question of the display of those results is attacked, and finally we address the problem of optimal treatment planning in hyperthermia. We first discuss inverse problems and the related mathematical questions. The general results are adapted to the above mentioned applications.

1 *Introduction*

The aim of medical imaging is to provide morphological information about an examined patient. There is an obvious possibility using surgical treatment but in the stage of diagnosis this is not applicable. In order to have a riskless and painless method a source of radiation , for example an x ray tube, is used, and changes are detected when the radiation is travelled through the patient. In those cases, where the searched-for quantity can not be directly measured, but where from an interaction of the subject to be studied and the external source, the information has to be determined ,we are talking about *inverse problems*.

In mathematical terms we have a set X of parameters, describing the examined quantity, and a set Y of possible results. The action of taking the measurement is described by a mapping A

$$A : X \to Y$$

and the problem to be solved is to find parameters f, which are mapped by A to the measured data g; i.e., we have to solve the operator equation of the first kind

$$Af = g.$$

Typically there are measurement errors, hence the set Y has to be chosen large enough, it does not only contain the image of X under the mapping A, but also the typical data noise. As an example we can think that $A(X)$ are differentiable functions, but in general the noise is far from being smooth, which means that Y is for example a L_2 space. That has the consequence that not for all $g \in Y$ the problem is solvable. Hadamard has called a problem *well posed*, if it is solvable for all $g \in Y$, if the solution is unique, and if, after the introduction of suitable topologies, the solution depends continuously on the data. If one of those conditions is not met, he calls the problem *ill posed* or *incorrectly posed*.

From the above discussions it is obvious that inverse problems are ill posed in that sense. What hurts more is the discontinuity of the solution operator as we shall see in the sequel. We first describe the mathematical difficulties related to ill-posed problems. As a remedy we discuss regularization methods which are used in the last sections to solve the above mentioned problems from the applications.

2   *Ill − Posed Problems*

In the following we assume that X, Y are Hilbert spaces and that the operator $A : X \to Y$ is linear and compact. For the selfadjoint, positive semidefinit operator A^*A we can find nonnegative real eigenvalues σ_n^2 and normalized eigenfunctions $v_n \in X$. We then define for $\sigma_n > 0$ the normalized functions u_n by

$$Av_n = \sigma_n u_n$$

and observe

$$A^* u_n = \sigma_n v_n.$$

The triple $\{v_n, u_n; \sigma_n\}_{n \in \mathbb{N}}$ is called a *complete singular system* of the compact operator A. If A is degenerated it has a finite dimensional range, if not the σ_n decay to zero; i.e.,

$$\sigma_n \to 0 \ for \ n \to \infty.$$

In the following we exclude the first mentioned trivial case. For nondegenerated compact operators A the range $\mathcal{R}(A)$ is not closed, hence we can not find for all $g \in Y$ a solution of the problem

$$Af = g.$$

We also have to consider the fact that the null space $\mathcal{N}(A^*)$ is not trivial. Hence we have the decomposition

$$Y = \overline{\mathcal{R}(A)} \oplus \mathcal{N}(A^*).$$

In order to define a solution for a larger set than just $\mathcal{R}(A)$ we consider

$$g \in \mathcal{R}(A) \oplus \mathcal{N}(A^*) =: \mathcal{D}(A^\dagger)$$

and we minimize the defect

$$\|Af - g\|.$$

If there are more than one minimizer we chose the unique one with minimal norm and define in that way a mapping

$$A^\dagger : \mathcal{D}(A^\dagger) \to X$$

which we call the *generalized inverse*. The solution f can then be characterized as the unique element in $\mathcal{N}(A)^\perp$ which solves the normal equation

$$A^*Af = A^*g.$$

Another possibility to determine the generalized solution is with the help of the singular value decomposition. We realize

$$A^\dagger g = \sum_{\sigma_n > 0} \frac{1}{\sigma_n} < g, u_n > v_n.$$

The v_n are in the range of A^*, hence orthogonal to the kernel of A, resulting in $A^\dagger g \in \mathcal{N}(A)^\perp$. It is straightforward to verify that $A^*AA^\dagger g = A^*g$.

As a consequence of this representation we observe that those components of the solution are strongly affected by noise where the singular values σ_n are small. Hence we use the decay of the σ_n to characterize the ill-posedness of the operator.

Definition 2.1. The operator A is called ill posed of the order α, if

$$\sigma_n = O(n^{-\alpha}).$$

If σ_n decays faster than polynomial, we call A exponentially ill posed.

Due to the fact, that the σ_n tend to zero we realize that for operators A with nonclosed range the operator $A^\dagger$ is not continuous. This means that small perturbations in the right-hand side g can cause large deviations in the solution. As substitute for continuity we introduce regularization. A regularization is a familiy of operators $\{T_\gamma\}_{\gamma > 0}$ with

$$T_\gamma : Y \to X.$$

Assuming that we have erroneous data g^ε with

$$\|g^\varepsilon - g\| \leq \varepsilon,$$

then we want that in the case of vanishing data errors the images $T_\gamma g^\varepsilon$ tend to $A^\dagger g$. This means that we have to chose γ in dependence of ε, and possibly of g^ε.

Definition 2.2. A family of operators $T_\gamma : Y \to X$ is called a regularization, if there exists a function

$$\gamma : \mathbb{R}_+ \times Y \to \mathbb{R}_+$$

such that for all $g^\varepsilon \in Y$ and $g \in \mathcal{D}(A^\dagger)$ with $\|g^\varepsilon - g\| \leq \varepsilon$

$$\lim_{\varepsilon \to 0} T_{\gamma(\varepsilon, g^\varepsilon)} g^\varepsilon = A^\dagger g.$$

280

In that case we call $\gamma(\varepsilon, g^\varepsilon)$ an $\underline{a-posteriori}$ parameter choice. If γ is independent of g^ε then it is an $\underline{a-priori}$ parameter choice.

A possibility of constructing regularizations is via the singular value decomposition and a function

$$F_\gamma : I\!R_+ \times Y \to I\!R_+$$

giving

$$T_\gamma g = \sum F_\gamma(\sigma_n, g)\sigma_n^{-1} < g, u_n > v_n.$$

The function F is called a *filter*. If F is independent of g then the method is linear and an a-priori parameter choice is sufficient for convergence.

$\underline{Theorem\ 2.3.}$ Assume that F_γ is independent of g. If

$$|F_\gamma(\sigma)\sigma^{-1}| \le c_\gamma \ for\ all\ \gamma,$$

$$|F_\gamma(\sigma)| \le c \ for\ all\ \gamma, \sigma,$$

$$\lim_{\sigma \to 0} F_\gamma(\sigma) = 1 \ for\ all\ \sigma > 0,$$

then the corresponding T_γ is a regularization with a-priori parameter choice.

In the following we give some examples.

1.

$$F_\gamma(\sigma) = \begin{cases} 1, & \sigma \ge \gamma, \\ 0, & \sigma < \gamma. \end{cases}$$

Here we get the *truncated singular value decomposition*

$$T_\gamma g = \sum_{\sigma_n > \gamma} \sigma_n^{-1} < g, u_n > v_n.$$

For the total error we estimate

$$\|T_\gamma g^\varepsilon - A^\dagger g\| \le \Big(\sum_{\sigma_n \ge \gamma} \sigma_n^{-2} | < g^\varepsilon - g, u_n > |^2 \Big)^{1/2} + \Big(\sum_{\sigma_n < \gamma} \sigma_n^{-2} | < g, u_n > |^2 \Big)^{1/2}.$$

We observe that the second term, the *filter error*, tends to zero for $\gamma \to 0$. But the first term, the *data error* grows unboundedly for $g^\varepsilon \in \overline{\mathcal{R}(A)} \setminus \mathcal{R}(A)$. The typical picture in treating ill-posed problems shows up, the opposite behaviour of the two error terms.

2.

$$F_\gamma(\sigma) = \frac{\sigma}{\sigma^2 + \gamma^2}.$$

This filter corresponds to the $Tikhonov - Phillips$ regularization, where we start from the minimization of

$$\|Af - g\|^2 + \gamma^2\|f\|^2$$

and where we have to solve the regularized normal equation

$$(A^*A + \gamma^2 I)f = A^*g.$$

Also some stochastical methods like best linear estimator or Bayes estimation can be viewed as generalized Tikhonov-Phillips method, and vice versa.

3. Let $\gamma = \frac{1}{m}$ for $m \in I\!N$ and consider for $0 < \beta < 2\|A\|^{-2}$ the filter

$$F_m(\sigma) = 1 - (1 - \beta\sigma^2)^m.$$

This filter corresponds to the $Landweber\ iteration$

$$f^{m+1} = f^m + \beta A^*(g - Af^m)$$

with starting vector $f^0 = 0$.

The two error terms show here that too many iterations destroy the result.

4. We mention the $conjugate\ gradient\ method$ as an example for a nonlinear regularization. Here

$$F_m(\sigma, g) = P_{m-1}(\sigma^2, g)\sigma$$

where the polynomials P_{m-1} of degree $m - 1$ generate the iteration

$$f^m = P_{m-1}(A^*A, g)A^*g.$$

It can be shown that for no a priori parameter choice this is regularization method.

For more details on regularization methods we refer to [1] and [5] and the references cited there.

3   *Medical Imaging*

The most wide – spread technique in medical imaging is x-ray computerized tomography (CT). As radiation source serves an x-ray tube, and the x-ray attenuation coefficient, which is proportional to the density of the tissue, has to be identified from the data.

In magnetic resonance imaging (*MRI*) a homogeneous magnetic field is used, and gradient fields for different directions are applied. The measured data pertain to the Fourier transform of the distribution of hydrogen nuclei in the body.

In emission CT (*SPECT* and *PET*) radiopharmaceuticals are injected and their distribution in the body has to be computed from the measured radiation. Here an integral transform related to the Radon transform, which is discussed in the following section, describes the mathematical model.

Besides the x-ray CT which serves as an example for presently used imaging technology we briefly mention the problems in ultrasound CT.

3.1   *X − Ray CT*

The standard technique in medical imaging is *x − ray computer tomography*. Here slices through the patient are studied. From an x-ray tube x rays are sent through the patient and on the opposite side the arriving photons are counted. Under the physical assumption that the rays are travelling on straight lines, that the attenuation of the intensity ΔI is proportional to the intensity I itself and the travelled path Δt we get after introducing the proportionality factor f

$$\Delta I = -If\Delta t.$$

Letting $\Delta t \to 0$ we find for each ray L the ordinary differential equation

$$\frac{\dot{I}}{I} = -f \ \ with \ \ I(-\infty) = I_0 \ , \ \ I(\infty) = I_L$$

which has the solution

$$-\ln \frac{I_L}{I_0} = \int_L f dt.$$

Parametrizing the rays by the unit vector $\omega \in S^1$, $\omega(\varphi) = \left(\cos\varphi, \sin\varphi\right)^\top$ we get with $\omega^\perp = \omega(\varphi + \frac{\pi}{2})$ the line

$$L = \{s\omega + t\omega^\perp : t \in I\!\!R\}$$

and hence the integral transform

$$Rf(\omega, s) = \int_{I\!\!R} f(s\omega + t\omega^\perp) \, dt$$

which is called the *Radon transform*. In higher dimensions the Radon transform associates with a function its integrals over all hyperplanes. If, as in x-ray CT, only line integrals are involved, we call the corresponding transform the x-ray transform.

For those transforms we can compute a singular value decomposition. For the sake of generality we extend the considerations here to arbitrary dimensions. We define for $\omega \in S^{N-1}$

$$Rf(\omega, s) = \int_{\boldsymbol{R}^N} f(x)\delta(s - x \cdot \omega)\, dx.$$

As function space X we use $L_2(\Omega)$, where Ω is the unit ball in $\boldsymbol{R}^N$, hence we assume that we determine compactly supported functions. We denote by

$$Z = S^{N-1} \times \boldsymbol{R} \subset \boldsymbol{R}^{N+1}$$

the unit cylinder, and by w the weight function

$$w(s) = (1 - s^2)^{(N-1)/2}.$$

As Y we use

$$Y = L_2(Z, w^{-1}).$$

The adjoint operator of R as mapping from $L_2(\Omega) \to L_2(Z, w^{-1})$ is

$$R^*g(x) = \int_{S^{N-1}} w^{-1}(x \cdot \omega)g(\omega, x \cdot \omega)\, d\omega.$$

If we denote by D^U a unitary representation of the orthogonal group $O(N)$ defined by

$$D^U f(x) = f(U^{-1}x)$$

we observe that D^U and R^*R commute. Hence R^*R is invariant under the action of the orthogonal group and we can find invariant subspaces with the help of the spherical harmonics. We can state the following result, describing a singular value decomposition for the Radon transform, with the help of the Gegenbauer polynomials C_m^ν, the Jacobi – polynomials $P_m^{(\alpha, \beta)}$ and a basis $Y_{\ell, k}$, $k = 1, \cdots, M(N, \ell)$, of the spherical harmonics of degree ℓ, see [3].

Theorem 3.1. *Let*

$$v_{m,\ell,k} = |x|^\ell\, P_{(m-\ell)/2}^{(0,\ell+N/2-1)}(2|x|^2 - 1)\, Y_{\ell,k}\left(\frac{x}{|x|}\right),$$

$$u_{m,\ell,k}(\omega, s) = w(s)C_m^{N/2}(s)Y_{\ell,k}(\omega),$$

$$\sigma_{m,\ell,k}^2 = 2^N \pi^{n-1} \frac{m!}{(m + N)!}.$$

Then

$$\{(v_{m,\ell,k},\ u_{m,\ell,k};\ \sigma_{m,\ell,k}) : m \geq 0,\ m + \ell\ \text{even},\ k = 1, \cdots, M(N, \ell)\}$$

is a complete singular system for $R : L_2(\Omega) \to L_2(Z, w^{-1})$.

We observe that the $\sigma_{m,\ell,k}$ decay like $O(m^{(1-N)/2})$, hence the ill-posedness increases with the dimension N.

In any practical application there are only a finite number of data available. It is obvious that the searched-for density distribution cannot be uniquely determined by those data. It is shown in [4] that the functions in the null space, the so-called ghosts, consist essentially of high frequency components, which means a restriction in the possible resolution, because high frequency components correspond to small details in the picture. For more results and reconstruction algorithms the reader is referred to [9].

3.2 *Ultrasound CT*

By far less advanced in the technical realization than x-ray CT is ultrasound CT, because the mathematical problems are much more difficult. For deriving a mathematical model we assume that time harmonic waves are sent to the patient and that the scattered waves are recorded. If we denote by

$$u^i(x) = e^{ik\theta x}, \ \theta \in S^2$$

the incoming plane wave, then in $I\!R^3$ the scattered wave u^s has the following asymptotic behaviour

$$u^s(x) = \frac{e^{ik|x|}}{|x|} f(k, \theta, \frac{x}{|x|}) + o(|x|^{-1}).$$

This means that u^s is essentially a spherical wave, modified by the complex-valued function f, the so-called far-field pattern or scattering amplitude. The total field $u = u^i + u^s$ is the solution of the Helmholtz equation which we write down in the form of the Schrödinger equation

$$(\Delta + k^2)u = Vu$$

where V denotes the potential which has to be determined. A simple linearization, the so-called first Born approximation, is found if we replace $V(u^i + u^s)$ by Vu^i. The function u^i solves the homogeneous Helmholtz equation which results in

$$(\Delta + k^2)u^s = Vu^i.$$

Of course we cannot measure u^s where V is supported hence we have to solve the inverse problem where for example the far field pattern is given.

A nonlinear approximation for the mapping of the potential V to the far field f is the so-called Eikonal approximation which was developed in [8] for treating the forward problem. In order to incorporate absorption in our model we allow V to be complex-valued with $\Im V \leq 0$. An approximation of the far-field pattern can be described by

$$f \approx A_k V = -ikB_{2k}F_{\theta\perp}^2 EPV,$$

see [6]. The operator

$$PV(\theta, b) = \int_{\mathbb{R}} V(b + t\theta)\, dt, \ b \in \theta^{\perp},$$

is the $3D$ - x-ray transform, see [7]. The nonlinear operator E is defined for complex numbers

$$Ez = \exp(\frac{-\imath}{2k}z) - 1.$$

It obviously preserves the support of PV. Then $F_{\theta\perp}^2$ is the $2D$ Fourier transform on the plane perpendicular to the direction θ of the incoming plane wave. Finally B_{2k} is the bandlimiting operator restricting the Fourier transformed function to frequencies smaller than $2k$ in modulus. The theoretical results from the study of the forward problem; i.e., computing an approximation of f for given potential V, indicate that this method also gives reasonable results in the case of multiple scattering.

4   *Vision*

The results in medical imaging typically are provided in form of images of planes through the patient. In order to get a vision of bone structures or the heart for example $3D$ pictures are displayed. This leads to the $3D$ representation of objects. In mathematical terms we can formulate the problem in the following way. Given are points on the boundary $\partial\Omega$, and we want to find Ω. Of course the mapping which associates to an object points on its boundary is far from being injective, and there is no stability in any sensible topology.

In order to find a generalized solution additional information has to be used. In contrast to CAD, where Ω consists only of a finite number of objects with simple geometry, we have to describe " natural " objects. We assume that $\Omega \in \mathbb{R}^3$ is compact with nonempty interior, $\partial\Omega$ is orientable and has finite measure. When we assume that the data are stemming from a stack of x-ray pictures we can conclude that parallel planes cutting the object perpendicular to image planes give polyeders with special structure, see [11].

Further restrictions like the search for objects of smallest surface or of smallest volume reduce the indeterminacy in the problem.

Finally the images have to be represented. One can use *image spaces*, where $\mathbb{R}^3$ is decomposed into voxels V_k and an approximation to Ω is found by

$$\Omega_h = \bigcup \{V_k : meas(V_k \cap \Omega) \geq c\},$$

see [2]. On the other hand we can use the *object space* where the surface to be displayed is described as

$$(\partial\Omega)_h = \bigcup \{triangles \ (a_{k1}, a_{k2}, a_{k3}) \ : \ a_{kj} \in \partial\Omega\},$$

see [11]. The decision which of the representation is preferred depends on the operations performed on the images.

5   *Therapy*

As an example for an inverse problem in therapy we discuss hyperthermia treatment planning. In hyperthermia the tumor is heated to more than 42.5° Celcius. Then it is better treatable by radiation therapy. Of course the healthy parts in the body must not be overheated, see [12].

The heating is realized by magnetic fields and the questions are the following. What is the smallest detail in the heating pattern which can be achieved by a given finite number of antennas. And of course what is the optimal control of the antennas. The mathematical model is formulated by the Helmholtz equation for the field E

$$(\Delta + k^2)E = 0 \ in \ \Omega,$$
$$E = g \ on \ \partial\Omega$$

and the bioheat equation which describes the changes in the temperature due to blood flow

$$T_t + aT = |E|^2.$$

The inverse problem on hand is the determination of the control g when a temperature T is prescribed. We attack the problem in two steps. First we determine dependent on T the necessary incoming field in the water bolus surrounding the patient using the Lippmann-Schwinger equation. Then we compute the control of the antennas; i.e., phase and amplitude, via least squares approximation.

The ill-conditioned nature of the problem is obvious if we consider the mapping form the given control g to the filed in the water bolus for an infinte number of antennas. We get an integral transform from $L_2(S^1) \to L_2(V(0,1))$ where $V(0,1)$ is the unit ball in $\mathbb{R}^2$. It is

$$Tg(x) = \int_{S^1} \frac{e^{ik|x-R\omega|}}{|x - R\omega|} \ g(\omega) \ d\omega$$

where $R > 1$ is the radius of the antenna circle. Because of $|x| \leq 1 < R$ the kernel is smooth and therefore we again face an extremely ill-posed problem.

6   *References*

[1] Bertero, M., De Mol, C., Viano, G.A.: The stability of inverse problems, in Baltes, H.P.(ed.) Inverse Scattering Problems, Springer, 1980

[2] Herman, G.T., Udupa, J.K.: Display of 3D discrete surfaces, Proc. Spie **283** (1981) 90-97

[3] Louis, A.K.: Tikhonov-Phillips regularization of the Radon transform, in Hämmerlin, G., Hoffmann, K.H. (eds.) Constructive Methods for the Practical Treatment of Integral Equations, ISNM 73, 211-223, 1985

[4] Louis, A.K.: Nonuniqueness of inverse Radon problems : the frequency distribution of the ghosts. Math. Z. **185** (1984) 429-440

[5] Louis, A.K.: Inverse und schlecht gestellte Probleme, Stuttgart :Teubner, 1989

[6] Louis, A.K.: The Eikonal approximation in ultrasound CT, to appear in IMA Proceedings, Springer 1989

[7] Maaß ,P.: The x-ray transform : singular value decomposition and resolution, Inverse Problems **3** (1987) 729-741

[8] Molière, G, Z. für Naturforschung, **2A** 133, 1947

[9] Natterer, F.: The mathematics of computerized tomography. Teubner-Wiley, 1986

[10] Newton, R.G.: Scattering theory of waves and particles, 2nd ed. Springer, 1986

[11] Tönnies, K.D.: 3D-Repräsentation der Morphologie von anatomischen Objekten durch Approximation ihrer Oberfläche, Dissertation, TU Berlin, 1987

[12] Wust, P., Nadobny, J., Felix, R., Deuflhard, P., John, W., Louis, A.K.: Numerical approaches to treatment planning in deep RF hyperthermia, Strahlentherapie, to appear

Prof. Dr. Alfred Louis, Fachbereich Mathematik der Technischen Universität Berlin, Straße des 17. Juni 136, D-1000 Berlin 12, FRG

DYNAMICAL SYSTEM IDENTIFICATION

FROM NOISY DATA

S. Beghelli   R.P. Guidorzi   U. Soverini

Summary

Some classical schemes in algebraic system identification are first recalled and compared. It is shown that, in most cases, the solution is obtained thanks to additional assumptions which are not deducible from the available data. The identification problem for linear dynamic systems is then solved on the basis of the Frisch scheme, in order to obtain the whole set of models compatible with noisy input–output sequences. The main result here proposed concerns the unicity of the solution when the data are affected by additive white noise.

1. Introduction

Identification of mathematical models from measured data is an important problem in many scientific disciplines. In most part of statistic theory the fundamental idea consists in assuming that there exist true linear laws which can completely describe the behaviour of the system under study. Starting from this true linear system and postulating some characteristic features on the disturbances, the problem is then reduced to determine how these disturbances affect the linear behaviour. In most cases the model is validated on the basis of assumptions which are *a priori* unverifiable and not deducible from the available data.

These a priori knowledges or assumptions on the noise characteristics are what Kalman calls *prejudices*, [12], [13], [14], [15]. The central idea in his works, is the already well-known *Uncertainty Principle* which states that the solution of a noisy identification problem is not unique or, with different words, that uncertain data imply uncertain models.

If this viewpoint is accepted, then the identification problem can be stated as the definition of a certain model class based on some physical or mathematical conveniences (e.g. linear model, additive noise, etc...) and of a criterion which specifies how the data are misfitted by the model. Of course, in absence of noise, the criterion should give the exact model when the data are generated by a system belonging to the specified model class. This idea can be found in the *Uniqueness Principle* stated by Kalman: if the data are exact and complete there is one and only one minimal system which reproduces the data.

In general, however, the model will never fit the data exactly and what is not explained by it will be declared to be noise. "Noise" looses in this context any stochastic meaning since it can be simply considered as any deviating factor which is not compatible with the model. It is also important to observe that fixing the model and the identification scheme involves at the same time the definition of what is considered noise, and the identification of the system implies the identification of the noise environment. If we deal with linear systems, it is natural to consider as noise any deviation from linear relations.

2. Identification schemes for algebraic linear systems

In agreement with previous considerations, the identification of linear relations from noisy data can be mathematically stated as follows.

Let us consider a finite sequence of n variables $x_1, x_2, ..., x_n$ observed at N different times (with $N > n$). If linear relations exist among these variables, they are described by models of the type

$$(2.1a) \qquad a_1 x_1 + a_2 x_2 + ... + a_n x_n = 0 .$$

Let X be the $(N \times n)$ matrix storing the previous measures. Models of the type (2.1a) are described by the columns of a matrix A such that

$$(2.1b) \qquad XA = 0$$

or, equivalently,

$$(2.1c) \qquad X^T XA = 0$$

where $\Sigma = X^T X$ is a sample covariance matrix, under a zero-mean assumption for all variables. The number of independent linear relations is indicated by the rank of Σ. When the data are corrupted by noise then $rank\,[\Sigma] = n$, so that no relation can be obtained unless the data are modified. A classical scheme, reproposed by Kalman, which introduces a minimum number of additional assumptions is the following.

The Frisch scheme

1 — All variables are treated symmetrically and each variable is affected by an unknown amount of additive noise;

2 — Each noise component is independent of every other noise component and of every variable.

Note that the second assumption is in accordance with the principle that noise should not be modelable by linear relations. Each variable x_i $(i = 1, 2, ..., n)$ is thus defined as

$$(2.2) \qquad x_i = \hat{x}_i + \tilde{x}_i$$

where the unknown right–hand terms represent the true value of the i–th variable, $\hat{x}_i$, and the additive noise on this variable, $\tilde{x}_i$. The identification problem can thus be formulated as follows.

Problem 2.1 — Given an $(n \times n)$ symmetric positive definite covariance matrix Σ, find all diagonal matrices $\tilde{\Sigma}$ with non–negative elements such that

$$(2.3) \qquad \hat{\Sigma} = \Sigma - \tilde{\Sigma} \geq 0$$

The analysis of this problem will be carried out in next section, with particular attention to the case of $rank[\hat{\Sigma}] = n - 1$.

In the following some other classical identification schemes are considered. It will be shown that such methods do not verify all previous requirements and force the solution to be unique introducing some additional assumptions (prejudices).

The Linear Least–Squares scheme

A classical formulation of the Linear Least–Squares problem can be stated as follows.

Problem 2.2 — Denote with X^i the $(N \times (n-1))$ matrix obtained from the data matrix X by deleting its i–th column x^i. Determine the $(n-1)$–dimensional vector a^i that minimizes $\|x^i - X^i a^i\|_2$.

It is quite evident that this approach does not treat the variables in a symmetric way. In fact, since all the elements of X^i and of x^i belong to the same set of data, there is no reason to assume all the noise on x^i and X^i noiseless.

Regressing the variable x_i on the remaining ones $x_1, ..., x_{i-1}, x_{i+1}, ..., x_n$ is equivalent to limit, for each i, the solutions of condition (2.3) to those matrices $\tilde{\Sigma}$ having only the i–th element $\tilde{\sigma}_i$ different from zero. Note that this corresponds to assume that only the i–th variable is noisy and the remaining ones are noisefree. It can be easily verified [4] that the variance of the additive noise on x_i is given by

$$(2.4) \qquad \tilde{\sigma}_i = det\,[\Sigma] \,/\, det\,[\Sigma_i] \;.$$

where Σ_i is obtained from Σ by deleting its i–th row and column.

The Linear Least–Squares approach therefore cannot provide an admissible solution if all variables are noisy. Moreover it must be noted that the noise environment considered by this scheme is very structured $(rank\,[\tilde{\Sigma}] = 1)$.

The least eigenvalue filtering scheme

One of the most used filtering schemes in identification problems is based on the following result.

Theorem 2.1 — For any $(n \times n)$ symmetric positive definite covariance matrix Σ, the matrix $\hat{\Sigma} = \Sigma - \sigma I_n$ is non–negative definite if and only if σ is equal to the least eigenvalue of Σ.

Theorem 2.1 suggests to take as noise covariance matrix $\tilde{\Sigma} = \sigma I_n$, where σ is the least eigenvalue of Σ. This scheme provides a realistic, even if not general, noise model in which all variables are considered to be affected by the same amount of disturbance. Note that conditions one and two stated in the Frisch scheme are both fulfilled so that the associated solution belongs to the solution set of this scheme.

The Total Linear Least–Squares scheme

Unlike the Linear Least–Squares approach, this scheme provides a unique solution compatible with the assumption that all data are noisy. In the noise space the problem formulation can be the following.

Problem 2.3 — Given an $(n \times n)$ symmetric positive definite covariance matrix Σ, determine the non–negative definite matrix $\tilde{\Sigma}$ such that $\hat{\Sigma} = \Sigma - \tilde{\Sigma}$ is non–negative definite and $\|\hat{\Sigma}\|_F$ is minimum ($\| \cdot \|_F$ = Frobenius norm).

The phylosophy of this approach has received remarkable attention in the literature even if under different names (errors–in–variables, orthogonal regression). The solutions can be found on the basis of the classical singular value decomposition technique. Under the assumption that the least eigenvalue λ_m of Σ is simple, it can be proved [6] that the unique solution of the problem is $\tilde{\Sigma} = \lambda_m h_m h_m^T$, where h_m is the eigenvector associated to λ_m.

It can be easily verified that the solution h_m in the parameter space is exactly the same given by the least eigenvalue scheme. The Total Linear Least–Squares approach, in spite of its numerical robustness, does not provide a realistic noise model because $(n-1)$ independent linear relations in the noise covariance matrix are assumed $(rank[\tilde{\Sigma}] = 1)$.

3. The Frisch scheme for algebraic linear systems

In this section the main results regarding the Frisch scheme are briefly recalled. The identification of algebraic linear systems with this scheme is mathematically equivalent to find all diagonal matrices $\tilde{\Sigma}$ with non–negative elements such that condition (2.3) is fulfilled. Note that the rank of $\hat{\Sigma}$ may change by varying $\tilde{\Sigma}$ and, consequently, the same set of data may be linked by different numbers of linear relations.

A procedure to find all the solutions of the Frisch scheme corresponding to matrices $\hat{\Sigma}$ with rank equal to $(n-1)$ will now be considered.

Algorithm 3.1 — Starting from the first variable, x_1, we can assume that the associated noise variance $\sigma_1^* = k_1 \tilde{\sigma}_1^0$ is a fraction $(0 \leq k_1 \leq 1)$ of the maximum allowable value $\tilde{\sigma}_1^0 = det\,[\Sigma]\,/\,det\,[\Sigma_1]$, where Σ_1 is the matrix obtained from Σ by deleting its first row and column. Let us now define $\hat{\Sigma}^1$ as follows

$$(3.1) \qquad \hat{\Sigma}^1 = \Sigma - diag\,[\sigma_1^*, 0, ..., 0]\ .$$

If $\hat{\Sigma}^1$ is singular $(k_1 = 1)$ a solution has been obtained and the procedure goes on with a new value of k_1. Once an allowable value for σ_1^* has been selected the maximal allowable noise on the second variable, x_2, is given by

$$(3.2) \qquad \tilde{\sigma}_2^1 = det\,[\hat{\Sigma}^1]\,/\,det\,[\hat{\Sigma}_2^1]$$

where $\hat{\Sigma}_2^1$ is obtained from $\hat{\Sigma}^1$ by deleting its second row and column. Let us now consider a fraction of the maximum allowable noise on x_2, $\sigma_2^* = k_2 \tilde{\sigma}_2^1$ $(0 \leq k_2 \leq 1)$ and define the corresponding matrix

$$(3.3) \qquad \hat{\Sigma}^2 = \Sigma - diag\,[\sigma_1^*, \sigma_2^*, 0, ..., 0]\ .$$

If $\hat{\Sigma}^2$ is singular $(k_2 = 1)$ another solution has been obtained and the procedure goes on with a new value of k_2 until the noise covariance matrix $\tilde{\Sigma} = diag\,[\sigma_1^*, \sigma_2^*, ..., \sigma_n^*]$ is completed. It should be noted that σ_i^* must be a fraction of $\tilde{\sigma}_i^{i-1}$ when $i \leq n-1$ while at the last step $\sigma_n^* = \tilde{\sigma}_n^{n-1} = det\,[\hat{\Sigma}^{n-1}]\,/\,det\,[\hat{\Sigma}_n^{n-1}]$.

Remark 3.1 — Note that every allowable noise covariance matrix $\tilde{\Sigma}$ (i.e. such that $\hat{\Sigma} = \Sigma - \tilde{\Sigma} \geq 0$) defines a point $(\sigma_1, ..., \sigma_n)$ belonging to the first orthant of the noise space $\mathcal{R}^n$, which is mapped into one and only one point $(a_1, ..., a_n)$ of the solution space $\mathcal{R}^n$.

Moreover it is possible to prove the following result [4]:

Theorem 3.1 — The solution set defined by Algorithm 3.1 is a convex hypersurface belonging to the first orthant of the noise space whose section, with a plane parallel to a coordinate one, is a hyperbola segment. In particular cases the hyperbola degenerates into a pair of straightlines or into a single point.

Remark 3.2 — The hypersurface defined by Theorem 3.1 partitions the first orthant of the noise space into two regions. The points over the hypersurface correspond to non definite matrices $\hat{\Sigma}$, those under the hypersurface to positive definite matrices.

The previous results give the whole solution set and describe its characteristic properties when the Frisch problem is analysed in the noise space. Different and more complex techniques are required when the analysis is carried out in the parameter space. An important result in this direction has been obtained by Kalman [13] using the Perron–Frobenius theorem for matrices with all positive elements.

Theorem 3.2 — The solution set of equation $(\Sigma - \tilde{\Sigma})a = 0$ is the convex simplex whose n vertices are the n least squares solutions. For every noise matrix $\tilde{\Sigma}$, $rank[\Sigma - \tilde{\Sigma}] = n - 1$ if and only if the covariance matrix Σ is positive definite and Σ^{-1} is sign similar to a positive matrix.

(A matrix Σ is sign similar to a matrix with all positive entries (positive matrix) if there exists a matrix $T = diag[\pm 1, \ldots, \pm 1]$ such that $T\Sigma T$ is a positive matrix.)

Remark 3.3 — Loosely speaking it can be said that the area of the simplex in the parameter space is linked to the amount of noise on the data and gives a measure of uncertainty in the solution set.

Note also that the solution given by the least eigenvalue scheme is *in the middle* of such simplex and this may explain the well known statistical consistency of this approach.
If we add an increasing amount of noise to the considered variables, then the least-squares solutions reach and trepass the hyperplanes delimiting the orthant which contains the initial simplex and in this situation the inverse positiveness condition of Theorem 3.2 is lost. Hence there exist noise matrices $\tilde{\Sigma}$ for which $rank[\Sigma - \tilde{\Sigma}] < n - 1$ and consequently the data may be linked by more than one relation. In such conditions a further requirement is added to the Frisch scheme, the determination of the value of $corank[\hat{\Sigma}]$ defined as

$$(3.4) \qquad corank[\hat{\Sigma}] = max_{\tilde{\Sigma}}(n - rank[\Sigma - \tilde{\Sigma}])$$

The maximality in the corank definition is very important since it corresponds to the maximum number of independent linear relations that may link the data. The corank is the only invariant of the problem that can be uniquely identified from the data. The determination of this value for matrices of non trivial dimension is an important but unsolved mathematical problem. Recently new geometrical concepts have been introduced [6], [7], [8] to solve this problem; the alghorithm proposed is based on testing a finite number of vectors satisfying particular properties. Starting from these geometrical considerations a conjecture has been advanced that, in the parameter space, the solution set might be a collection of convex polyhedral sets lying in the orthants.

Remark 3.4 — Observe that the case of $corank[\hat{\Sigma}] > 1$ does not modify the hypersurface of the admissible solutions in the noise space. In fact if noise values corresponding to a rank of $\hat{\Sigma}$ lower than $(n-1)$ exist, they belong to the hypersurface defined by Theorem 2.2.

4. The Frisch scheme for linear dynamic systems

Let us consider a finite sequence of the variables $y(\cdot)$ and $u(\cdot)$ observed with a constant sampling interval. If dynamic linear relations exist among these variables, they can be described by models of the type

$$(4.1) \qquad \hat{y}(t+n) = \sum_{i=0}^{n-1} \alpha_i\, \hat{y}(t+i) + \sum_{i=0}^{n} \beta_i\, \hat{u}(t+i)$$

which describe linear single–input single–output discrete–time systems whose order is n and whose parameters are α_i and β_i.

Let us consider at first the following problem.

Problem 4.1 (realization) — Given a noiseless input–output sequence $\hat{u}(\cdot)$, $\hat{y}(\cdot)$ generated by a system of type (4.1), determine the order n and the parameters α_i, β_i of the system.

Let us define the following vectors and matrices

$$(4.2a) \qquad \hat{u}^N(t+k) = [\hat{u}(t+k) \ldots \hat{u}(t+k+N-1)]^T$$

$$(4.2b) \qquad \hat{y}^N(t+k) = [\hat{y}(t+k) \ldots \hat{y}(t+k+N-1)]^T$$

$$(4.2c) \qquad X_k(\hat{u}) = [\hat{u}^N(t) \ldots \hat{u}^N(t+k-1)]$$

$$(4.2d) \qquad X_k(\hat{y}) = [\hat{y}^N(t) \ldots \hat{y}^N(t+k-1)]$$

$$(4.2e) \qquad \hat{\Sigma}_k(\hat{u}\hat{u}) = X_k^T(\hat{u})\, X_k(\hat{u})$$

$$(4.2f) \qquad \hat{\Sigma}_k(\hat{y}\hat{y}) = X_k^T(\hat{y})\, X_k(\hat{y})$$

$$(4.2g) \qquad \hat{\Sigma}_k(\hat{y}\hat{u}) = X_k^T(\hat{y})\, X_k(\hat{u}) = \hat{\Sigma}_k^T(\hat{u}\hat{y})$$

where N is assumed large enough to solve the problem considered. Let us partition now the matrix $\hat{\Sigma}_k$ as follows

$$(4.3) \qquad \hat{\Sigma}_k = \begin{bmatrix} \hat{\Sigma}_k(\hat{y}\hat{y}) & \hat{\Sigma}_k(\hat{y}\hat{u}) \\ \hat{\Sigma}_k(\hat{u}\hat{y}) & \hat{\Sigma}_k(\hat{u}\hat{u}) \end{bmatrix}.$$

To solve the realization problem it is possible to consider the sequence of increasing–dimension matrices

$$(4.4) \qquad \hat{\Sigma}_2\ \hat{\Sigma}_3\ \ldots\ldots\ \hat{\Sigma}_k\ \ldots$$

testing their singularity. As soon as a singular matrix $\hat{\Sigma}_k$ is found then

$$(4.5) \qquad n = k - 1$$

and the parameters $\alpha_0, \ldots, \alpha_{n-1}, \beta_0, \ldots, \beta_n$ describe the dependence relationship of the $(n+1)$-th vector of $\hat{\Sigma}_{n+1}$ on the remaining ones.

Remark 4.1 — In Problem 4.1 it has been assumed that N is large enough to avoid unwanted linear dependence relationships due to limitations in the dimension of the involved vector spaces; this means $N \geq 2(n+1)$. The minimal number of samples must be therefore equal to $3n+2$. If a lower number of samples is available then only a partial realization problem can be solved.

In the noisy case the following identification problem can be proposed.

Problem 4.2 (identification) — Given a noisy input–output sequence $u(\cdot)$, $y(\cdot)$ univocally determine, if possible, the order n and the parameters α_i, β_i of a model (4.1) of the system which has generated the noiseless sequences $\hat{u}(\cdot)$, $\hat{y}(\cdot)$.

Note that in presence of noise the procedure described for the solution of Problem 4.1 would obviously be useless since matrices $\hat{\Sigma}_k$ would always be non–singular. As in the algebraic case, a very natural assumption concerns the definition of the input–output variables as

$$(4.6a) \qquad u(t) = \hat{u}(t) + \tilde{u}(t)$$

$$(4.6b) \qquad y(t) = \hat{y}(t) + \tilde{y}(t)$$

where every noise term $\tilde{u}(t)$, $\tilde{y}(t)$ is independent of every other term and only $u(t)$ and $y(t)$ are known. Without loss of generality, all the variables may be assumed as having null mean value. Consequently the generic positive definite matrix Σ_k associated with the input–output noise–corrupted sequences may always be expressed as the sum of two terms

$$(4.7) \qquad \Sigma_k = \hat{\Sigma}_k + \tilde{\Sigma}_k$$

where

$$(4.8) \qquad \tilde{\Sigma}_k = diag\,[\tilde{\sigma}_y I_k, \tilde{\sigma}_u I_k] \geq 0$$

since no correlation has been assumed among the noise samples at different times. This condition is verified for additive white noise with variance $\tilde{\sigma}_y$ and $\tilde{\sigma}_u$ on the input–output sequences.

Under the previous assumptions the identification problem has been reconducted to the Frisch scheme. For its solution it is useful to analyse first how Problem 2.1 can be extended to the dynamic case.

Problem 4.3 — Given a sequence of increasing–dimension $(2k \times 2k)$ symmetric positive definite covariance matrices

$$(4.9) \qquad \Sigma_2 \ \ \Sigma_3 \ \ \ldots\ldots \ \ \Sigma_k \ \ \ldots.$$

find, for each k, all diagonal non–negative definite matrices $\tilde{\Sigma}_k = diag\,[\tilde{\sigma}_y I_k, \tilde{\sigma}_u I_k]$ such that

$$(4.10) \qquad \hat{\Sigma}_k = \Sigma_k - diag\,[\tilde{\sigma}_y I_k, \tilde{\sigma}_u I_k] \geq 0 \ .$$

Remark 4.2 — It is worth observing now that, unlike the algebraic case, for each k the noise space is always $\mathcal{R}_+^2$, while the parameter space is $\mathcal{R}^{2k}$.

First let us determine in the noise space the solutions $(\tilde{\sigma}_y',0)$ and $(0,\tilde{\sigma}_u')$ corresponding to the limit cases of noise affecting only the output or input sequences. This case can be considered as the natural extension to the dynamic case of the Least–Squares scheme. If we consider the same partition given for $\hat{\Sigma}_k$ in (4.3) on matrix Σ_k, then the following result can be proved.

Theorem 4.1 — The maximal admissible value for the output noise variance $\tilde{\sigma}_y'$ is the least eigenvalue of the matrix

$$(4.11a) \qquad \Sigma_k(yy) - \Sigma_k^T(uy)\,\Sigma_k^{-1}(uu)\,\Sigma_k(uy)$$

and, similarly, the maximal admissible value for the input noise variance $\tilde{\sigma}_u'$ is the least eigenvalue of the matrix

$$(4.11b) \qquad \Sigma_k(uu) - \Sigma_k(uy)\,\Sigma_k^{-1}(yy)\,\Sigma_k^T(uy)\ .$$

Proof — Since $\Sigma_k(uu)$ is nonsingular, the symmetric matrix $\Sigma_k - diag[\tilde{\sigma}_y I_k, 0 I_k]$ is equivalent [16] to

$$(4.12) \qquad diag[\Sigma_k(yy) - \Sigma_k^T(uy)\,\Sigma_k^{-1}(uu)\,\Sigma_k(uy) - \tilde{\sigma}_y I_k,\ \Sigma_k(uu)]\ .$$

Since $\Sigma_k(uu)$ is positive definite then condition (4.10) is satisfied if and only if

$$(4.13) \qquad \Sigma_k(yy) - \Sigma_k^T(uy)\,\Sigma_k^{-1}(uu)\,\Sigma_k(uy) - \tilde{\sigma}_y I_k \geq 0$$

i.e. only when $\tilde{\sigma}_y$ is the least eigenvalue of matrix (4.11a). $\tilde{\sigma}_u'$ can be obtained in similar fashion.

Remark 4.3 — Every matrix $\tilde{\Sigma}_k$ solving Problem 4.3 is characterized by parameters $\tilde{\sigma}_y$ and $\tilde{\sigma}_u$, whose values cannot exceed $\tilde{\sigma}_y'$ and $\tilde{\sigma}_u'$. With reference to Theorem 3.1, we can note that for each k the solution set of relation (4.10) describes, in the first quadrant of the $(\tilde{\sigma}_y, \tilde{\sigma}_u)$-plane, a curve whose concavity faces the origin. These curves cross the coordinate axes in the previously computed points.

The proof of the previous theorem suggests a way to solve Problem 4.3 as follows.

Algorithm 4.1 — We can assume that the input noise variance $\tilde{\sigma}_u = k\tilde{\sigma}_u'$ is a fraction $(0 \leq k \leq 1)$ of the maximal admissible value given by Theorem 4.1. Then the corresponding value $\tilde{\sigma}_y$ is the least eigenvalue of the matrix

$$(4.14) \qquad \Sigma_k(yy) - \Sigma_k^T(uy)\,[\Sigma_k(uu) - \tilde{\sigma}_u I_k]^{-1}\,\Sigma_k(uy)\ .$$

This procedure can be considered as an extension of Algorithm 3.1 to the dynamic case.

It has been noted, in the algebraic case, that different matrices can model the covariance of the noise corrupting the data and, consequently, different structural and/or parametric solutions can be found. The same problem arises in the dynamic case.

Previous results hold for every value of k. Since determination of the system order requires the increasing values of k to be tested, it is relevant to analyse the behaviour of the associated curves when k varies. This corresponds to a comparison of the admissible solution sets for different model orders. In this context the following result can be proved.

Theorem 4.2 — The solution sets of condition (4.10) for different values of k are non-crossing curves.

Proof — Denote with h and j two different values of the model order with $h > j$. Every point $(\tilde{\sigma}_y^h, \tilde{\sigma}_u^h)$ belonging to a curve associated to h corresponds to a non-negative definite matrix $\hat{\Sigma}_j$ since $\hat{\Sigma}_j$ can be obtained from $\hat{\Sigma}_h$ by deleting $2(h - j)$ rows and columns crossing on the main diagonal. Then, from the property stated in Remark 3.2, it follows that points $(\tilde{\sigma}_y^h, \tilde{\sigma}_u^h)$ will lie under or on the curve associated to j.

It is also important to observe that, since we assume that a system (4.1) has generated the noiseless data, for $k > n$ all the curves of type (4.10) have necessarily at least one common point, i.e. point $(\tilde{\sigma}_y^*, \tilde{\sigma}_u^*)$ corresponding to the true variances $\tilde{\sigma}_y^*$ and $\tilde{\sigma}_u^*$ of the noise affecting the output and the input of the system. The search for a solution for the identification problem can thus start from the determination in the noise space of this point. The following considerations can now be stated.

Remark 4.4 — With reference to the diagonal non-negative definite matrices $\tilde{\Sigma}_k = diag\,[\tilde{\sigma}_y^* I_k, \tilde{\sigma}_u^* I_k]$, where $\tilde{\sigma}_y^*$ and $\tilde{\sigma}_u^*$ are the actual variances of the noise affecting the output and the input of the system, the following properties hold:

— If $k \leq n$ the matrices $\hat{\Sigma}_k$ are positive definite.

— If $k > n$ the dimension of the null space of $\hat{\Sigma}_k$ and, consequently, the multiplicity of its least eigenvalue, is equal to $(k - n)$.

— For $k = (n + 1)$ matrix $\hat{\Sigma}_k$ is characterized by a linear dependence relation among its $2k$ vectors and the coefficients which link the k-th vector of $\hat{\Sigma}_k$ to the remaining ones are the system parameters.

— For $k > (n + 1)$ all linear dependence relations among the vectors of the matrix $\hat{\Sigma}_k$ are characterized by the same $(2n + 1)$ coefficients α_i, β_i.

Previous results do not exclude coincident curves associated to different values of k. If this situation can be observed for $k > n$ the noise variances cannot be univocally estimated and, consequently, the actual system is unidentifiable. The reader can easily evaluate what happens with the simple system $\hat{y}(t + 3) = \alpha \hat{u}(t)$ when the input $\hat{u}(t)$ is white noise with unknown variance $\hat{\sigma}_u$ and the additive noises are characterized by unknown variances $\tilde{\sigma}_y^*$ and $\tilde{\sigma}_u^*$. In this case variances $\tilde{\sigma}_y^*$, $\tilde{\sigma}_u^*$ and the parameter α cannot be univocally determined. On the other hand the same system can be identified if a suitable input sequence is applied.

The identification of dynamical systems with the Frisch scheme has been considered by other Authors [1], [2], [5], [9]. In their works it has been proved that when all measured variables include noise, it is not possible to uniquely identify a model but only a class of possible candidates. On the other hand, the results of this paper show that when increasing orders for the considered model are tested, a unique solution can be obtained. Nevertheless, this approach cannot be applied *sic et simpliciter* in the identification of real processes, since the hypotheses on the linearity, finite dimensionality and time independence of the system and on the additivity and whiteness of the noise are not usually verified. In such conditions the data cannot be explained by a model belonging to the considered class and the unicity of the solution is lost. The definition of a suitable criterion of selection in such cases is currently under investigation.

Conclusions

It has been shown that, unlike the algebraic case, the Frisch scheme in the identification of single–input single–output dynamical systems corrupted by additive white noise leads to a unique solution for both the order and the parameters if the input sequence is properly selected.
The extension of the Frisch scheme to the identification of multi–input, multi–output linear dynamic systems might lead to similar results. If the noise variances are univocally determined, the identification problem can be reduced also in this case to a realization one and then the associated algorithms to obtain a canonical or overlapping model of the system can be used [10], [11]. This will be the subject of future investigations.

References

[1] Anderson, B.D.O. *Identification of scalar errors–in–variables models with dynamics*, Automatica, vol. 21, pp. 709–716, 1985.

[2] Anderson, B.D.O. and M. Deistler *Dynamic errors–in–variables systems with three variables*, Automatica, vol. 23, pp. 611–616, 1987.

[3] Beghelli, S. and R.P. Guidorzi *Transformation between input–output multistructural models: properties and applications*, Int. J. Control, vol. 37, no. 6, pp. 1385–1400, 1983.

[4] Beghelli, S. and R.P. Guidorzi *Problemi di stima da dati affetti da rumore*, in Sistemi Dinamici 85, atti dell'incontro nazionale dei ricercatori del progetto nazionale M.P.I., Como, Villa Olmo, 1985.

[5] Deistler, M. *Linear errors–in–variables systems*, in Time Series and Linear Systems, ed. S. Bittanti, Springer–Verlag, Berlin, pp. 37–68, 1986.

[6] De Moor, B. *Mathematical concepts and techniques for modelling of static and dynamic systems*, Doctoral Thesis, Katholieke Universiteit Leuven, 1988.

[7] De Moor, B. and J. Vandewalle *A geometric approach to the maximal corank problem in the analysis of linear relations*, Proceedings of the 25th Conference on Decision and Control, Athens, Greece, pp. 1900–1995, 1986.

[8] De Moor, B. and J. Vandewalle *The uniqueness versus the non–uniqueness principle in the identification of linear relations from noisy data*, Proceedings of the 25th Conference on Decision and Control, Athens, Greece, pp. 1663–1665, 1986.

[9] Green, M. and B.D.O. Anderson *Identification of multivariable errors–in–variables models with dynamics*, IEEE Trans. on Aut. Control, vol. AC–31, pp. 467–471, 1986.

[10] Guidorzi, R.P. *Invariants and canonical forms for systems structural and parametric identification*, Automatica, vol. 17, pp. 117–133, 1981.

[11] Guidorzi, R.P. and S. Beghelli *Input–output multistructural models in multivariable systems identification*, Preprints of the 6th IFAC Symposium on Identification and System Parameter Estimation, Washington, D.C., pp. 461–465, 1981.

[12] Kalman, R.E. *Identification from real data*, Current Developments in the Interface: Economics, Econometrics, Mathematics, (edited by M. Hazewinkel and A.H.G. Rinnoy Kan), D. Riedel, Dordrecht, pp. 161–196, 1982.

[13] Kalman, R.E. *System identification from noisy data*, Dynamical Systems II, (edited by A.R.Bednarek and L. Cesari), Academic Press, pp. 135–164, 1982.

[14] Kalman, R.E. *Identification of noisy systems*, 50th Anniversary Symposium, Steklov Institute of Mathematics, USSR Academy of Sciences, Moskva, 1984

[15] Kalman, R.E. *The problem of prejudice in scientific modeling*, European Econometric Meeting, Budapest, Hungary, 1986.

[16] Satake, I. *Linear Algebra*, Marcel Dekker, Inc., New York, 1975.

The authors are with
Dipartimento di Elettronica, Informatica e Sistemistica
Università di Bologna
Viale del Risorgimento 2, 40136 Bologna, Italy

LINEARIZATION BASED ON EIGENVALUE ESTIMATES

Bernd Aulbach, Augsburg

Abstract: For the linearization of a system $\dot{x} = f(x)$ of nonlinear ordinary differential equations near a stationary state x_0 there exists a method from invariant manifold theory having the property that in any case the linearized system has the qualitatively same phase portrait near x_0 as the given nonlinear system. This method is based on the assumption that for each eigenvalue of the Jacobian of f at x_0 one is able to decide whether the real part is exactly zero or not. For critical systems with eigenvalues very close to the imaginary axis this generally cannot be done in practice since only estimates for the eigenvalues are available. In this paper we present a modification of the above-mentioned method to the case where the location of eigenvalues is known only approximately.

1   Introduction

First of all we pose a problem which is the starting point of this paper and which frequently arises in applications. A real-world phenomenon is mathematically modelled and the model turns out to be a system of nonlinear ordinary differential equations $\dot{x} = f(x)$. One is interested in the behavior of its solutions near a stationary state x_0. Since the equation is nonlinear in general it cannot be approached directly and therefore the straightforward idea is to linearize the equation near x_0, i.e. to compute the Jacobian $Df(x_0)$ of f at x_0 and to study the linear system $\dot{x} = Df(x_0)x$ first with the aim of carrying over the information to the original nonlinear system. The phase portrait of the approximating linear system now depends on the location of the eigenvalues of $Df(x_0)$ in the complex plane, in fact if there are eigenvalues on the imaginary axis it is well known that the linear system alone cannot describe, not even qualitatively, the phase portrait of the nonlinear system. Thus it is vital to find out if $Df(x_0)$ has eigenvalues with zero real-parts. This is where usually the trouble begins because the eigenvalues cannot be determined exactly, one rather depends on numerical methods for their determination which in turn give only estimates. Perhaps one is able to show in a realiable way that there are no eigenvalues close to the imaginary axis, then the linearization problem is trivial. However, if the numerical computations indicate eigenvalues with zero real-parts one cannot be sure whether the real-parts are indeed zero or whether they are nonzero but just too small to be detected by the numerical method. This being-zero-or-not however is the crucial point for the validity of an accurate linearization method. Under this "vague" assumption that there are eigenvalues with very small real-parts, perhaps zero, perhaps not, there has not been any mathematically proven linearization procedure yet. It is the aim of this note to present one.

To some extend this paper may be viewed as a continuation of our previous paper "Trouble with Linearization" [1] where we have cast a critical view on some of the common linearization techniques for systems of ordinary differential equations near a stationary state. In [1] we have done essentially two things. Firstly we have explained the invariant manifold linearization technique and indicated that the thus linearized system in any case exhibits the qualitatively same phase portrait as the original nonlinear system near the equilibrium under consideration. Secondly we have compared three frequently used linearization techniques with the invariant manifold technique and by means of counterexamples we have demonstrated that each of them may yield incorrect information about a given nonlinear system. Thus, in a sense, the paper [1] sent out the message that in a general case one should only use what we called the "center manifold procedure" because this is the only method which is mathematically secured in any situation including the critical cases. However, in the concluding remarks in [1] we have indicated that from a practical point of view the center manifold procedure has one essential drawback namely, apart from academic examples, its assumptions usually cannot be verified. This is because the method requires a decomposition of the underlying equation for which it is necessary to determine which of the eigenvalues of the linear part of the right-hand side lie exactly on the imaginary axis and which don't. Of course, in a mathematical model of a real phenomenon this cannot be done since the system contains uncertainties leading to parameters of the equation which are known only approximately. Based on eigenvalue estimates the only thing that can be said in general is whether the real parts of the eigenvalues are small (in some sense) or not. Our paper [1] ended: "For this "realistic" spectral decomposition there is no mathematically secured device for a linearization yet. A center manifold theory based on such a spectral assumption, however, is about to come". It is the aim of this paper to pick up this loose end. We will present that much of the generalized center manifold theory as it is relevant in the context of linearization.

2   The Problem

In order to deal with the above-mentioned problem rigorously we have to provide a mathematical framework. To this end let us consider throughout a system of autonomous ordinary differential equations

$$\dot{x} = f(x) \tag{1}$$

and suppose it has the coordinate origin $x = 0$ as a stationary solution, i.e. $f(0) = 0$. Furthermore suppose f is sufficiently smooth, e.g. continuously differentiable, to guarantee unique solutions and the validity of the decomposition

$$f(x) = Df(0)x + r(x) \quad , lim_{x \to 0} \frac{r(x)}{|x|} = 0$$

where $Df(0)$ is the Jacobian of f at 0.

The problem is to study the phase portrait of (1) in a neighborhood of the equilibrium point 0. The key idea is to replace the system (1) by a simpler, possibly linear system which exhibits near 0 the qualitatively same solution behavior as (1). But what does this mean precisely? In this paper we say two systems are **qualitatively the same** near 0 if there exists a bijective mapping h from an neighborhood M of 0 onto a neighborhood N of 0 such that both the mapping h and its inverse h^{-1} are continuous and h transforms trajectories (including orientation) of one equation onto trajectories of the other and vice versa. In the literature this is called **topological equivalence**.

By means of the concept of topological equivalence the above idea reads as follows: For (1) find a simpler, possibly linear system which is topologically equivalent near 0 to (1). A meanwhile classical result in this direction is the Hartmann-Grobman Theorem (see Hartmann [3], IX. Theroem 7.1) which says that (1) is topologically equivalent near 0 to the completely linearized system

$$(2) \qquad\qquad\qquad \dot{x} = Df(0)x$$

if $Df(0)$ has no eigenvalue with zero real part. On the other hand if $Df(0)$ does have eigenvalues with vanishing real parts then it is well known that the variety of nonlinear local phase portraits cannot be described by linear systems alone and this means that a complete linearization like that above cannot be excepted. The aim in this more general case must be then to reduce the given nonlinear system to a topologically equivalent system which has as many linear parts as possible. The standard result in this direction is the generalized Hartman-Grobman Theorem due to Palmer [4] which provided the basis for the center manifold procedure described in [1]. This method need not be repeated here because it will be presented in a more genereal setting below.

Instead we want to emphasize the problem associated with the set-up of this procedure. The differential systems has to be given in a form where the eigenvalues of $Df(0)$ on the imaginary axis are separated from those off the imaginary axis regardless of their distance from this axis. This means that if for instance $Df(0)$ has eigenvalues $-10^{20}, -10^{-20}, 0, 10^{-20}, 10^{20}$ the system has to be decomposed according to the three groups $\{-10^{20}, -10^{-20}\}, \{0\}$ and $\{10^{-20}, 10^{20}\}$. This might be reasonable from a theoretical point of view, from a practical point of view this is nonsense. A resulution of the cluster of three small eigenvalues will be impossible in general and even if it is possible the discrepancy between fast and slow "modes" urgently suggests the splitting $\{-10^{20}\}, \{-10^{-20}, 0, 10^{-20}\} and \{10^{20}\}$. Therefore a "realistic" assumption on the spectrum of $Df(0)$ is that it lies in certain vertical strips in the complex plane. If non of those strips contains the imaginary axis then one is in the classical case and the Hartman-Grobman Theorem which provides a complete linearization as above. The interesting situation arises when the imaginary axis is contained in one of the spectral strips. Without further information on the eigenvalues so far there has not been any mathematically proven linearization procedure. We will describe one in the course of this paper.

3   <u>The Theorem</u>

In order to present the mathematical background we first formulate the theorem providing the basis for the linearization procedure. We suppose it is known that the Jacobian $Df(0)$ has a group of eigenvalues close to the imaginary axis and that this group can be separated by vertical straight lines from the remaining eigenvalues. Then we change coordinates so as to bring $Df(0)$ into block-diagonal form $diag(A^-, A, A^+)$ where A contains all eigenvalues with small real-parts and A^-, A^+ the remaining ones in the left or right half-plane, respectively. Finally, according to this decomposition we split the state vector x into three components u, v, w. Altogether we suppose to have the given differential system in the form

$$\dot{u} = A^- u + r^-(u, v, w)$$
$$(3) \qquad \dot{v} = Av + r(u, v, w)$$
$$\dot{w} = A^+ w + r^+(u, v, w)$$

where, say, $u \in \mathbb{R}^k, v \in \mathbb{R}^m, w \in \mathbb{R}^n$. For the matrices A^-, A, A^+ and the functions r^-, r, r^+ we require the following respective conditions:

(H1)   There exist real constants $\alpha < \beta \leq 0 \leq \gamma < \delta$ such that for the real-part of each eigenvalue $\lambda^-, \lambda, \lambda^+$ of A^-, A, A^+, respectively, the estimate $Re\lambda^- < \alpha < \beta < Re\lambda < \gamma < \delta < Re\lambda^+$ holds true.

(H2)   The function $r = (r^-, r, r^+)$ is continuously differentiable in a neighborhood N of $(0, 0, 0)$ and vanishes together with its Jacobian at $(0, 0, 0)$.

We wish to emphasize that those assumptions (H1) and (H2) virtually impose no restrictions on system (3), at least if smoothness is of no concern. They just fomalize the natural set-up for the problem posed above. Nevertheless we get the following useful result.

<u>Theorem</u>: Under the above assumptions (H1) and (H2) there exist neighborhoods U, V, W of the coordinate origin in $\mathbb{R}^k, \mathbb{R}^m, \mathbb{R}^n$, respectively, and continuously differentiable functions

$$f^- : V \to U \quad , \quad f^+ : V \to W$$

which vanish together with their Jacobians at $O \in \mathbb{R}^m$ such that the following is true:

a) the graph of the function $(f^-, f^+) : V \to U \times W$ is an invariant manifold with respect to system (3), i.e. each solution of (3) starting on

$$G := \{(f^-(v), v, f^+(v) : v \in V\}$$

remains on G (in both time directions) as long as its v-coordinate remains in V,

b) System (3) is near $(0, 0, 0)$ topologically equivalent to the partially linearized system

$$\begin{aligned}
\dot{u} &= A^- u \\
\dot{v} &= Av + r(f^-(v), v, f^+(v)) \\
\dot{w} &= A^+ w
\end{aligned}$$

(4)

i.e. there exists a continuous mapping h from $U \times V \times W$ onto a neighborhood X of $(0, 0, 0)$ with continuous inverse $h^{-1} : X \to U \times V \times W$ such that the following is true: i) If $(u(t), v(t), w(t))$ is a solution of (3) in $U \times V \times W$ then $h(u(t), v(t), w(t))$ is a solution of (4) in X, and conversely, ii) if $(u(t), v(t), w(t))$ is a solution of (4) in X, then $h^{-1}(u(t), v(t), w(t))$ is a solution of (3) in $U \times V \times W$.

<u>Remarks</u>: 1. The statement of the theorem is a local one and in general the local neighborhoods U, V and W depend on the data of the system. Roughly speaking one can say the following. The size of U, V and W is directly proportional to the "widths" $\beta - \alpha$ and $\delta - \gamma$ of the gaps between the spectral strips and indirectly proportional to the norm of the Jacobian of the nonlinearity r near $(0, 0, 0)$.

2. As in the "classical" case of the Hartman-Grobman Theorem the mapping h establishing the qualitative equivalence cannot be expected to be differentiable (cf. Hartman [3,Ch.IX]) even with analytic right-hand side of (3).

3. The special case $\beta = 0 = \gamma$ corresponds to the previously studied situation. In this case the manifold G is a local center manifold and the theorem is a local version of the generalized Hartman-Grobman Theorem (cf. Palmer [4]).

4. The theorem says that the differential system which is relevant for the linearization has the dimension of the matrix A of critical eigenvalues. The criticality of eigenvalues, however, depends on the given problem; to be precise, it depends on which eigenvalues

close to the imaginary axis can be clearly and reliably separated from the remaining ones. This, of course, is a matter of methods (i.e. numerical) outside the theory of differential equations. In any case, when applying the theorem to a given system one should try to make the dimension of A as low as possible by shifting β and γ as close to 0 as possible.

4   The Procedure

The theorem in the previous section in theoretical in the sense that it does not explain how to find the linearized system (4), i.e. the function $(f^-, f^+) : V \to U \times W$. The linearization procedure for $\dot{x} = f(x), f(0) = 0$ based on this theorem works as follows:

1st step: Determine the eigenvalues of $Df(0)$. If none of them are zero the linear system $\dot{x} = Df(0)x$ describes the local phase portrait of $\dot{x} = f(x)$ near 0. If some of the eigenvalues have real-parts which cannot be proved to be non-zero then separate those from the remaining eigenvalues by vertical straight lines in the complex plane and make a linear change of coordinates in order to put system $\dot{x} = f(x)$ into the form (3).

2nd step: Determine the couple of functions $(f^-, f^+) : V \to U \times W$ as a local solution of the Cauchy-problem

$$
\begin{aligned}
&Df^-(v)[Av + r(f^-(v), v, f^+(v))] = A^- f^-(v) + r^-(f^-(v), v, f^+(v)) \\
&Df^+(v)[Av + r(f^-(v), v, f^+(v))] = A^+ f^+(v) + r^+(f^-(v), v, f^+(v))
\end{aligned}
\tag{5}
$$

$$
f^-(0) = 0 \in \mathbb{R}^k, f^+(0) = 0 \in \mathbb{R}^n, Df^-(0) = 0 \in \mathbb{R}^{k \times m}, Df^+(0) = 0 \in \mathbb{R}^{n \times m}. \tag{6}
$$

3rd step: Study the local phase portrait of the "reduced" equation

$$
\dot{v} = Av + r(f^-(v), v, f^+(v)).
$$

The system (4) then describes the original equation near $(0,0,0)$ in a qualitatively correct way.

Remarks: The first step needs no further mention, the second one, however, does. It consists of an initial value problem for a system of partial differential equations which in general cannot be solved in closed form. The standard way of dealing with it is to make an Ansatz for the unknown functions f^-, f^+ in form of a Taylor series in order to determine successively its coefficients. Once an approximation of f^-, f^+ has been obtained this way one has an approximation of the reduced equation appearing in the third step. All of this will be demonstrated by means of the following example.

5   An Example

In order to avoid the cases which are too special we have chosen a four-dimensional system where the first step of the linearization procedure already has been done. We take u and w one-dimensional with $A^- = -1, A^+ = 1$ and $v = (v_1, v_2)$ two-dimensional with A being in canonical form with eigenvalues $\epsilon \pm i\alpha$. Our assumptions require that the real part ϵ is a quantity which is only known to be small relative to the eigenvalues -1 and 1 of A^- and A^+, respectively. The imaginary part α may be arbitrary.

We write the given system (formally) expanded with arbitrary coefficients

$$\dot{u} = -u + \sum_{p+q+r+s=2}^{\infty} r^-_{pqrs} u^p v_1^q v_2^r w^s$$

$$\dot{v}_1 = \epsilon v_1 - \alpha v_2 + \sum_{p+q+r+s=2}^{\infty} r^1_{pqrs} u^p v_1^q v_2^r w^s$$

$$\dot{v}_2 = \alpha v_1 + \epsilon v_2 + \sum_{p+q+r+s=2}^{\infty} r^2_{pqrs} u^p v_1^q v_2^r w^s$$

$$\dot{w} = w + \sum_{p+q+r+s=2}^{\infty} r^+_{pqrs} u^p v_1^q v_2^r w^s$$

where the indices p, q, r, s are non-negative integers. For $f^- : \mathbb{R}^2 \to \mathbb{R}$ and $f^+ : \mathbb{R}^2 \to \mathbb{R}$ we make the Ansatz

$$f^-(v_1, v_2) = \sum_{i+j=2}^{\infty} f^-_{ij} v_1^i v_2^j$$

$$f^+(v_1, v_2) = \sum_{i+j=2}^{\infty} f^+_{ij} v_1^i v_2^j$$

where the first three terms with indices $00, 10$ and 01 need not appear because of the initial condition (6). Inserting this Ansatz into the system of equations (5) yields from the first equation

(7)

$$\sum_{i+j=2}^{\infty} i f_{ij}^- v_1^{i-1} v_2^j \cdot \Big[\epsilon v_1 - \alpha v_2 + \sum_{p+q+r+s=2}^{\infty} r_{pqrs}^1 \Big(\sum_{i+j=2}^{\infty} f_{ij}^- v_1^i v_2^j \Big)^p v_1^q v_2^r \Big(\sum_{i+j=2}^{\infty} f_{ij}^+ v_1^i v_2^j \Big)^s \Big] +$$

$$\sum_{i+j=2}^{\infty} j f_{ij}^- v_1^i v_2^{j-1} \cdot \Big[\alpha v_1 + \epsilon v_2 + \sum_{p+q+r+s=2}^{\infty} r_{pqrs}^2 \Big(\sum_{i+j=2}^{\infty} f_{ij}^- v_1^i v_2^j \Big)^p v_1^q v_2^r \Big(\sum_{i+j=2}^{\infty} f_{ij}^+ v_1^i v_2^j \Big) \Big] =$$

$$-\sum_{i+j=2}^{\infty} f_{ij}^- v_1^i v_2^j + \sum_{p+q+r+s=2}^{\infty} r_{pqrs}^- \Big(\sum_{i+j=2}^{\infty} f_{ij}^- v_1^i v_2^j \Big)^p v_1^q v_2^r \Big(\sum_{i+j=2}^{\infty} f_{ij}^+ v_1^i v_2^j \Big)^s$$

and a similar relation (7') from the second equation. Next we compare coefficients of like powers beginning with 2nd degree (of homogenuity), i.e. $v_1^2, v_1 v_2$ and v_2^2. This gives three equations for the unknown coefficients $f_{20}^-, f_{11}^-, f_{02}^-$ which written as a linear system of algebraic equations reads

$$(8) \qquad (f_{20}^-, f_{11}^-, f_{02}^-) \begin{pmatrix} 2\epsilon + 1 & -2\alpha & 0 \\ \alpha & 2\epsilon + 1 & -\alpha \\ 0 & 2\alpha & 2\epsilon + 1 \end{pmatrix} = (r_{0200}^-, r_{0110}^-, r_{0020}^-).$$

The determinant of the coefficient matrix is $(2\epsilon+1)\left[(2\epsilon+1)^2 + 4\alpha^2\right]$ and this is different from 0 if $\epsilon \neq -\frac{1}{2}$ regardless of the value of α. Thus, if ϵ is known to be smaller than $\frac{1}{2}$ in absolute value the first three coefficients of f^- can be determined uniquely. Similarly, the first three coefficients of f^+ can be determined uniquely from the equation (7') we have not written out explicitly. In fact, the determining equation is the same as (8) when replacing all "minus-indices" by "plus-indices".

Comparing coefficients of the four 3rd-degree terms $v_1^3, v_1^2 v_2, v_1 v_2^2$ and v_2^3 in (7) yields a linear system of equations

$$(f_{30}^-, f_{21}^-, f_{12}^- f_{03}^-) \begin{pmatrix} 3\epsilon + 1 & -3\alpha & 0 & 0 \\ \alpha & 3\epsilon + 1 & -2\alpha & 0 \\ 0 & 2\alpha & 3\epsilon + 1 & -\alpha \\ 0 & 0 & 3\alpha & 3\epsilon + 1 \end{pmatrix} = (r_{0300}^-, r_{0210}^-, r_{0120}^- r_{0030}^-) +$$

(9)

$$(r_{1100}^- f_{20}^- + r_{0101}^- f_{20}^+, r_{1010}^- f_{20}^- + r_{0011}^- f_{20}^+ + r_{0101}^- f_{11}^+ + r_{1100}^- f_{11}^-,$$

$$r_{1100}^- f_{02}^- + r_{0101}^- f_{02}^+ + r_{1010}^- f_{11}^- + r_{0011}^- f_{11}^+, r_{1010}^- f_{02}^- + r_{0011}^- f_{20}^+)$$

where it should be noted that the right-hand side contains apart from the given r-coefficients only f-coefficients which have been determined before. The determinant of the matrix in (9) is

$$(3\epsilon + 1)^2 \left[(3\epsilon + 1)^2 + 7\alpha^2\right] + 3\alpha^2(3\epsilon + 1)^2 + 9\alpha^4$$

and this is positive if $|\epsilon| < \frac{1}{3}$ regardless of the value of α. Thus the 3rd-degree coefficients of f^- can be computed if ϵ is known to be smaller than $\frac{1}{3}$ in modulus. Accordingly the 3rd-degree terms of f^+ can be found and, in fact, also the determination of the higher-degree terms of f^- and f^+ follows this pattern. The previously determined coefficients enter the right-hand side, the inhomogeneous part of the determining equation. It should be noticed that as seen in the example all coefficients with the same degree have to be handled simultaneously. This leads to systems a linear algebraic equations of higher and higher dimensions. It is also worth noting that the unique solvability of the determining equation for the f^-- and f^+-coefficients, as seen in the above cases, depends on ϵ. In fact, it is to be expected that the n-th-degree coefficients of f^- and f^+ can be determined uniquely only if $|\epsilon| < \frac{1}{n}$. This demonstrates in which way the approximability of the functions f^-, f^+ and herewith the reduced equation depends on the width of the strip around the imaginary axis which is known to contain the critical eigenvalues. This effect is an examplification of the theoretical fact that - in contrast to the "classical" center manifold theory - the functions f^- and f^+ may not be as smooth as the right-hand side of the underlying differential equation.

6   The Proof

The proof of the above theorem has not been published yet but it is too long and technical to be presented here. It is based on a rather general theorem on invariant manifolds in [2].

References

[1] Aulbach, B.: Trouble with linearization, in "Mathematics in Industry", 229-246, Stuttgart: Teubner 1984.

[2] Aulbach, B.: Hierarchies of invariant manifolds. J. Nig. Math. Soc., to appear.

[3] Hartman, P.: Ordinary differential equations. New York: Wiley 1964

[4] Palmer, K.J.: Qualitative behavior of a system of ODE near an equilibrium point - A generalization of the Hartman-Grobman theorem. Preprint, Inst. für Angew. Math. Univ. Bonn 1980

Bernd Aulbach

Institut für Mathematik

Universität Augsburg

Memminger Str. 6

D-8900 Augsburg

AN ADAPTIVE LINEAR APPROACH
TO NONLINEAR FILTERING

by

Giovanni B. Di Masi and *Wolfgang J. Runggaldier*

Dipartimento di Matematica Pura ed Applicata

Università di Padova, Padova, Italy

Abstract : A nonlinear filtering problem is considered for a dynamic model with piecewise linear coefficient and with initial condition and disturbances distributed according to finite mixtures of normal distributions.
It is shown that, for vanishing variances of the normal distributions relative to the signal process, the infinite-dimensional optimal filter coincides with the finite dimensional filter for a suitably defined adaptive linear model.
The results obtained illustrate, in the particular situation considered here, the relevance of adaptive linear techniques for the approximation of optimal nonlinear filters.

1. INTRODUCTION AND PROBLEM STATEMENT

Let $\{(x_t, y_t), t = 1,2,...\}$ be a partially observed, discrete-time stochastic process, with $x_t \in R^n$ the state or signal process and $y_t \in R^m$ the observation or measurement process. The filtering problem consists in estimating the signal x_t on the basis of the actually observed measurements $y^t := \{y_1, y_2, ... y_t\}$. The most complete solution to this problem consists in providing the conditional law of x_t given the history y^t of the observations. In what follows this conditional law will admit a probability density function which will be denoted by $p(x_t | y_t)$.

It is well known that in general the solution to the filtering problem is infinite-dimensional and that one of the few situations in which a finite-dimensinal solution is available is when the state and observation processes are described by a linear stochastic system, namely

(1.a) $\qquad x_{t+1} = A_t x_t + B_t + \Sigma^x v_{t+1}$

(1.b) $\qquad y_t \quad = C_t x_t + D_t + \Sigma^y w_t$

where $\{v_t\}$ and $\{w_t\}$ are independent standard white noise processes, i.e. sequences of indeddpendent random variables with standard normal law and the initial condition x_0 is normal with mean $\hat{x}_0$ and covariance matrix P_0, and is furthermore independent of $\{(x_t, y_t)\}$. In fact, such model describes a Gaussian stochastic process $\{(v_t, w_t)\}$, so that the conditional density can be characterized in terms of its mean $\hat{x}_t$ and covariance matrix P_t. The algorithm which provides recursive relations for the computation of these quantities is the widely used Kalman filter (KF)[2].

For a nonlinear model such as

$$(2.a) \qquad x_{t+1} = a_t(x_t) + n^x_{t+1}$$

$$(2.b) \qquad y_t \;\; = c_t(x_t) + n^y_t$$

it is necessary in general to resort to approximate solutions. The most popular of such solutions, especially in engineering applications, is the so-called extended Kalman filter (EKF)[2], which essentially consists, at each time t, in a linearization of a_t and c_t in (2) around the most recent estimate $\hat{x}_t$ obtained and in the use of the KF algorithm for the derivation of $\hat{x}_{t+1}$ and P_{t+1}.

The EKF algorithm has been highly appreciated because of its very simple structure. Nevertheless it has some drawbacks which make its use rather critical in several situations. In fact the underlying idea of the algorithm is the approximation of the a posteriori density $p(x_t \mid y^t)$ by a normal density ; furthermore, in order to exploit the results for the linear case, the EKF assumes normally distributed noises and initial conditions. It is however well known that in the nonlinear situation both these facts are far from being realistic. Finally the EKF possesses only local asymptotic (in the sense of vanishing noise) properties [2 Ch.8 Th.2.1].

In the present paper we shall consider a reasonably simple nonlinear model, namely one with a_t and c_t in (2) continuous and piecewise affine, i.e. given by

$$(3.a) \qquad a_t(x) = \sum_{i=1}^{N} [A_t(i)x + B_t(i)]\, I_{\pi_i}(x)$$

$$(3.b) \qquad c_t(x) = \sum_{i=1}^{N} [C_t(i)x + D_t(i)]\, I_{\pi_i}(x)$$

where $\{\pi_i\colon i=1,\dots,N\}$ is a finite partition of R^n. Such problems have recently received increasing attention in the context of nonlinear filtering theory [3,4,8,9,10].

We shall also assume noises and initial condition distributed according to mixtures of normal random variables i.e., denoting by $\mathfrak{N}(\cdot;m,S)$ the normal density with mean vector m and covariance matrix S and writing ~ for " is distributed according to ",

$$(4.a) \qquad x_0 \sim \sum_{i=1}^{N^0} \alpha^0_i \mathfrak{N}(\cdot;\mu^0_i,\varepsilon\Sigma^0_i)$$

$$(4.b) \qquad n^x_t \sim \sum_{i=1}^{N^x} \alpha^x_i \mathfrak{N}(\cdot;\mu^x_i,\varepsilon\Sigma^x_i)$$

$$(4.c) \qquad n^y_t \sim \sum_{i=1}^{N^y} \alpha^y_i \mathfrak{N}(\cdot;\mu^y_i,\Sigma^y_i)$$

Our goal will be to study the connection between the optimal (infinite-dimensional) solution to such problem and the (finite dimensional) solutions to the linear problems corresponding to the various linear behaviours of a_t and c_t in (3), i.e. problems corresponding to models of the form

$$(5.a) \qquad x^i_{t+1} = A_t(i)x^i_t + B_t(i) + \varepsilon\Sigma^x v_{t+1} \; ; \qquad\qquad x_0 \sim \mathfrak{N}(\cdot;\mu,\varepsilon\Sigma)$$

$$(5.b) \qquad y^i_t \; = C_t(i)x^i_t + D_t(i) + \Sigma^y w_t$$

In particular it will be shown that for vanishing ε, namely when x_0 and n^x_t in (4) tend to discrete random variables, the optimal filter can be approximated in terms of a suitable mixture of filters for models as in (5).

2.CONSTRUCTION OF AN ADAPTIVE LINEAR MODEL

We shall be concerned here with the derivation of an adaptive linear model which, for each value of a suitably chosen parameter θ, represents a system of the form (5). The nonlinear filtering problem for this model will be explicitly computable in terms of a mixture of linear filters and will provide the asymptotic (in the sense of vanishing ε) approximate solution to the original filtering problem for model (2)-(4). To this end, notice first that it is possible to give a representation of the initial condition x_0 in terms of a finite-valued random parameter θ_0 in the following way . Let θ^0 be distributed according to

$$(6) \qquad P(\theta^0 = \theta^0_i) = \alpha^0_i \qquad\qquad (i = 1,2,...,N^0)$$

and let μ^0 and Σ^0 be mappings such that

$$(7) \qquad \mu^0(\theta^0_i) = \mu^0_i \quad , \quad \Sigma^0(\theta^0_i) = \Sigma^0_i \qquad (i = 1,2,...,N^0)$$

and let $v_0 \sim \mathfrak{N}(\cdot;0,I)$. Then a representation of x_0 in (4) is given by

$$(8) \qquad x_0 = \mu^0(\theta^0) + \varepsilon\Sigma^0(\theta^0)v^0$$

In an analogous way we have for the noises the representation s

$$(9.a) \qquad n^x_t = \mu^x(\theta^x) + \varepsilon\Sigma^x(\theta^x)v_t$$

$$(9.b) \qquad n^y_t = \mu^y(\theta^y) + \Sigma^y(\theta^y)w_t$$

where θ^x and θ^y are finite-valued random variables distributed according to

$$(10.a) \qquad P(\theta^x = \theta^x_i) = \alpha^x_i \qquad\qquad (i = 1,2,...,N^x)$$

$$(10.b) \qquad P(\theta^y = \theta^y_i) = \alpha^y_i \qquad\qquad (i = 1,2,...,N^y)$$

the mappings $\mu^x, \Sigma^x, \mu^y, \Sigma^y$ are defined by

$$(11.a) \qquad \mu^x(\theta^x_i) = \mu^x_i \quad , \quad \Sigma^x(\theta^x_i) = \Sigma^x_i \qquad (i = 1,2,...,N^x)$$

$$(11.b) \qquad \mu^y(\theta^y_i) = \mu^y_i \quad , \quad \Sigma^y(\theta^y_i) = \Sigma^y_i \qquad (i = 1,2,...,N^y)$$

and $\{v_t : t = 0,1,...\}$ and $\{w_t : t = 0,1,...\}$ are independent standard white Gaussian noises. Notice that the mutual dependence of the random variables x_0, n^x_t, n^y_t in (8),(9) is related to the mutual dependence of the random variables $\theta^0, \theta^x, \theta^y$ so that various dependence structures are possible according to the various choices of the joint distribution $p(\theta^0, \theta^x, \theta^y)$. In particular it is worthwhile to note that in the finite horizon situation , allowing a time dependence of the θ variables , it is possible, using finite-valued θ's, to cover also the case when noises n^x_t and n^y_t are sequences of independent random variables. In what follows we shall assume that for the given nonlinear filtering problem (2)-(4), initial condition and disturbances are exactly described by (8) and (9) with a given joint distribution $p(\theta^0,\theta^x,\theta^y)$. Furthermore we shall let $\theta'=[\theta^{0'},\theta^{x'},\theta^{y'}]'$ and use in our notation the argument θ regardless of which components of θ are actually involved.

For the construction of the adaptive linear model we consider the processes ξ_t and η_t defined by

$$(12.a) \qquad \xi_{t+1} = a_t(\xi_t) + \mu^x(\theta) \qquad \xi_0 = \mu^0(\theta)$$

$$(12.b) \qquad \eta_t = \sum_{i=1}^{N} i\, I_{\pi_i}(\xi_t)$$

Notice that ξ_t is an asymptotic (for vanishing ε)approximation of the state process x_t, and η_t selects the element of the partition $\{\pi_i\}$ where ξ_t lies.

We now consider the model

$$(13.a) \qquad \bar{x}_{t+1} = A_t(\eta_t)\bar{x}_t + B_t(\eta_\tau) + \mu^x(\theta) + \varepsilon\Sigma^x(\theta)v_t$$

$$\bar{x}_0 = \mu^0(\theta) + \varepsilon\Sigma^0(\theta)$$

$$(13.b) \qquad \bar{y}_t = C_t(\eta_t)\bar{x}_t + D_t(\eta_t) + \mu^y(\theta) + \Sigma^y(\theta)w_t$$

Taking into account that η_t is a deterministic function of θ and combining terms, (13) can be written with abuse of notation as

$$(14.a) \qquad \bar{x}_{t+1} = A_t(\theta)x_t + B_t(\theta) + \varepsilon\Sigma^x(\theta)v_t$$

$$\bar{x}_0 = \mu^0(\theta) + \varepsilon\Sigma^0(\theta)$$

$$(14.b) \qquad \bar{y}_t = C_t(\theta)x_t + D_t(\theta) + \Sigma^y(\theta)w_t$$

which is an adaptive linear model , namely a linear model depending on a (random) parameter θ. The way in which such model has been constructed suggests that it can be considered as a reasonable candidate for the asymptotic approximation of the optimal filter for the original problem (2)-(4). It will be shown in the next section that this asymptotic approximation actually holds.

3. ASYMPTOTIC RESULTS

In this section we shall first discuss the solution to the nonlinear filtering problem corresponding to the adaptive model (14). Then we shall show that for vanishing ε this filter coincides with optimal filter for the original model (2)-(4). In what follows we shall denote by $\bar{p}^{\varepsilon}_t(\cdot|y^t)$ and $p^{\varepsilon}_t(\cdot|y^t)$ the filtering densities corresponding to the original model (2)-(4) and to the adaptive model (14) respectively.

The existence of an explicitly computable finite-dimensional solution to the filtering problem for the adaptive linear model (14) can be easily realized. In fact we have, with obvious notation

$$(15) \qquad \bar{p}^{\varepsilon}_t(x|y^t) = \sum_{\theta} \bar{p}^{\varepsilon}_t(x|y^t,\theta) \, \bar{p}^{\varepsilon}_t(\theta|y^t)$$

The term $\bar{p}^{\varepsilon}_t(x|\theta,y^t)$ corresponds to the linear model (14) with fixed θ and can therefore be computed using a Kalman filter ; the term $\bar{p}^{\varepsilon}_t(\theta|y^t)$ is a distribution for a finite-valued variable and can be computed via classical Bayesian techniques. An algorithm for the effective computation of the filter, based on resuts in [5] , is given in [7] and for convenience is stated here without proof. In the sequel $\propto$ will denote proportionality and for a nonsingular matrix M we let $M^{-2} := (M^{-1})'M^{-1}.$

Theorem 1: The conditional joint distribution $\bar{p}^{\varepsilon}_t(x,\theta|y^t)$ for the adaptive model (14) satisfies

$$(16) \qquad \bar{p}^{\varepsilon}_t(x,\theta|y^t) \propto [\, \det\Sigma^0(\theta) \, (\det \Sigma^x(\theta) \det \Sigma^y(\theta))^t]^{-1} \cdot$$

$$p(\theta) \left[\prod_{s=0}^{t} \det N_s(\theta) \right]^{-1/2} \exp\left[-\tfrac{1}{2} x'M_t(\theta)x + x'h^y_t(\theta) + k^y_t(\theta) \right]$$

where N_t, M_t, h^y_t and k^y_t satisfy the recursive relations

$$N_t(\theta) = A'_{t-1}(\theta)(\Sigma^x(\theta))^{-2}A_{t-1}(\theta) + M_{t-1}(\theta); \quad N_0(\theta) = I$$

$$M_t(\theta) = C'_t(\theta)(\Sigma^y(\theta))^{-2}C_t(\theta) + (\Sigma^x(\theta))^{-2}$$

$$\qquad - (\Sigma^x(\theta))^{-2}A_{t-1}(\theta)N_t^{-1}(\theta)A'_{t-1}(\theta)(\Sigma^x(\theta))^{-2}; \quad M_0(\theta) = (\Sigma^0(\theta))^{-2}$$

$$h^y_t(\theta) = C'_t(\theta)(\Sigma^y(\theta))^{-2}[y_t - D_t(\theta)] + (\Sigma^x(\theta))^{-2}B_{t-1}(\theta)$$

$$+(\Sigma^x(\theta))^{-2}A_{t-1}(\theta)N_t^{-1}(\theta)[h^y_{t-1}(\theta) - A'_{t-1}(\theta)(\Sigma^x(\theta))^{-2}B_{t-1}(\theta)];$$

$$h^y_0(\theta) = (\Sigma^0(\theta))^{-2}\mu^0(\theta)$$

$$k^y_t(\theta) = k^y_{t-1}(\theta) + \frac{1}{2}[h^y_{t-1}(\theta)' - B'_{t-1}(\theta)(\Sigma^x(\theta))^{-2}A_{t-1}(\theta)]N_t^{-1}(\theta) \cdot$$

$$\cdot [h^y_{t-1}(\theta) - A'_{t-1}(\theta)(\Sigma^x(\theta))^{-2}B_{t-1}(\theta)] - \frac{1}{2}B'_{t-1}(\theta)(\Sigma^x(\theta))^{-2}B_{t-1}(\theta)$$

$$-\frac{1}{2}[y_t - D_t(\theta)]'(\Sigma^y(\theta))^{-2}[y_t - D_t(\theta)];$$

$$k^y_0 = -\frac{1}{2}(\mu^0(\theta))'(\Sigma^0(\theta))^{-2}\mu^0(\theta)$$

$\blacklozenge$

The following theorem, whose proof is in [7], shows that the solution to the adaptive linear problem given in Theorem 1 can be considered as an asymptotic approximation to the original filtering problem.

Theorem 2: Assume that for $i = 1,...,N$ and all t

$$\lim_{\|x\| \to +\infty} \|C_t(i)\,x\| = +\infty$$

Then for every function f continuous and with polynomial growth we have a.s.

$$\lim_{\varepsilon \to 0} \int f(x)\,\bar{p}^\varepsilon_t(x|y^t)\,dx = \lim_{\varepsilon \to 0} \int f(x)\,p^\varepsilon_t(x|y^t)\,dx$$

In particular , denoting by $\overset{\Delta\varepsilon}{x}_t$ and $\hat{x}^\varepsilon_t$ the respective conditional means and by $\bar{e}^\varepsilon_t$ and e^ε_t the conditional mean square errors, both computed with respect to the conditional distribution for the original model, i.e.

$$\bar{e}^\varepsilon_t = \int (x-\overset{\Delta\varepsilon}{x}_t)\,(x-\overset{\Delta\varepsilon}{x}_t)'p^\varepsilon_t(x|y^t)\,dx$$

$$e^\varepsilon_t = \int (x-\hat{x}^\varepsilon_t)\,(x-\hat{x}^\varepsilon_t)'p^\varepsilon_t(x|y^t)\,dx$$

we have

$$\lim_{\varepsilon \to 0} \overset{\Delta}{x}{}^{\varepsilon}_{t} = \lim_{\varepsilon \to 0} \hat{x}{}^{\varepsilon}_{t}$$

$$\lim_{\varepsilon \to 0} \bar{e}^{\varepsilon}_{t} = \lim_{\varepsilon \to 0} e^{\varepsilon}_{t}$$

♦

4.CONCLUSIONS

We have considered here a nonlinear filtering problem relative to a model with piecewise affine coefficients and with initial condition and disturbances distributed according to finite mixtures of Gaussian densities. Starting with such model we have constructed an adaptive linear model that is parametrized parametrized by a finite valued random parameter and whose dynamics is described in terms of the various linear dynamics appearing in the original model. It is shown that, when the variances of the Gaussian components of the state disturbances vanish, the optimal filter for the original model coincides with the optimal filter for the adaptive model. Furthermore, the latter filter can be explicitly computed by a finite-dimensional procedure. The results obtained have connections with an heuristic approach to nonlinear filtering known as Gaussian Sum Estimator (GSE) [1,11] and, for the special model considered here, they provide formal justification to the GSE technique. Adaptive linear methods have been studied also in connection with approximations of nonlinear stochastic control problems [6].

REFERENCES

1.   Alspach, D.L.; Sorenson, H.W.: Nonlinear Bayesian Estimation Using Gaussian Sum Approximations.*IEEE Trans. Aut.Control* AC 17 (1972) 439-448

2.   Anderson, B.D.O.; Moore, J.B.: *Optimal Filtering*.Prentice-Hall, 1979

3.   Benes, V.; Karatzas I.:Filtering for Piecewise Linear Drift and Observation. *Proc. 20th Conf.on Dec. and Control* (1981) 583 - 589

4.   Fleming, W.H.; Ji D.; Pardoux E.: Piecewise Linear Filtering with Small Observation Noise.In *Analysis and Optimization of Systems* (A. Bensoussan ,J.L. Lions eds.) Springer-Verlag L.N.in Controland Info. Sci. 111 (1988), 725 - 739

5.   Di Masi, G.B.; Runggaldier, W.J.: On Measure Tranformations for Combined Filtering and Parameter Estimation in Discrete Time. *Sys.and Control Letters* 2 (1982) 57-62

6.   Di Masi, G.B.; Runggaldier, W.J. : Small Noise Analysis for Piecewise Linear Stochastic Control problems. *IIASA Tech. Rept.* WP 87-124, 1987

7.   Di Masi, G.B.; Runggaldier, W.J.: Asymptotic Analysis for Piecewise Linear Filtering.In *Analysis and Optimization of Systems* (A. Bensoussan ,J.L. Lions eds.) Springer-erlag L.N.in Controland Info. Sci. 111 (1988) 752-759

8.   Kolessa A.E. Recursive Filtering Algorithms for Systems with Piecewise Linear Nonlinearities. *Avtom. Telemekh.* 5 (1986) 48 -55 (English translation :480 -486)

9.   Pardoux E.;Savona C.: Piecewise Linear Filtering. In *Stochastic Differential Systems, Stochastic Control Theory and Applications* (W. Fleming and P.L. Lions eds.), IMA Volume in Mathematics and its Applications 10, Springer-Verlag, 1987

10.   Savona C.: Approximate Nonlinear Filtering for Piecewise Linear Systems.*Sys. and Control Letters* (1988) .To appear.

11.   Sorenson, H.W.; Alspach, D.L.: Recursive Bayesian Estimation using Gaussian Sums.*Automatica* 7 (1971) 465-479

Authors' address:
Università di Padova
Dipartimento di Matematica Pura ed Applicata
Via Belzoni, 7
I - 35131 Padova , Italy

STOCHASTIC FILTERING IN A RELIABILITY FRAME

C.A. Clarotti,   W.J. Runggaldier

Summary:
   Fundamentals of reliability are briefly surveyed.
   The role of stochastic filtering in reliability is enlightened.
   The major contribution of filtering in reliability assessment is
shown to be simplifying the related numerical analysis.

1.   Reliability Theory & Practice. A Glossary

Reliability is concerned with assesing the probability that a given enegineering equipment will correctly perform a given task, for a pre-established time under specified operating conditions. This probability is referred to as "equipment-reliability".

If an equipment successfully performs its task until time t and is not capable of performing it any more from time t on, we shortly say that "the equipment fails" or "an equipment failure occurred" at time t. Equipment life is synonym with equipment failure time. From what precedes, it readily follows that the equipment reliability is nothing but the equipment failure time survival function ($\bar{F}(t)$). In all the reliability applications equipment-life has a density ($f(t)$). This makes it possible to characterize the distribution of the equipment time-to-failure (T_f) via the hazard rate $\lambda(t;\theta)$ of the distribution (failure rate)

$$(1.1) \quad \lambda(t;\theta) = \frac{f(t;\theta)}{\bar{F}(t;\theta)}$$

In equation (1.1) the dependence of $\lambda(\cdot)$ on a possibly unknown "parameter" θ is evidentiated (θ may be a vector).

Reliability is concerned with the assessment of equipment-survival-time probabilities for the purpose of making decisions such as:
- different design solutions are available for a given equipment;
- whichever solution is adopted, the equipment results to be prone to failure;
- correct operation of the equipment will produce benefits;

- equipment failure will cause a loss to be incurred;
- implementations of different design solutions have different costs;
- which is the best solution?

The above decision must be statistical, i.e. if θ in eq. (1.1) is unknown, it must be "estimated" by making use of observed failure and/or survival times relative to equipments "similar" to the one of interest. If these data are statistically relevant, the equipment is classed as "a component", otherwise "it is" a system.

For the purpose of reliability assessment, a system is subdivided into components, that is, system task is subdivided into subtasks each to be accomplished by a component. The underlying heurstic idea is to assess on statistical data component-reliabilities and then to calculate via probability calculus the reliability of the system. This being theoretically correct ot not will be discussed in the next section, for the moment let us survey the properties of "well designed" systems [1].

Let n be the number of system components. The binary indicator of component i is the variable x_i such that

$$(1.2) \quad x_i = \begin{cases} 1 \text{ , if component i is successfully performing its task} \\ \quad i = 1, \ldots, n \\ \\ 0 \text{ , otherwise} \end{cases}$$

ϕ the binary indicator of the system is defined likewise.

Well designed systems (coherent systems) are such that:

1) $\phi = \phi \ (x_1, \ldots, x_n)$;

2) $\phi(\cdot)$ is increasing in the natural partial ordering in $\{0,1\}^n$;

3) in a frequentist frame, under the hypothesis of independence of component failure times, it results

$$P_r\{\phi = 1\} = h(p_1, \ldots, p_n)$$

$$p_i = P_r\{x_i = 1\} \ , \ i = 1, \ldots, n$$

4) $h(p_1, \ldots, p_n)$ increasing in each p_i

2.   <u>Statistics for Reliability Decisions</u>

2.1   <u>The Frequentist Approach</u>

The use of the MLE (Maximum Likelihood Estimator) is a

mandatory choice in a frequentist setting.

Indeed, for the decomposition into components to lead to coherent decisions, the estimator of system reliability must be related to system-component-reliability estimators via the function $h(\cdot)$ defined in the previous section.

This is the case if the adopted estimators possess the invariance property with respect to an increasing transform. This invariance property is possessed only by the MLE. The inevitability of using the MLE makes classical statistics unsuitable for the sake of reliability decisions. This is so because:

- Most of reliability data come from the operative life of components in the plants where they are installed (field data), ad-hoc life-testing campaigns being impossible due to .cost and duration constraints.

- Field sample are characterized by nonhomogeneous stopping rules in the sense that:
  1. Both random and non-random withdrawals of units under observation can occur.
  2. The mission-time, and consequently the on-test time, of the unit need not be the same all over the plants which are providing data.

- Even if component lives are exponential [2], the distribution of the MLE of the unknown parameter θ is not calculable in an explicit form under such general sampling plans. Furthermore, even in the exponential case the typical sample size and the typical values of reliabilities to be estimated are such that the MLE of concern is very far from asymptotical normality. As a consequence, confidence intervals are not available for supporting decisions.

2.2 The Bayesian Approach and the Coming-In of Filtering

Field data are treatable by Bayes statistics. In a Bayes framework the stopping rules are not required to be the same for the components on test as long as they are not informative with respect to the parameters of the distribution being investigated [4]. In most cases, the stopping rules of field data are not contingent on the performance and so they are non-informative as to the parameter of interest.

For simplicity consider the case of decision relative to a component. Let u be the safety goal for component reliability, that

320

is the minimal value of the component reliability that is judged to be satisfactory.

The steps for making Bayesian decisions as to using or not the component for a time t_M (mission-time) are [3]:

i.   Use Bayes theorem and derive $\Pi(\theta)$, your posterior on θ.

ii.   Deliver the component to mission or not, according to whether or not it is

$$(2.1) \quad R(t_M) = \int \exp\left(-\int_0^{t_M} \lambda(x,\theta)\,dx\right) \Pi(\theta)\,d\theta \; > u$$

Having stated this, let us consider another source of difficulty with field data: contamination due to maintenance;

In operating plants, components routinely undergo preventive maintenance operations which perturb the evolution towards the failure of the components.

Our inference problem is then:

- We have a file of maintained-component-histories.

- We want to assess the reliability of a component "of the same family" in the case of no maintenance (motivation for this could be: we would like to relax the maintenance policy and to know the maintenance-free-probability of survival).

Remember that "of the same family" in a Bayesian reliability frame means components whose lives are independent given the value of an unknown parameter θ.

First note that if maintenance is foreseen, $\lambda(\cdot,\theta)$ must be increasing in the time-argument. Non-increasing $\lambda(\cdot,\theta)$ entails that the residual life survival function is non-decreasing in the actual component age, which makes the maintenance meaningless (If the component survived up to now it has more chance of surviving than a fresh component, maintenance aims to take the component back to the fresh status).

A reasonable model for the effect of maintenance is [4],[5] to think of the age of the component in terms of a stochastic process $\tau(t)$ defined according to eq. (2.2).

$$(2.2) \quad \tau(t) = \begin{cases} t, \; t < T \\[4pt] t - \sum_{k=1}^{[t/T]} W_k; \; t \geq T \end{cases}$$

where

kT: time of k-th maintenance operation, k=1, 2,…

W_k: random variable of conditional pdf $g(\cdot|\tau)$

Thinking of the effect of maintenance in terms of age reduction makes sense because $\lambda(\tau,\theta)$ is increasing in τ.

We may assume that the file of observations consists of the history of just one component because if, as we are assuming, component lives are independent given θ, the case of more than one component can be treated by processing component-records sequentially as shown in fig. 1. [4],[5]

Define the jump process

$$Y_t = \begin{cases} 1 < \text{if the component under observation is working at time t} \\ \\ 0: \text{otherwise} \end{cases}$$

Keeping in mind eq. (2.1), the inference problem we have to solve for the sake of decision making can be summarized as follows.

Calculate:

$$(2.3) \quad E\left\{\exp\left(-\int_t^{t+t_M} \lambda(s-t\,;\,\theta)\,ds\right) \mid Y^t\right\}$$

where:

Y_t is the jump process of the maintened component under observation;

$E\{\cdot|Y^t\}$ means conditional expectation given the history of Y_t up to time t;

t: is the time at which the observation of the maintened component terminated;

t_M is the mission time of the non-maintained component whose reliability we want to assess; the latter component is new at time t, i.e. when it starts operating.

Note that, it is

$$t = T_f \wedge T'$$

T_f: failure-time of the component under observation

T': a non-informative censoring time.

This formulation suggests to regard the problem as a stochastic

INFERENCE FLOW CHART

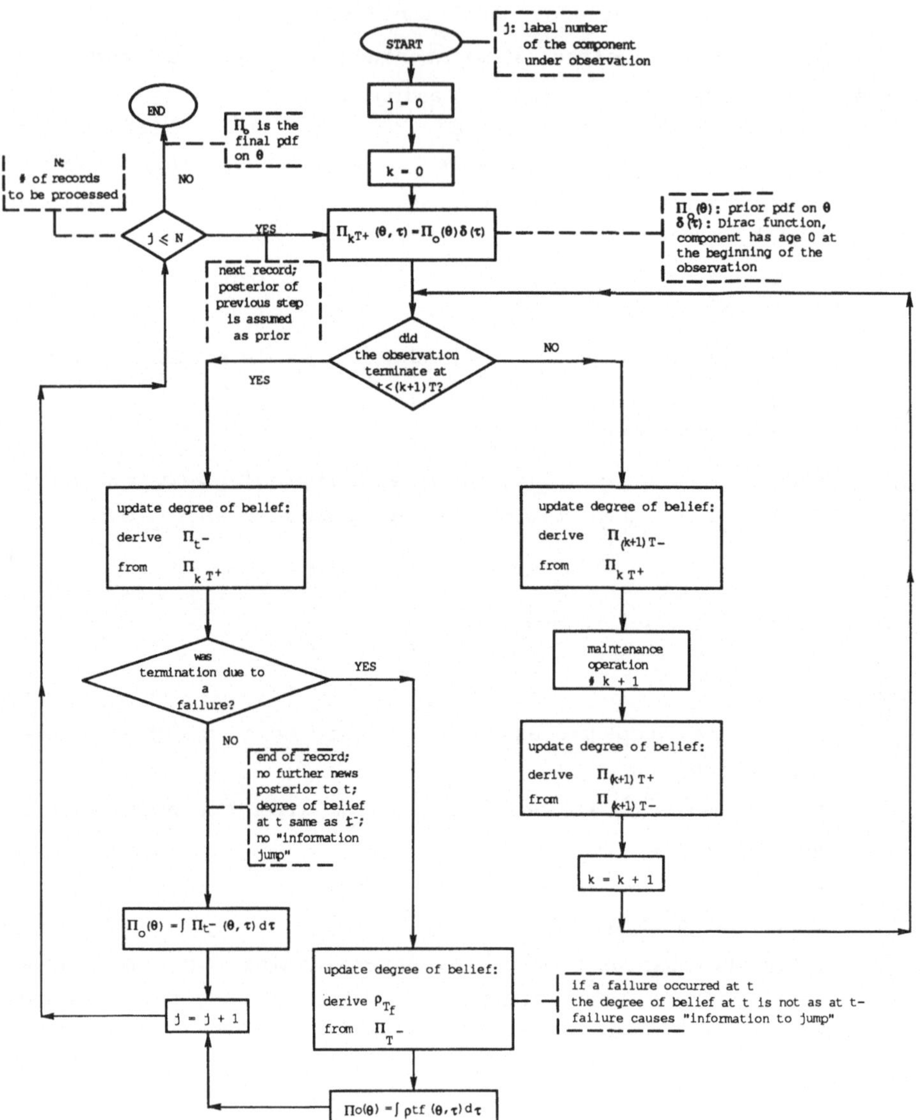

filtering problem, i.e. as the problem of calculating

$$(2.4) \quad E\{f(\theta, \tau(t)) | Y^t\}$$

for any arbitrary integrable function of the non-observable process $(\theta, \tau(t))$.

For any integrable function means also for the function exp $(iu\theta_t + iv\tau_t)$ which specializes (2.4) into the characteristic function of the conditional joint distribution of θ and τ at time t. Antitransforming, when feasable, then yields the conditional joint pdf of θ and τ given Y^t, that is the Bayesian solution of the inference problem.

This was done in [5] and yielded $(\Pi_s(\tau, \theta))$ and $\rho_s(\tau, \theta)$ are the joint pdf's conditional on respectively $s<T_f$ and $s=T_f$, and $\propto$ means: proportional to)

$$\Pi_s(\tau;\theta) \propto \exp\left(\int_0^{s-kT} \lambda(\tau-s+kT+x,\theta)\,dx\right) \Pi_{kT+}(\tau-s+kT;\theta)$$

$$kT<s\leq(k+1)T$$

$$(2.5)$$

$$\Pi_{kT+}(\tau;\theta) = \int_\tau^{kT} \Pi_{kT^-}(x;\theta)\,g(x-\tau|x)\,dx$$

$$\rho_s(\tau,\theta) \propto \lambda(\tau,\theta)\,\Pi_{s^-}(\tau,\theta)$$

This is how stochastic filtering came into reliability business [5], but this is not the very contribution that stochastic filtering can give to reliability.

As a matter of fact eqs.(2.5) are obtainable by making use of just Bayes theorem [4]. To become convinced of this note that, after formal multiplication by $d\tau d\theta$ the first of eqs.(2.5), for instance reads:

"probability that the component is of age τ at time s is equal to the probability that the component was of age $(\tau-s+kT)$ $(s-kT)$-time-units-earlier, times the conditional probability of a survival of length s-kT", which is nothing else but Bayes formula.

The value of stochastic filtering seems rather to be in its potential as numerical computation tool, as shown in next section.

3.   Stochastic Filtering as Computational Tool

After processing observed data according to eqs. (2.5) what we have is not our posterior pdf on τ and θ but rather a numerical approximation of its. Of course our approximation can be made as "close" as desired to the "true" posterior but if "close" is understood according to the metric

$$\sup_{(\theta,\tau)} \ | \ f_1(\theta,\tau) \ - \ f_2(\theta,\tau|$$

this is not enough for making sure that the expected value of the function of interest (see eqs.(2.1),(2.3)) with respect to the approximating pdf converge to the expected value with respect to the "true" posterior. A well known example of the converse being true is the fact that a normal distribution can be made as close as desired (in the above sense) to a Cauchy distribution, yet the former has a finite expectation and the latter has not.

Loosely speaking, we can say that stochastic filtering shows us what kind of metric to introduce on the set of all possible unnormalized pdf's on τ and θ in order to be sure that "a pdf close enough" to the "true" posterior means a pdf which yields an accurate numerical estimate of the filter (2.3). We say loosely speaking because an unnormalized pdf will appear in the numerical estimation of (2.3), but its role will be just formal.

In [6] an accurate numerical estimate of (2.3) is derived. An outline of how this can be done is following.

The starting point is: instead of the original problem we consider an approximation of its such that:

i.   The approximation concerns only the state process $(\theta,\tau(t))$ and not the observation Y_t.

In this way no information is lost.

ii. If $(\theta,\tau(t))_n$ is the approximation of $(\theta,\tau(t))$ then

$$E\left\{ \exp\left(-\int_t^{t+t_M} \lambda(s-t,\theta_n) \ \Big| \ Y^t \right\} \xrightarrow[\forall t]{a.s.}$$

(3.1)
$$E\left\{ \exp\left(-\int_t^{t+t_M} \lambda(s-t,\theta)\,ds\right) \ \Big| Y^t \right\}$$

iii. The lhs of (3.1) is explicitly calculable.

Note that did we have simply

(3.2) $E\left\{ \exp\left(-\int_t^{t+t_M} \lambda(s-t,\theta)\,ds\right) \right\}$

in place of the conditional expectation on the rhs of (3.1), then, since the function in (3.2) is continuous and bounded, from

$$(\theta,\tau(\cdot))_n \xrightarrow{d} (\theta,\tau(\cdot))$$

it would follow

$$(3.3) \quad E\left\{ \exp\left(-\int_t^{t+t_M} \lambda(s-t,\theta_n)\,ds\right)\right\} \longrightarrow E\left\{ \exp\left(-\int_t^{t+t_M} \lambda(s-t,\theta)\,ds\right)\right\}$$

We then reconduct the case of *conditional* expectations to the case of simple expectations by means of a probability-measure-transform such that for any integrable function $f(\cdot)$ it results

$$(3.4) \quad E\{f(\theta,\tau(t))\,|\,Y^t\} = \frac{V_t(N,f)}{V_t(N,1)}$$

where

the argument N on the rhs of (3.4) evidentiates the dependence on $N_t=1-Y_t$; $V_t(N,f)$ and $V_t(N,1)$ are *simple expectations* with respect to θ and τ_t for fixed t and fixed N_t, namely

$$(3.5) \quad V_t(N,f) = E\left\{ f(\theta,\tau(t))\,\lambda^{N_t}(\theta,\tau(t))\exp\left(\int_0^t (1-\lambda(\theta,\tau(t))\,ds\right)\right\}$$

The searched approximation of filter (2.3) is then obtained by approximating V_t via a discretization

$$(\theta,\tau(\cdot))_n \xrightarrow{d} (\theta,\tau(\cdot))$$

which entails

$$(3.6) \quad V_t^{(n)}(N,f) \xrightarrow[\forall t]{a.s.} V_t(N,f)$$

as both $\lambda(\cdot)$ and $f(\cdot)$ (this in the case of filter (2.3)) are continuous and bounded and $t<T'$. The procedure for calculating $V_t^{(n)}$ can be found in [6] and is not reproduced here. We just cite the fact that $V_t^{(n)}$ can be put in the form of an expectation of $f(\cdot)$ with respect to a suitable, unnormalized pdf on θ,τ that can be computed recursively. In this sense stochastic filtering "singles out" a particular pdf such that the convergence of the expectation of interest is insured.

References

[1]   Barlow, R.E.; Proschan, F.: Statistical Theory of Reliability
        and Life Testing, Holt,Rinehart and Winston, New York, 1975,
        Chapts 1,2.

[2]   Bartholomew, D.J.: The Sampling Distribution of an Estimate
        Arising in Life Testing. Technometrics, $\underline{5}$ (1963) 361-374.

[3]   Clarotti, C.A.: Failure Rate Estimation. A Dangerous Nonsense
        in a Bayesian View. Rel. Eng. & Syst. Safety $\underline{20}$ (1988) 117-126.

[4]   Clarotti, C.A.; Koch, G.; Spizzichino, F.: Bayes Inference from
        Failure Data Contaminated by Maintenance. IEEE Trans. on Rel.
        $\underline{34}$ n°4 (1985) 377-381.

[5]   Clarotti, C.A.; Koch, G.; Spizzichino, F.:A Filtering Model for
        Bayesian Analysis of Failure Data Contaminated by Maintenance.
        To Appear in Stat. & Prob. Letters.

[6]   Runggaldier, W.J.; Clarotti, C.A.: On Approximations for
        Stochastic Filtering with Applications to Reliability, in
        Quantitative Methoden in den Wirtschaftswissenschaften;
        Kohlas, Kall, Zehnder, Topp, Eds; Springer-Verlag (1989) to
        appear.

C.A.Clarotti. ENEA TIB-AQ, CRE Casaccia, S.P. Anguillarese 301, 00100,
Rome, Italy.

W.J. Runggaldier. Dept. of Mathematics, University of Padova, Via
Belzoni 7, 35131 Padova, Italy.

Numerical Solution of
Asymptotic Two-Point Boundary Value Problems
with Application to the Swirling Flow over a Plane Disk [†]

by

Hans Josef Pesch and **Peter Rentrop**

Mathematisches Institut, Technische Universität München

Abstract. A technique is presented for the numerical solution of asymptotic two-point boundary value problems. Thereby the boundary layer part of the solution, defined over a finite interval, is splitted from the asymptotic part of the solution. By a linearization technique one obtains surrogate boundary conditions so that the infinite problem can be efficiently approximated by a finite problem. This allows the convenient application of standard software for the solution of two-point boundary value problems. As a main example, the swirling flow of a viscous incompressible fluid over an infinite plane disk is investigated. Limitations of the procedure are discussed.

1   Introduction and Survey

In applications, symmetry assumptions are a justified and useful tool for the treatment of partial differential equations. The symmetries reduce the number of independent variables. In many cases the partial differential equations degenerate to ordinary differential equations which can be handled easier. A typical disadvantage, one has to pay for, is that the reduced problem is defined on an infinite interval. One obtains a so-called *asymptotic two-point boundary value problem.*

For two examples from hydrodynamics and quantum mechanics we propose a linearization technique to treat the infinite interval. Under certain assumptions, this technique can be easily applied to other asymptotic two-point boundary value

---

[†] This paper is in final form and no version of it will be submitted for publication.

problems. The basic idea is to separate the boundary layer part of the solution, defined by the given nonlinear system on a *finite* interval, from the asymptotic part, defined on an *infinite* interval. This asymptotic part is obtained by linearization and yields approximatate boundary values for the nonlinear system. Both parts can be matched together, so that standard software for the numerical solution of two-point boundary value problems, e.g., the *multiple shooting* method, can be applied. The limitations of the procedure presented are discussed for the simple Thomas-Fermi model of an atom.

2   Similarity Solutions for a Swirling Flow Problem

The swirling flow of a viscous incompressible fluid over an infinite plane disk is described by the Navier-Stokes equations. In Fig. 1 the radial velocity $u(r,z)$, the tangential velocity $v(r,z)$ and the vertical velocity $w(r,z)$ of the fluid particles are explained. One can imagine that the rotating flow is generated by a second rotating disk above the fixed infinite disk.

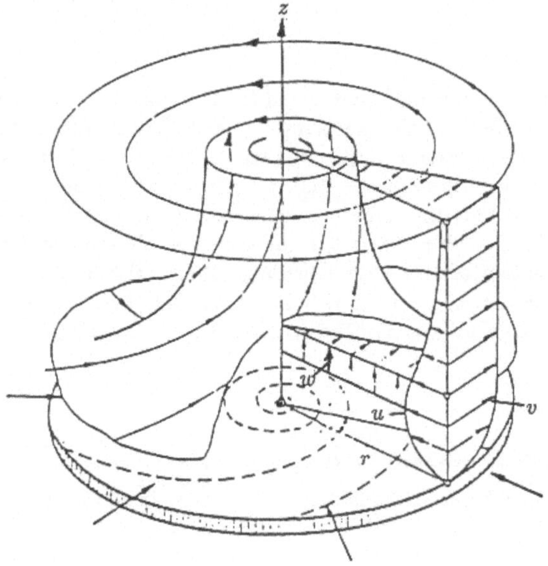

Fig. 1: Swirling Flow Model

In order to simplify the Navier Stokes equations, one relates the radial distance r and the height z by the new variable

$$(2.1) \qquad \eta = \frac{z}{r^{(n+1)/2}}$$

with the similarity parameter n where

$$-1 \leq n \leq 1 \quad \text{with} \quad n = \begin{cases} -1 & \text{for a solid body rotation} \\ +1 & \text{for a potential vortex} \end{cases} .$$

Following Troesch [13] and the cited literature there, this leads to an asymptotic two-point boundary value problem:

$$
\begin{aligned}
& f''' + \frac{1}{2}(3-n)\,f\,f'' + n\,f'^2 + g^2 - 1 - s\,f' = 0 \;, \\
& g'' + \frac{1}{2}(3-n)\,f\,g' + (n-1)\,g\,f' - s\,(g-1) = 0 \;, \\
& f(0) = 0 \;, \quad f'(0) = 0 \;, \quad g(0) = 0 \;, \\
& \lim_{x \to \infty} f'(x) = 0 \;, \qquad \lim_{x \to \infty} g(x) = 1 \;.
\end{aligned}
\tag{2.2}
$$

Related rotating disk problems have been treated, e.g., by Baindl [2], Lentini and Keller [6], and in Ref. 9. The parameter s in eq. (2.2) characterizes the magnitude of an applied magnetic field if the fluid possesses a conductivity.

3 Asymptotic Two-Point Boundary Value Problems

The swirling flow problem (2.2) represents a typical asymptotic two-point boundary value problem of the form

$$
\begin{aligned}
& y'(x) = f(x, y(x)) \quad \text{where} \quad y \colon [a, \infty] \to \mathbb{R}^n \;, \\
& r(y(a), \lim_{x \to \infty} y(x)) = 0 \;.
\end{aligned}
\tag{3.1}
$$

Numerical standard methods must fail because of the infinite interval length. The modern codes can be only applied to two-point boundary value problems in the so-called *standard form*, see, e.g., Ascher and Russel [1]:

$$
\begin{aligned}
& y'(x) = f(x, y(x)) \;, \\
& r(y(a), y(L)) = 0 \;.
\end{aligned}
\tag{3.2}
$$

In praxis, one often tries to cut the infinite interval by replacing infinity by an arbitrary large value of L. Then we obtain the standard form (3.2). If the solutions due to different values of L change below a prescribed accuracy only, one is satisfied with these results.

Obviously, this heuristic procedure has several shortcomings:

- How is the value of L to be chosen?

- What happens if L is too short? Are there boundary layers at infinity?

A correct analysis of each asymptotic problem would be based on a singular perturbation technique as it was worked out, e.g., by Lentini and Keller [6] and Markowich [7]. However, it is possible to avoid this cumbersome analysis and transfer most of the work to the computer. The main idea of Refs. 6 and 7 is to split the solution into a boundary layer part on a finite interval $[0, L]$—a two-point boundary value problem in standard form (3.2)—and into an asymptotic part. The asymptotic part is essentially described by the linearized system on an infinite interval $[L, \infty)$. In our strategy, we pick up this idea and we use the solution of the linearized system to obtain boundary values at L for the boundary layer solution. Hereby, the heuristic approach of Ref. 9, where the asymptotic boundary value problem also was divided into those two parts, can be continued to a more general technique. The idea will be now explained by several examples in the following section.

4   Linearization Technique

4.1   The Klein-Gordon Equation

The nonlinear Klein-Gordon equation is one of the simplest nonlinear relativistic equations in mathematical physics. The radial steady state solutions are defined by the following asymptotic two-point boundary value problem (see, e.g., Stenger [10] and Strauss, Vazquez [12]):

$$(4.1) \qquad \begin{aligned} &u'' + \frac{2}{x} u' - u = u^3 \;, \\ &u'(0) = 0 \;, \quad u(\infty) = 0 \;. \end{aligned}$$

In this equation two different numerical difficulties are hidden:

- A singularity at $x = 0$: Usual numerical integration procedures will fail. As a remedy, power series techniques must be applied (see, e.g., Refs. 5 and 11).

- A singularity at $x = \infty$: A linearization technique which is the subject of this paper will help to overcome this numerical hurdle.

  A linearization of eq. (4.1) leads to the system

$$(4.2) \qquad v'' + \frac{2}{x} v' - v = 0 \;.$$

Introducing the new variable $w(x) = x\,v(x)$, eq. (4.2) is transformed to

$$(4.3) \qquad w'' - w = 0 \ .$$

Therefore the general solution of eq. (4.2) is

$$(4.4) \qquad v(x) = A\,x^{-1}\,e^{-x} + B\,x^{-1}\,e^{x} \ .$$

However, the solution (4.4) of the linearized problem and the solution of the nonlinear problem (4.1) must have the same asymptotic behavior because of the boundary condition at infinity. Therefore, we obtain $B = 0$ requiring that $\lim_{x \to \infty} v(x) = 0$ must hold. Hence, the solution of eq. (4.2) with the required asymptotic behavior can be written as

$$(4.5) \qquad v(x) = A\,x^{-1}\,e^{-x}$$

where A describes an unknown integration constant.

In the following step, the linearized solution (4.5) and the solution of (4.1) are matched to determine the unknown value of A. For this purpose, the boundary conditions of the nonlinear Klein-Gordon equation (4.1) at $x = L$ are defined by $u(L) = v(L)$ and $u'(L) = v'(L)$ where L is to be chosen large enough so that the missing term of the linearized system, namely $u(x)^3 = A^3\,x^{-3}\,e^{-3x}$ is small compared to u. With $u'(0) = 0$ we have three boundary conditions for only a second order differential equation. Therefore, a third artificial differential equation is introduced for the unknown integration constant. Hence, $A' = 0$. Moreover, this choice of surrogate boundary conditions allows to estimate L much less than for the heuristic technique of cutting off the interval as mentioned in section 3. It need not be explicitly mentioned that, as a consequence, also considerably less computing time is needed.

Using the abbreviations $y_1 = u$, $y_2 = u'$ and $y_3 = A$ we have a well defined two-point boundary value problem in standard form

$$
(4.6) \qquad
\begin{aligned}
&y_1' = y_2 \ , \\[2mm]
&y_2' = -\frac{2}{x}\,y_2 + y_1 + y_1^3 \ , \\[2mm]
&y_3 = 0 \ , \\[2mm]
&y_2(0) = 0 \ , \\[2mm]
&y_1(L) = y_3(L)\,L^{-1}\,e^{-L} \ , \\[2mm]
&y_2(L) = -y_3(L)\,(L^{-1} + L^{-2})\,e^{-L}
\end{aligned}
$$

where L is chosen appropriately. Additionally, the power series technique at $x = 0$ has to be included.

Without any further numerical difficulties, it was possible to compute several solutions of (4.6) using the multiple shooting method (cf. Stoer and Bulirsch [11] and Deuflhard and Bader [3]). Figs. 2 and 3 show five solutions obtained. Obviously, there may exist infinitely many solutions satisfying an oscillation theory.

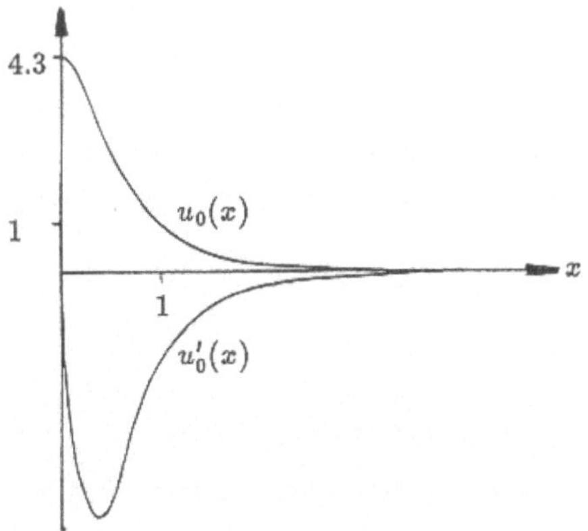

Fig. 2: Solution $u_0(x)$ of the Klein-Gordon Equation

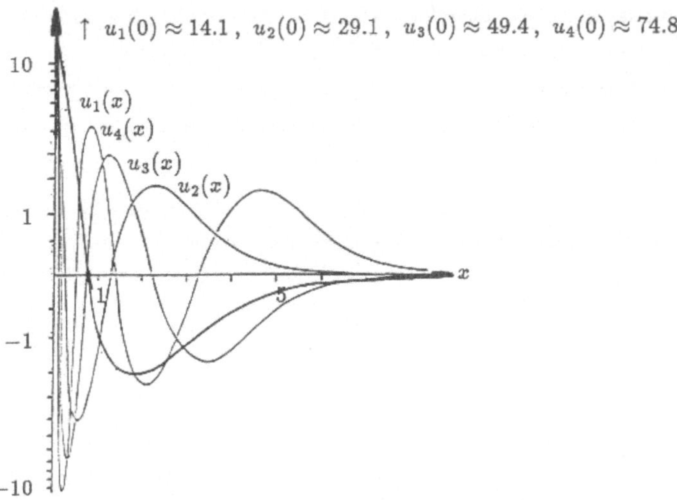

Fig. 3: Solutions $u_1(x)$ to $u_4(x)$ of the Klein-Gordon Equation

4.2 The Swirling Flow Equation

The technique presented for the treatment of the Klein-Gordon equation is now transfered to the more complicated eqs. (2.2). Firstly, we transform this boundary value problem to a problem with homogeneous boundary conditions by

$$(4.7) \qquad f(x) = a + F(x) \, , \quad g(x) = 1 + G(x) \, .$$

The linearization of the transformed problem results in

$$(4.8) \qquad Y' = C\,Y \quad \text{with} \quad C = \begin{pmatrix} 0 & 1 & 0 & 0 & 0 \\ 0 & 0 & 1 & 0 & 0 \\ 0 & s & \frac{1}{2}(3-n) & -2 & 0 \\ 0 & 0 & 0 & 0 & 1 \\ 0 & 1-n & 0 & s & \frac{1}{2}(3-n) \end{pmatrix} \, .$$

In the case of distinct eigenvalues of C —in eq. (4.8), there is one eigenvalue equal to zero, the others are two complex pairs—the solution is given by

$$(4.9) \qquad Y(x) = \sum_{j=1}^{5} c_j\, e^{\lambda_j x}\, v_j$$

where λ_j and v_j denote the eigenvalues and eigenvectors, resp. The c_j's are integration constants.

Only the eigenvalues with $\Re\,\lambda_j < 0$ satisfy the boundary conditions at infinity. The solution parts with $\Re\,\lambda_j \geq 0$ are cut out. Into this pattern, the heuristic derivation of the asymptotic in Ref. 9 fits, too.

Now, the following procedure is recommended for the numerical solution of asymptotic homogeneous two-point boundary value problems:

General Algorithm for the Treatment of
Asymptotic Homogeneous Two-Point Boundary Value Problems

- Linearize the differential equations. This requires the only additional analytical work.

- Compute the eigenvalues and eigenvectors of the linearized system using standard software, e.g., from the libraries IMSL or NAG.

- Apply a projector to cut out the eigenvalues with $\Re\,\lambda_j \geq 0$.

- Introduce as many artificial differential equations for the remaining integration constants as free boundary conditions at $x = L$ are given by the asymptotic.

- Replace the boundary conditions of the original system at infinity by boundary conditions at $x = L$ matching the asymptotic.

- Use a 'standard form' algorithm for two-point boundary value problems, e.g., the multiple shooting method.

- Compute solutions due to different values of L to estimate the accuracy of the asymptotical Ansatz by the size of the truncation error.

The system of the swirling flow problem to be solved now, consists of nine differential equations, namely
–five first order differential equations resulting from the original problem,
–one artificial differential equation for the unknown parameter a of the transformation to homogeneous boundary conditions; see eq. (4.7),
–two artificial differential equations for the integration constants associated with the complex eigenvalue pair with $\Re \lambda_{2,3} < 0$,
–one artificial differential equation for either the interval length L or the similarity parameter n or the magnetic field parameter s. As in Ref. 4 this differential equation is introduced for homotopy purposes.

Fig. 4 presents a typical solution showing the correspondence between the asymptotic and boundary layer solution in the neighborhood of $x = L$. Obviously, the solution $F(x)$ does not coincide with its asymptotic for $x \ll L$. Moreover, the solutions shown in Fig. 5 are very difficult to obtain because of the high numerical sensitivity of the problem for the required parameter values $n = 0.1027$ and $s = 0$. In order to reduce the sensitivity of this problem, it is necessary to reverse the direction of integration. A detailed analysis of the sensitivity of shooting methods has been given by Mattheij [8]. As a rule of thumb one may say, that increasing the value of n makes the computations more difficult whereas increasing the value of s $(s > 0)$ simplifies the computations. Following Troesch [13], there exists no solution for $n > 0.1218$ and $s = 0$.

The $f'(x)$-$g(x)$ plane, i.e., the projection of the $u(r,z)$-$v(r,z)$ plane into the $z = 0$ plane, as shown in Fig. 6, allows the physical interpretation of the swirling flow (cp. also Fig. 1). The regions of inward ($u > 0$ or $f' > 0$) and outward ($u < 0$ or $f' < 0$) flows can be clearly identified.

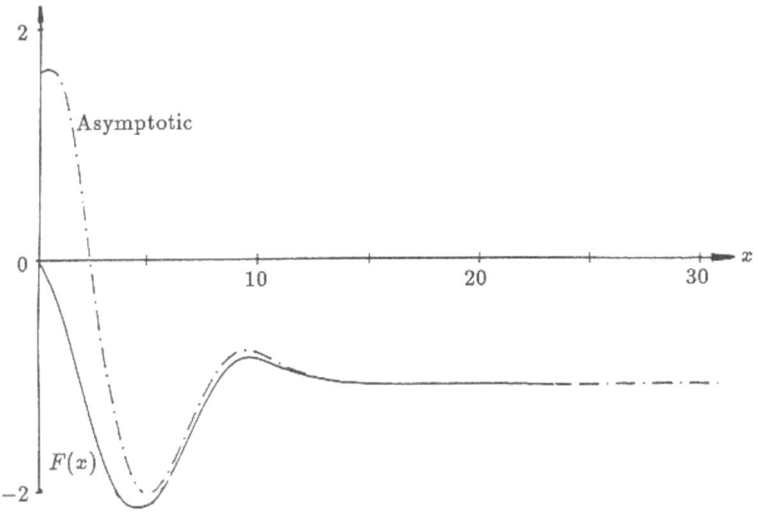

Fig. 4: Correspondence Between the Nonlinear System and Its Asymptotic
$(\, n = -0.175\,,\ s = 0.011\,,\ L = 24\,)$

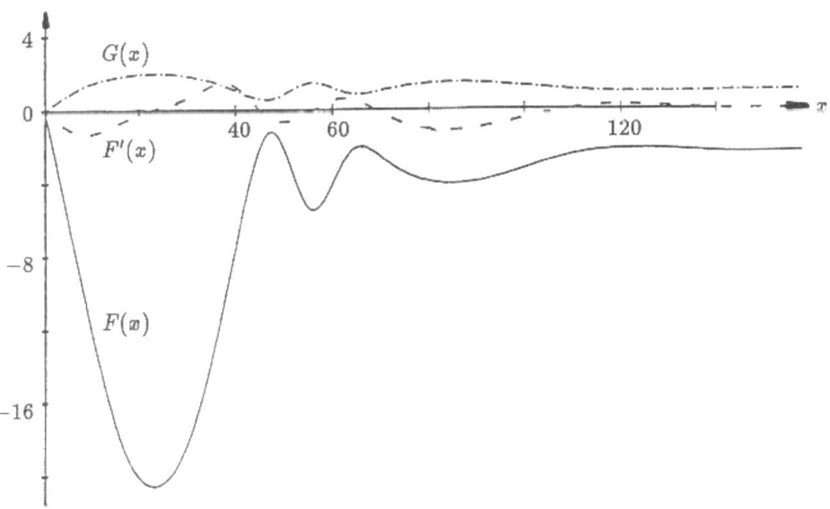

Fig. 5: Solution for $n = 0.1027$ and $s = 0$ $(\, L = 158\,)$

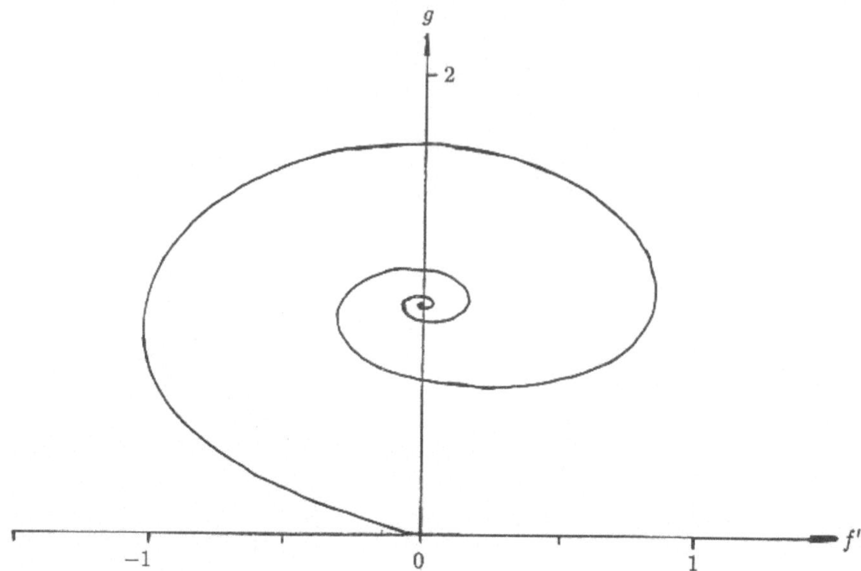

Fig. 6: Solution in the $f'(x)$-$g(x)$ Plane Showing the Swirl of the Flow

4.3   The Thomas-Fermi Model

The linearization technique depends on a particular property of the linearized system, namely that this system yields additional nontrivial information. Consequently, the technique presented may fail. This is demonstrated for the following Thomas-Fermi model. This problem describes the electric field of an atom or ion:

$$y''(x) = y(x)^{3/2}/x^{1/2} \, ,$$

(4.10)

$$y(0) = 1 \, , \quad \lim_{x \to \infty} y(x) = 0 \ \text{(for an atom)} \quad \text{or} \quad y(b) = 0 \ \text{(for an ion)}.$$

The linearization leads to $v(x) \equiv 0$ and yields no nontrivial information. However, there exists a nontrivial asymptotic solution

(4.11) $v(x) = 144/x^3$

which satisfies the differential equation and the boundary condition at infinity. Fig. 7 shows the solutions for $b = 2.75$ in the non-asymptotic case and for the asymptotic case the value of L is chosen to be $L = 5$.

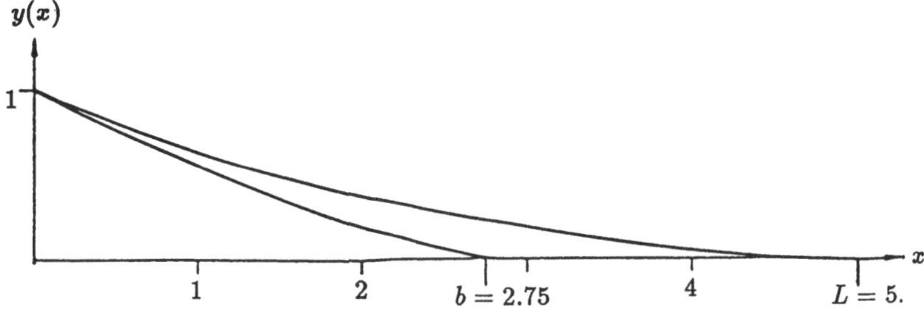

Fig. 7: Solution of the Thomas-Fermi Model

Conclusion

A method for an efficient numerical solution of asymptotic two-point boundary value problems is presented. Hereby, the original problem is splitted into a nonlinear problem on a finite interval and into a linearized problem on an infinite interval. In order to transfer analytical work to the computer, the number of differential equations is enlarged by introducing unknown integration constants as artificial differential equations. Via these additional differential equations and associated boundary conditions, the linearized part is coupled with the nonlinear system so that standard software for the solution of two-point boundary value problems, e.g. multiple shooting, can be used for the augmented problem.

References

[1.] Ascher, U.; Russel, R. D.: Reformulation of Boundary Value Problems into 'Standard Form'. *SIAM Review* **23** (1981) 238–254

[2.] Baindl, G.: Numerische Berechnung asymptotischer Randwertprobleme der Hydromechanik mit Hilfe der Mehrzielmethode. Department of Mathematics, Munich University of Technology: Diploma Thesis 1983.

[3.] Deuflhard, P.; Bader, G.: Multiple Shooting Techniques Revisited. In: Deuflhard, P.; Hairer, E. (eds.): Numerical Treatment of Inverse Problems in Differential and Integral Equations. *Progress in Scientific Computing* **2** (1983) 74–94

[4.] Deuflhard, P.; Pesch, H. J.; Rentrop, P.: A Modified Continuation Method for the Numerical Solution of Nonlinear Two-Point Boundary Value Problems by Shooting Techniques. *Numer. Math.* **26** (1976) 327–343

[5.] Diekhoff, H. J.; Lory, P.; Oberle, H. J.; Pesch, H. J.; Rentrop, P.; Seydel, R.: Comparing Routines for the Numerical Solution of Initial Value Problems of Ordinary Differential Equations in Multiple Shooting. *Numer. Math.* **27** (1977) 449–469

[6.] Lentini, M.; Keller, H. B.: Boundary Value Problems Over Semi-Infinite Intervals and Their Numerical Solution. *SIAM J. Numer. Anal.* **17** (1980) 577–604

[7.] Markowich, P. A.: Analysis of Boundary Value Problems on Infinite Intervals. *SIAM J. Math. Anal.* **14** (1983) 11-37

[8.] Mattheij, R. M. M.: Decoupling and Stability of Algorithms for Boundary Value Problems. *SIAM Review* **27** (1985) 1–44

[9.] Pesch, H. J.; Rentrop, P.: Numerical Solution of the Flow Between Two Counter-Rotating Infinite Plane Disks by Multiple Shooting. *ZAMM* **58** (1978) 23–28

[10.] Stenger, F.: Numerical Methods Based on Whittaker Cardinal or SINC Functions. *SIAM Review* **23** (1981) 165–224

[11.] Stoer, J.; Bulirsch, R.: *Introduction to Numerical Analysis*. Berlin, New York, Heidelberg: Springer 1980.

[12.] Strauss, W.; Vazquez, L.: Numerical Solution of a Nonlinear Klein-Gordon Equation. *J. Comp. Phys.* **28** (1978) 271–278

[13.] Troesch, B. A.: The Limiting Vortex in the Similarity Solution of a Swirling Flow. *Appl. Math. & Comp.* **6** (1980) 133–144

Address: Dr. H. J. Pesch, Prof. Dr. P. Rentrop
Mathematisches Institut
Technische Universität München
Arcisstraße 21
Postfach 20 24 20
D 8000 München 2

ON EQUILIBRIUM POINTS OF THE VARIATIONAL ADAPTIVE CONTROL SCHEME

G. Casalino, R. Minciardi

Summary: Within the framework of adaptive control schemes based on implicit model identification, a special position is held by the so-called variational adaptive control schemes. Their particular feature is the fact that the adaptation step does not consist in a simple redesign of the control law on the basis of the information provided by the implicit model, but in a modification of the pre-existing regulator, according to a specified control objective. Variational adaptive control schemes based on finite or infinite- horizon quadratic control objectives are considered. It is shown that the·test of simple conditions on a generic equilibrium point can tell us whether the optimal control law has been attained or not.

1. Introduction

Within the wide research field related to the theory of adaptive control (see, for instance, the recent survey by Astrom [1]), one of the main streams has been that of the so-called certainty-equivalence adaptive control schemes. Such schemes are characterized by the separation of the adaptive controller structure into: 1) the proper control law; ii) a recursive identification algorithm giving a model of the process to be controlled; iii) an adaptation algorithm which, at each control step, adjusts the parameters of the control law on the basis of the identified model. It has been readily recognized that the use of efficient identification algorithms is of fundamental interest in such schemes. In this connection, Åström and Wittenmark [2] noted, already in the early seventies, that, for a certain control objective, even in presence of a true process model having an ARMAX structure, the use of Recursive Least Squares was allowed (in lieu of the more cumbersome Extended Least Squares). More specifically, they developed an algorithm admitting the optimal control law as an equilibrium point, which, under certain conditions, always converges to the "optimal" equilibrium point.

Later on, a certain number of authors have devoted attention to the use of Recursive Least Squares in an adaptive control context, with the assumption of an underlying ARMAX structure of the true process model (see the references in [3], for instance). Essentially, the problem is whether a "simple" identification procedure (namely, Recursive Least Squares) can be used in order to identify process models which, even structurally different from the true ARMAX process model, can be used in order to derive the optimal control law for the true system. A general theoretical framework regarding the possibility of this use has been provided in [3], where two types of control design criteria were explicity considered, namely: i) pole placement criteria, and ii) quadratic optimization criteria. In particular, for the case of quadratic optimization, it has been shown [3] that this use is allowed under certain conditions. More specifically, let us designate by the term <u>implicit models</u> those ARX models (i.e., compatible with the use of Recursive Least Squares) which can be used in order to correctly predict the system's output behaviour in a certain closed-loop condition. Then, in [3] it has been proved that, with specific reference to finite-horizon quadratic control objectives, the identification of implicit models (instead of the true ARMAX process models) is allowed provided that their structure is properly defined, and a certain number of parameters a-priori fixed. Actually, the fulfilment of the last condition requires a certain amount of a-priori knowledge about the true system, which besides increases as the control horizon increases. On the counterpart, no use of (ARX) implicit models is generally allowed in connection with an infinite-horizon control objective.

The introduction of <u>extended implicit models</u> [4,5] has given the possibility to overcome the above cited drawbacks. By extended implicit models, we simply intend those implicit models which take into account the presence of a stochastic measurable additional signal ("dither noise") in the structure of the control law. The introduction of such models has: i) removed the necessity of the a-priori knowledge about the true system, with reference to the finite-horizon control objective; ii) allowed the correct definition of an algorithm for the adaptive infinite-horizon quadratic optimization, based on (extended) implicit model identification.

This paper deals with this last algorithm, denoted as <u>Variational Adaptive Control Algorithm</u>. Its main feature is that the adaptation step does not consist in a simple redesign of the control law on the basis of the information provided by the (extended) implicit model, but in a modification of the pre-existing regulator. After the definition of the algorithm, and the mention of a preliminary basic result about its equilibrium points, a new result is provided concerning the possibility of testing in a simple way whether a generic equilibrium point corresponds to the optimal control law or not.

2. The <u>variational adaptive control algorithm</u>

In this section, only the basic structure and motivations of the variational adaptive control algorithm will be reported; a detailed derivation and justification of this structure can be found in [4,5]. It is supposed that the structure of the (SISO) system to be controlled is of ARMAX type

$$Ay_i = Bu_i + Ce_i \tag{2.1}$$

where $A(q^{-1})$, $B(q^{-1})$, $C(q^{-1})$ are polynomials of order n, with no unstable common factors, being A and C monic, B with first coefficient zero, $\partial(A,B,C) = n$ (i.e., at least one of the last coefficients of A,B,C is $\neq 0$), C(z) with zeros outside the unit circle, and $\{e_i\}$ stationary white sequence. It is assumed that the system is initially governed by the (stabilizing) control law

$$Ru_i = Py_i + T\eta_i \qquad (2.2)$$

where $R(q^{-1})$, $P(q^{-1})$, $T(q^{-1})$ are polynomials of order n, being R and T monic, and $\{\eta_i\}$ is an arbitrary (measurable) stochastic sequence ecting as a "dither noise", independent from $\{e_i\}$. The control objectives that are considered (with respect to the true input-output process model (1)) are both of the quadratic type, namely:

a) minimize, at each time instant i, the finite-horizon quadratic cost functional

$$E\{ \sum_{t=1}^{t+m-1} (y_{t+1}^2 + p\,u_t^2) \,|\, I_i \} \qquad (2.3a)$$

b) minimize the infinite-horizon quadratic cost functional

$$\lim_{m \to \infty} \frac{1}{m} E [\sum_{t=i}^{t+m-1} (y_{t+1}^2 + pu_t^2) \,|\, I_i] \qquad (2.3b)$$

where $I_i \overset{\Delta}{=} \{y_i, y_{i-1}, \ldots, u_{i-1}, u_{i-2}, \ldots, \eta_{i-1}, \eta_{i-2}, \ldots\}$ by means of strategies of type $u_i = \bar{u}_i + \eta_i$, being $u_i = \gamma[I_i]$.

In connection with the above defined control problem b), we can define the following adaptive control algorithm [4,5].

<u>Basic Variational Adaptive Control Algorithm</u> (for the case of infinite-horizon quadratic control objective).

At each iteration step i, perform the following operations:

i) update (via Recursive Least Squares) the estimates of the parameters of an extended implicit model having structure

$$\mathcal{A}\,y_t = \mathcal{B}u_t + \mathcal{D}\eta_t + \xi_t \qquad (2.4)$$

where $\mathcal{A}(q^{-1})$, $\mathcal{B}(q^{-1})$, $\mathcal{C}(q^{-1})$ are polynomials of order $(n+1)$, $\mathcal{A}$ is monic, $\mathcal{B}$ and $\mathcal{C}$ are with first coefficient zero, and a single parameter is arbitrarily fixed; let this updating give

$$\hat{\mathcal{A}}_i, \hat{\mathcal{B}}_i, \hat{\mathcal{C}}_i;$$

ii) build a state-space representation of the equations representing the estimated extended implicit model and the pre-existing regulator, namely

$$\hat{\mathcal{A}}_i \, y_t = \hat{\mathcal{B}}_i \, u_t + \hat{\mathcal{C}}_i \, \eta_t + \xi_t \tag{2.5a}$$

$$R_{i-1} \, u_t = P_{i-1} \, y_t + T_{i-1} \, \eta_t \tag{2.5b}$$

where $\partial(R_{i-1}, P_{i-1}, T_{i-1}) \leq n$. Let this state-space representation be given by

$$x_{t+1} = F_i \, x_t + G \, \eta_t + K_i \, \xi_t \tag{2.6a}$$

$$y_{t+1} = H_i \, x_{t+1} + \xi_{t+1} \qquad u_t = S \, x_{t+1} \tag{2.6b}$$

being y_t and u_t outputs, η_t the input, and

$$x_t \triangleq \text{col } [y_{t-1}, \ldots, y_{t-n-1}, u_{t-1}, \ldots, u_{t-n-1}, \eta_{t-1}, \ldots, \eta_{t-n-1}]$$

iii) perform a single iteration of the matrix Riccati equation based on the state-space model obtained at the previous step, and an matrix M_i obtained a the previous iteration, namely

$$M_{i+1} = F_i^T [M_i - M_i G (G^T M_i G)^{-1} G^T M_i] F_i + W_i \tag{2. 7}$$

where $W_i \triangleq pSS^T + H^T H$;

iv) compute "the optimal control variation"

$$\delta u_i = -(G^T M_i G)^{-1} G^T M_i [(F_i - K_i H_i) x_i + K_i y_i] \tag{2.8}$$

and then apply the control action

$$u_i = \bar{u}_i + \delta u_i + \eta_i \tag{2.9}$$

where $\{\eta_i\}$ is a white (zero-mean) stationary sequence, and $\bar{u}_i$ is the control action generated on the basis of the pre-existing regulator $(R_{i-1}, P_{i-1}, T_{i-1})$, with $\eta_i = 0$;

v) update the polynomial expression of the regulator $(R_{i-1}, P_{i-1}, T_{i-1})$, on the basis of (8, 9), thus obtaining the "new" regulator (R_i, P_i, T_i), again of maximum order n. ◁

For the above algorithm a first basic result has been established in [5].

Theorem 2.1. Let (R^o, P^o, T^o) be the regulator obtainable on the basis of the true system (1) knowledge. Then, assume that the following conditions hold:

i) no common factors in (R^o, P^o);

ii) $\partial(R^o, P^o) = n$.

Then the algorithm admits a unique whitening equilibrium point. Moreover, this equilibrium point is characterized by the optimal control law (R^o, P^o, T^o). □

In the above Theorem, the term "whitening equilibrium point" has been used to designate equilibrium points characterized by white stationary prediction error sequences $\{\xi_t\}$ (i.e., corresponding to the attainment of an extended implicit model).

Finally, it is just worth mentioning that a Basic Variational Adaptive Control Algorithm can also be defined in connection with a m-step quadratic optimization criterion (cost (3a)). In this case, (m-1) iterations (instead of a single one) of the matrix Riccati equation must be performed at step iii) of the algorithm, with initialization at W. The same basic result in Theorem 2.1 still holds even for this case.

3. Conditions for the optimality of equilibrium points

The result of the previous section does not remove the possibility of the existence of non-whitening equilibrium points of the considered algorithm. In this section, we want to provide a result concerning the possibility of assessing the optimality of a generic equlibrium point of the algorithm on the basis of a simple condition on the parameters of the model and the regulator characterizing the equilibrium point.
The result under concern can be stated as follows.

Theorem 3.1. Consider the Basic Variational Adaptive Control Algorithm, and assume that the conditions in Theorem (2.1) hold. Furtherly assume:

i) (A,B) without common factors, and $\partial(A,B) = n$;
and that the identification-adaptation algorithm has reached a
convergence point $(\mathcal{A}, \mathcal{B}, \mathcal{C}, R, P, T)$ such that

ii) the regulator (R,P,T) is asymptotically stabilizing for
both the true system (A,B,C) and the estimated model $(\mathcal{A}, \mathcal{B}, \mathcal{C})$;

iii) (R,P) without common factors and $\partial(R,P)=n$;

iv) $\partial(\mathcal{A}R - \mathcal{B}P) \leq n$.

Then $(\mathcal{A}, \mathcal{B}, \mathcal{C}, R, P, T)$ individuates a whitening equilibrium
point, with (R,P,T) coinciding with the optimal regulator. $\square$

<u>Proof</u>. Before entering into the details of the proof, let us
mention three basic implications of the assumptions in the
Theorem. First of all, the assumption that the algorithm has
reached a convergence point characterized by

$$\mathcal{A} y_i = \mathcal{B} u_i + \mathcal{C} \eta_i + \xi_i \tag{3.1}$$

where $\{\xi_i\}$ is the prediction error sequence, imply

$$S_{\xi y}(\tau) \overset{\Delta}{=} E(\xi_i z_{i-\tau}) = 0 \qquad \tau = 1,2,\ldots,(n+1) \tag{3.2}$$

for z equal to y, u and η. Actually, for $z=\eta$, $\tau=1$, (2)
does not hold if we assume that the single coefficient which is
arbitrarily fixed in the extended implicit model to be
identified is δ_1. This choice does not affect the generality of
our discussion at all.

Moreover, it is a well known result that assumption i) is
equivalent to a matricial condition, i.e., rank $Z=2n$, and that
assumption iii) is similarly equivalent to rank $V=2n$, being

$$
Z \overset{\Delta}{=}
\left[
\begin{array}{c}
\left.\begin{array}{ccccccc}
1 & a & \ldots & a_n & & & \emptyset \\
  & & 1 & & & & \\
\emptyset & & & \ddots & & & \\
  & & & 1 & a & \ldots & a_n \\
\end{array}\right\} n \\
\left.\begin{array}{ccccccc}
0 & b & \ldots & b_n & & & \\
  & 1 & & & & & \\
\emptyset & & \ddots & & & & \emptyset \\
  & & & 0 & b & \ldots & b_n \\
  & & & & 1 & & \\
\end{array}\right\} n
\end{array}
\right]
\qquad
V \overset{\Delta}{=}
\left[
\begin{array}{c}
\left.\begin{array}{ccccccc}
1 & r & \ldots & r_n & & & \\
  & 1 & & & & & \emptyset \\
\emptyset & & \ddots & & & & \\
  & & & 1 & r & \ldots & r_n \\
  & & & & 1 & & \\
\end{array}\right\} n \\
\left.\begin{array}{ccccccc}
P_0 & P_1 & \ldots & P_n & & & \\
  & 0 & 1 & & & & \emptyset \\
\emptyset & & & \ddots & & & \\
  & & & P_0 & P_1 & \ldots & P_n \\
  & & & & 0 & 1 & \\
\end{array}\right\} n
\end{array}
\right]
\tag{3.3}
$$

<u>Step</u> <u>1</u> (Decomposition). The prediction error sequence is given by the dynamic equation (1). Then, by applying superposition, one can write

$$\xi_i = \xi_i^e + \xi_i^\eta \quad ; \quad y_i = y_i^e + y_i^\eta \quad ; \quad u_i = u_i^e + u_i^\eta \qquad (3.4)$$

where superscripts e, η denote the contributions to the global signals due to $\{e_i\}$, $\{\eta_i\}$, respectively. Obviously, due to the independence of $\{e_i\}$ and $\{\eta_i\}$, it turns out that each term superscribed by e is independent from any term superscribed by η and vice versa. As a consequence of the above fact, it easily follows that conditions (2) become

$$S_{\xi^e y^e}(\tau) = 0 \quad ; \quad S_{\xi^e u^e}(\tau) = 0 \quad ; \quad S_{\xi^\eta y^\eta}(\tau) = 0; \quad S_{\xi^\eta u^\eta}(\tau) = 0$$

$$\tau = 1,2,\ldots,(n+1) \qquad (3.5a)$$

$$S_{\xi^\eta \xi^\eta}(\tau) = 0 \qquad \tau = 2,\ldots,(n+1) \qquad (3.5b)$$

<u>Step</u> <u>2</u> (Analysis of the contribution due to $\{e_i\}$). Throughout this analysis $\eta_i \equiv 0$ will be assumed. Let us define $Q = \mathcal{R}R - \mathcal{B}P$, which is characterized by $\partial(Q) \leq n$ (assumption iv)). Then, it is easily recognized that the following relations hold

$$y_i^e = Rw_i \quad ; \quad u_i^e - Pw_i \qquad (3.6)$$

where w_i is defined by $Qw_i = \xi_i^e$, and turns out to be a stationary stochastic sequence, due to the assumed asymptotic stability of Q.

Now, by keping into account the definition of matrix V given by (3) and relations from (5a), it follows that

$$V[S_{\xi^e w}(1),\ldots,S_{\xi^e w}(2n)]^T = 0 \qquad (3.7)$$

which implies, due to assumptions iii),

$$S_{\xi^e w}(\tau) = 0 \quad ; \quad \tau = 1,2,\ldots,2n \qquad (3.8)$$

Then, from the model equation (1), and from the system and regulator equations, it is possible to relate $\{\xi_i^e\}$ with $\{e_i\}$, namely

$$(AR-BP)\,\xi_i^e = QCe_i \tag{3.9}$$

where obviously

$$\partial(AR-BP) \leq 2n \tag{3.10}$$

and, due to assumption iv),

$$\partial(QC) \leq 2n \tag{3.11}$$

Now, let us multiply both sides of (9) by $w_{i-(2n+1)}$ and take expectation, obtaining

$$S_{\xi^e w}(2n+1) + l_1 S_{\xi^e w}(2n) + \ldots + l_{2n} S_{\xi^e w}(1) =$$
$$= S_{ew}(2n+1) + g_1 S_{ew}(2n) + \ldots + g_{2n} S_{ew}(1) \tag{3.12}$$

where $l_1, \ldots, l_{2n}$ are the coefficients of the (monic) polynomial (AR-BP), and $g_1, \ldots, g_{2n}$ are the coefficients of the (monic) polynomial QC. In (12) the l.h.s. reduces to $S_{\xi^e w}(2n+1)$, due to (8), whereas the r.h.s. is clearly zero. Applying the same reasoning to $S_{\xi^e w}(2n+2)$, etc. we can conclude that

$$S_{\xi^e w}(\tau) = 0 \qquad \tau \geq 1 \tag{3.13}$$

Now observe that, from the definition of w_i, we have

$$S_{\xi^e \xi^e}(\tau) = S_{\xi^e w}(\tau) + q_1 S_{\xi^e w}(\tau+1) + \ldots + q_n S_{\xi^e w}(\tau+n) \tag{3.14}$$

where $q_1, \ldots, q_n$ are the coefficients of the (monic) polynomial Q. Then, from (14), and due to (13), we have

$$S_{\xi^e \xi^e}(\tau) = 0 \qquad \tau \geq 1 \tag{3.15}$$

Then we can conclude that $\{\xi_i^e\}$ is a white sequence. But, from (9), ξ_i^e can be represented by a linear combination of $e_i, e_{i-1}, \ldots$. Thus, we can conclude that actually

$$\xi_i^e = e_i \tag{3.16}$$

which implies (from (9))

$$(AR-BP) = QC = (\mathcal{A}R - \mathcal{B}P)C \tag{3.17}$$

Step 3 (Analysis of the contribution due to $\{\eta_i\}$). Now we suppose $e_i \equiv 0$. Then, from the true system and controller equations,

$$(AR-BP)y_i = BT\eta_i \quad ; \quad (AR-BP)u_i = AT\eta_i \tag{3.18}$$

which, due to (7)) yield

$$y_i = Bv_i \quad ; \quad u_i = Av_i \tag{3.19}$$

where v_i is defined by $QCv_i = T\eta_i$ and turns out to be a stationary stochastic sequence, due to the asymptotic stability of Q.

Now, by keeping into account the definition of matrix Z, given by (3), and relations (19), from (5a) it follows that

$$Z[S_{\xi^\eta v}(1), \ldots, S_{\xi^\eta v}(2n)]^T = 0 \tag{3.20a}$$

$$Z[S_{\xi^\eta v}(2), \ldots, S_{\xi^\eta v}(2n+1)]^T = 0 \tag{3.20b}$$

Since Z is non singular (assumption i)), (20a,b) imply

$$S_{\xi^\eta v}(\tau) = 0 \qquad \tau = 1, \ldots, 2n+1 \tag{3.21}$$

Now let us relate $\{\xi_i^\eta\}$ with $\{\eta_i\}$. It is easy to show that

$$(AR-BP)\xi_i^\eta = [T(\mathcal{A}B - A\mathcal{B}) - \mathcal{Q}(AR-BP)]\eta_i \tag{3.22}$$

Then, observe that (17) implies

$$(A - \mathscr{A} C)/P = (B - \mathscr{B} C)/R \qquad (3.23)$$

which, being (R,P) without common factors implies that both sides are equal to the same polynomial. Let us call S this polynomial. Then, due to the identity (whose proof is immediate)

$$\mathscr{A} B - A \mathscr{B} = (B - \mathscr{B} C)(\mathscr{A} R - \mathscr{B} P)/R \qquad (3.24)$$

and due again to (17), (24) can be rewritten as

$$QC \, \xi^{\eta}_{i} = (TSQ - \mathscr{D} QC) \, \eta_{i} \qquad (3.25)$$

or even (due to the stability of Q)

$$c \, \xi^{\eta}_{i} = (TS - \mathscr{D}) c) \, \eta_{i} \qquad (3.26)$$

Note that $\partial(S) \leq n+1$. Then multiplying both sides of (26) by $v_{i-(2n+2)}$ and taking expectations, we have

$$S_{\xi^{\eta}v}(2n+2) + c_{1} S_{\xi^{\eta}v}(2n+1) + \ldots + c_{n} S_{\xi^{\eta}v}(n+2) =$$

$$= f_{1} S_{\eta v}(2n+1) + \ldots + f_{2n+1} S_{\eta v}(1) \qquad (3.27)$$

where $f_{1}, \ldots, f_{2n+1}$ are the coefficients of polynomial $(TS - \mathscr{D}) c)$ (which has null zero-th order coefficient like S and $\mathscr{D}$). In (27) the l.h.s reduces to $S_{\xi^{\eta}v}(2n+2)$, due to (21), and the r.h.s. is obviously zero. Applying the same reasoning to $S_{\xi^{\eta}v}(2n+3)$, etc. we can conclude that

$$S_{\xi^{\eta}v}(\tau) = 0 \qquad \tau \geq 1 \qquad (3.28)$$

Now observe that, from the definition of v_{i}

$$S_{\xi^{\eta}(QCv)}(\tau) = S_{\xi^{\eta}(T\eta)}(\tau) \qquad \tau \geq 1 \qquad (3.29)$$

where the l.h.s. is zero due to (28). Then, it is immediate to conclude that

$$S_{\xi^{\eta}\xi^{\eta}}(\tau) = 0 \qquad \tau \geq 1 \qquad (3.30)$$

Moreover, observe from (26) that ξ_i is influenced only by η_{i-1}, η_{i-2},... . Then, (30) simply implies $\xi_i^\eta = 0 \; \forall i$, which in turn imposes (again through (26))

$$ST - \textcircled{D}C = 0 \qquad (3.31)$$

or even

$$\textcircled{D} = ST/C \qquad (3.32)$$

But, due to (23) (and the definition of S), S and C are coprime (if they were not, then the common factor of S and C, would also be factor of A and B, which is excluded by assumption i)). Then, it must be

$$T=C \; ; \; \textcircled{D} = S = (B-\textcircled{B}C)/R = (A-\textcircled{R}C)/P \qquad (3.33)$$

<u>Step</u> <u>4</u> (Conclusion of the proof). Then, due to superposition, $\{\xi_i\}$ on the whole is a white sequence. This means that (1) is actually an extended implicit model, which is moreover consistent with the fact that $(\textcircled{R}, \textcircled{B}, \textcircled{D})$ satisfy (17) and (33) (see[5]). Then, due to the assumed convergence of the control law, and to Theorem 2.1, we can conclude that (R,P,T) coincides with the optimal regulator.

<u>References</u>

[1] Aström, K.J.: Adaptive Feedback Control. Proc. of the IEEE 75, (1988), pp. 185-217.

[2] Aström, K.J., Wittemark, B.: On Self-Tuning Regulators. Automatica 9, (1973), pp. 185-199.

[3] Casalino G., Davoli F., Minciardi R., Zappa G.: On Implicit Modelling Theory: Basic Concepts and Application to Adaptive Control. Automatica 29, (1987), pp. 189-201.

[4] Casalino G., Davoli F., Minciardi R.: Extended Implicit Models and Application to LQ Adaptive Optimization. Proc. 2nd IFAC Workshop on Adaptive Systems in Control and Signal Processing, Lund, Sweden (1986).

[5] Casalino G., Minciardi R.: On Implicit Modeling Theory: Further Results with Application to LQ Adaptive Optimization and Development of a Variational Adaptive Control Scheme. Submitted.

The authors are with the Department of Communications, Computer and System Science, University of Genoa - Via Opera Pia 11/A - 16145 - Genova, Italy.

Proceedings of the Conference

Mathematics in Industry

October 24—28, 1983 Oberwolfach

Edited by Prof. Dr. H. NEUNZERT, University of Kaiserslautern, W.-Germany

1984. 287 pages. 16,2 × 23,5 cm. ISBN 3-519-02610-4. Paper DM 52,—

Contents

ORGANIZED COOPERATION BETWEEN UNIVERSITY AND INDUSTRY

R. S. Anderssen and F. R. de Hoog: A Framework for Studying the Application of Mathematics in Industry / A. B. Tayler: Oxford Study Groups with Industry; 1967—1983 / H. Wacker: Hydro Energy Optimization / J. Spanier: Applied Mathematics Education at the Claremont Colleges / H. Neunzert: Mathematic in the University and Mathematics in Industry — Complement or Contrast? / K. Hoffmann: On Establishing Contacts with Industry / M. Schulz-Reese: A Report of the „Kaiserslauterer Modellversuch": Continuing Mathematical Education / H.-E. Gross and U. Knauer: University Education as Preparation for Professional Praxis / A. M. Kempf: Mathematical Modelling in the French Grandes Ecoles. The Particular Case of the E.S.I.E.A.

INDIVIDUAL PROJECTS AT THE UNIVERSITIES

C. Cercignani: Mathematics and Fluiddynamics / M. Primicerio: Sorption of Swelling Solvents by Glassy Polymers / M. Shinbrot: Icebreaking by Hovercraft / B. Rihtarsic, F. Krmelj and I. Kuscer: Oscillations in Pipelines of Hydroelectric Power Plants / A. K. Louis: The Limited Angle Problem in Computerized Tomography / H. Frank: Computer Aided Design in Piping of Chemical Plants / W. Krüger: The Trippstadt-Problem / B. Aulbach: Trouble with Linearization

PROBLEMS POSED BY INDUSTRY

J. Bukovics: Oscillations of a Gasbody with Absorbant Walls (A Problem Occuring in Structural Acoustics of Passenger Cars) / P. Causemann: Repuirements for a Calculating Program Regarding a Two-Mass Vibration System to Optimize Damping Force Characteristics for Vehicle Shock Absorbers / A. Gamst: Geometric Design of Mobile Radio Telephone Systems / U. Pallaske: Large Systems of Stiff Ordinary Differential Equations. Numerical Treatment by System Reduction / R. Zobel: Validation of a Vehicle Crash Model

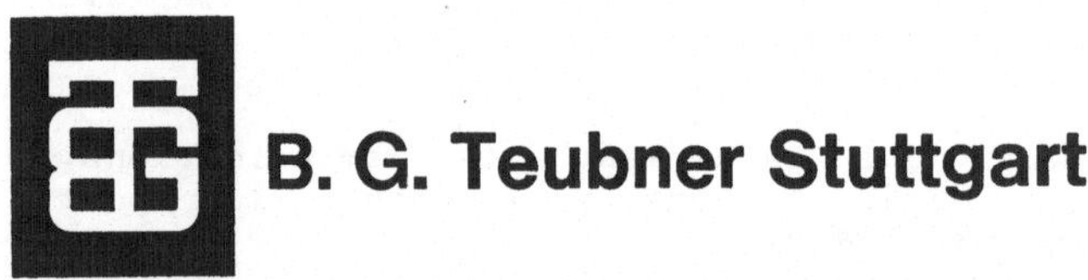

B. G. Teubner Stuttgart

ECMI VOL. 1

Proceedings of the

First European Symposium on Mathematics in Industry

ESMI I October 30–November 1, 1985, Amsterdam

Edited by Prof. Dr. M. Hazewinkel, Centrum voor Wiskunde en Informatica, Amsterdam, Prof. Dr. R. M. M. Mattheij, University of Eindhoven and Prof. Dr. E. W. C. von Groesen, University of Twente, Enschede

1988. XIV, 238 pages. 16,2 x 23,5 cm. Coproduction Kluwer-Teubner

ISBN 3-519-02170-6 (Teubner) Bound DM 72,–

ISBN 90-277-2730-9 (Kluwer) Bound Dfl 130.00/£ 39,50

These proceedings contain the lectures of the (invited) speakers of the European Symposium on Mathematics in Industry, held in Amsterdam in 1985. They represent an important account of (a variety of) case studies and problems posed by industry and solved by or in cooperation with academic institutions. At this symposium it was also discussed how to establish a firm contact between universities and industry, preferably on a European level. Both the introductions of several speakers to this topic, as well as information about the actually successful outcome of this, viz. ECMI (the European Consortium for Mathematics and Industry) are also contained in these proceedings.

Contents

A. Bensoussan: Application of stochastic control to the management of electricity production / Y. Cherruault, I. Karpouzas: Applications of mathematical models in the pharmaceutical industry: identification and optimal control applied to therapeutics / A. Fasano, M. Primicerio: Heat and mass transfer in quasi-steady ground freezing processes / M. Heilio: Making university- and industrial mathematics meet. Some experiences from Finland / S. D. Howison: The Oxford study group with industry, 1968-1985 / W. Krüger: Mathematically modelling in lifetime testing / J. K. Lenstra: Interface operations research and computer science / S. McKee: Academic-industrial collaboration in numerical analysis / J. Molenaar, G. W. M. Staarink: The optimal form of diamond heat sinks / J. D. Petersen: Accelerated fatigue-test with stochastic load-functions from the rainflow matrix / W. Schempp: Computational signal geometry and laser optics / G. Soderlind: Some industrial problems in dynamic simulation / C. B. Vreugdenhill: ISNAS: on information system for flow simulation based on the Navier-Stokes equations / Hj. Wacker: Mathematical research and industrial consulting in Linz: power plants – reactions columns – heating of slabs / A collection of capsule description of mathematics-in-industry problems / List of participants / List of the members of the committee for developing industrial mathematics in the Netherlands

 B. G. Teubner Stuttgart

 Kluwer Academic Publishers

ECMI VOL. 2

Case Studies in Industrial Mathematics

Edited by Prof. Dr.-Ing. H. W. Engl, Prof. Dr. Hj. Wacker and Dr. W. Zulehner, University of Linz

1988. X, 218 pages. 16,2 x 23,5 cm. Coproduction Kluwer-Teubner

ISBN 3-519-02171-4 (Teubner) Bound DM 72,–

ISBN 90-277-2731-7 (Kluwer) Bound Dfl 140.00 / £ 43,00

In the book eigth projects are presented, which were treated by groups of mathematicians from the Johannes Kepler Universität at Linz, Austria. As the title of the book indicates, the projects are "real world" problems from industry. They were studied in close contact with engineers. Numerical methods play a crucial role in all projects. The book concentrates on problems close to the mathematical expertise of the working groups, which includes nonlinear equations, optimization and control, inverse and ill-posed problems. Most of the presented problems come from a few areas of applications, namely hydro energy production, various phases of steel production and modelling of large-scale chemical systems.

Contents

Project 1 H. W. Engl/E. H. Lindner: Computing Eddy Current Losses in Reactor Coils

Project 2 W. Zulehner: Calculation of the Hydrodynamic Coefficients for Bodies of Revolution

Project 3 H. W. Engl/T. Langthaler: Control of the Solidification Front by Secondary Colling in Continuous Casting of Steel

Project 4 D. Auzinger/Hj. Wacker: Optimal Reheating of Slabs in a Pusher Type Reheating Furnace

Project 5 W. Zulehner: On the Design of the Volute of a Centrifugal Pump

Project 6 D. Auzinger/L. Peer/Hj. Wacker/W. Zulehner: Numerical Calculation of Separation Processes

Project 7 W. Bauer/E. H. Lindner/Hj. Wacker: Optimization of Systems of Hydro Energy Power Plants

Project 8 E. H. Lindner/Hj. Wacker: A Black Box Technique for Determining the Efficiency Function of a Hydroelectric Storage Power Plant

B. G. Teubner Stuttgart

Kluwer Academic Publishers

ECMI VOL. 3

Proceedings of the

Second European Symposium on Mathematics in Industry

ESMI II March 1–7, 1987, Oberwolfach

Edited by Prof. Dr. H. Neunzert, University of Kaiserslautern

1988. VIII, 359 pages. 16,2 x 23,5 cm. Coproduction Kluwer-Teubner

ISBN 3-519-02172-2 (Teubner) Bound DM 82,–

ISBN 90-277-2732-5 (Kluwer) Bound Dfl 180.00 / £ 56,00

Industrial mathematics is a special method to get interesting problems; a special attitude of curiosity for technical or economical questions; a general rather broad knowledge in all branches of mathematics. These proceedings include only examples of "how to do". On the one hand classical topics in industrial mathematics as questions of mathematical modelling, e.g. in chemical industry; numerics, e.g. process simulation, CAD; optimization problems, e.g. in steel industry; and finally image and signal processing are treated. On the other hand the applicability of "pure" mathematics to practical problems is demonstrated by examples in system theory (filtering) and in the application of the analysis of the wave equation to control problems in space technology.

Contents

R. V. Mendes, M. Faria, L. Streit: Map Dynamics in Gearbox Models / R. M. M. Mattheij, S. W. Rienstra: On an Off-shore Pipe Laying Problem / K. Merten: Mathematische Probleme des rechnergestützten Entwurfs höchstintegrierter Schaltungen / P. S. Teigen, Statoil: Stability Problems in Offshore Installation Operations / J. Struckmeier, F.-J. Pfreundt: Identification of Deterministic Nonlinear Dynamical Systems / M. Hazewinkel: Introduction to Nilpotent Approximation Filtering / M. Primicerio: Dynamics of "Slurries" / C. Bardos: Practical Applications of the Notions of Propagation of Singularities / S. D. Howison: Complex Variables in Industrial Mathematics / E. Cumberbatch: Mosfet Modelling / V. Capasso: Reaction-Diffusion Models for the Spread of a Class of Infectious Diseases / G. Stoyan: Numerical Solution of Pipeline System Problems by Monotone Difference Approximations / S. A. Halvorsen: Dynamic Model of a Metallurgical Shaft Reactor with Irreversible Chemical Kinetic and Moving Lower Boundary / P. F. Hodnett: Wave Induced Washout of Submerged Vegetation in Shallow Irish Lakes / J. Hoschek: Approximate Conversion of Spline Curves/ P. Rentrop, U. Wever: Interpolation Algorithms for the Control of a Sewing Machine/ M. Bercovier, E. Jankovich, M. Durand: The Development of a Mechanical Model for a Tire: A 15 Year Story to Replace Test Machines / D. Auzinger, L. Peer, Hj. Wacker: Steady State Simulation of Chemical Plants – Flowsheeting/ H. W. Engl, G. Landl: A Scheduling Problem in the Production Line "Steel Making – Continuous Casting – Hot Rolling" / W. Schempp: A Reference Model for Laser Opto-Electronics, Digital Signal Processing, and Holography / U. Eckhardt, G. Maderlechner: Application of the Projected Radon Transform in Picture Processing

 **B. G. Teubner Stuttgart**

 **Kluwer Academic Publishers**